Molecular Biology in Plant Pathogenesis and Disease Management

Molecular Biology in Plant Pathogenesis and Disease Management

Disease Development

Volume 2

P. Narayanasamy

Former Professor and Head,
Department of Plant Pathology,
Tamil Nadu Agricultural University,
Coimbatore, India

Springer

Author

Dr. P. Narayanasamy
32 D Thilagar Street
Coimbatore-641 002
India
pnsamy@dataone.in

ISBN 978-1-4020-8244-3 e-ISBN 978-1-4020-8245-0

Library of Congress Control Number: 2007943471

Cover Picture: Courtesy of R. Viswanathan, Sugarcane Breeding Institute, Coimbatore, India

Printed on acid-free paper.

9 8 7 6 5 4 3 2 1

springer.com

Dedicated to the Memory
of My Parents
for their Love and Affection

Contents

Table of Contents for Volumes 1 and 3

Volume 1

Volume 3

Preface

The discovery of the structure of DNA followed by the introduction of recombinant DNA technology has provided significant impetus to the applications of molecular techniques widely in biological sciences including plant pathology. Molecular techniques have been very useful in understanding the phenomenon of plant pathogenesis (disease development) and for elucidating the intricacies of the interactions between microbial plant pathogens and the host plants at cellular and molecular levels. Molecular genetic tools allow the identification and functional analysis of genes involved in the interplay of pathogens and their host plants.

The interaction of products of pathogenicity genes and host defense genes determine the progress of pathogenesis either leading to development of disease in susceptible plants or disease suppression in resistant plants that can perceive the presence of pathogen and activate their defense genes that encode various antimicrobial compounds or form formidable barriers arresting the invasion of host tissues by the pathogen concerned. Successful pathogens have mechanisms that specifically counteract and dismantle plant defense components, resulting in avoidance of host's surveillance system, prevention of activation of plant defenses and development of pathways to suppress manifestation of host defense responses. Various kinds of effector proteins secreted by fungal and bacterial pathogens create conditions favoring pathogen colonization of host tissues to varying degrees. On the other hand, viral pathogens even with such small genomes that carry information for the synthesis of viral nucleic acid and other structural components, are able to deplete efficiently the resources of infected plants to their advantage. Enormous genetic data required for epidemiological analysis for studying the patterns of disease occurrence and development in populations of plants under natural conditions have been gathered using molecular techniques. By determining pathogen gene frequencies, changes within and among populations, as a result of both natural selection of genes and population increases and spread, could be assessed. Molecular epidemiological approaches have been applied for tracking disease outbreaks, for development of prediction models to determine the sources of inoculum and for understanding evolution of pathogen virulence. It may be expected that novel methods of managing crop diseases more effectively may become available, by studying the signal pathways involved in plant pathogenesis and determining the vulnerable genetic targets for intervention in the life cycles of plant pathogens.

This book presents updated and comprehensive information in an easily understandable style, on the molecular biology of plant-pathogen interactions in three volumes: (1) Microbial plant pathogens, (2) Molecular biology of plant disease development and (3) Molecular biology in crop disease management. The usefulness and effectiveness of molecular techniques to establish the identity of pathogens precisely, to have a better understanding of the intricacies of the success or failure of pathogen infection respectively in compatible and incompatible plant species and to develop more effective crop disease management systems is highlighted with suitable examples. Appendices containing protocols included in appropriate chapters will be useful for students, teachers and researchers of various departments offering courses and pursuing research programs in molecular biology in general and plant pathology in particular.

Coimbatore
India

P. Narayanasamy

Acknowledgement

I wish to record my appreciation and thanks to my colleagues and graduate students of the Department of Plant Pathology, Tamil Nadu Agricultural University, Coimbatore, India, for the intellectual interaction and assistance in various ways. Dr. T. Ganapathy has provided all technical assistance with enormous patience. The secretarial assistance of Mrs. K. Mangayarkarasi for the preparation and presentation of the manuscript of this book has been significant. Permission granted by the copyright holders to reproduce the figures published in various journals is sincerely acknowledged.

The abundant affection and immense support of my wife Mrs. N. Rajakumari, made it possible to devote my undivided attention for the preparation of this book. To all my family members Mr. N. Kumar Perumal, Mrs. Nirmala Suresh, Mr. T. R. Suresh and Mr. Varun Karthik, I am glad to express my thanks for their affectionate encouragement to heighten the level of my involvement in academic endeavors.

Finally it is with a sense of gratitude, I extend my thanks to Mr. Pappa Vidyaakar, founder of Udavum Karangal, for his enduring encouragement and appreciation for my academic and humanistic activities.

Chapter 1
Introduction

When the microbial pathogens reach an appropriate host plant species and if favorable environmental conditions prevail, they develop rapidly resulting in the derangement of various physiological functions of the host, followed by the development of characteristic macroscopic symptoms. The pathogen suppresses innate natural disease resistance (NDR) mechanisms by producing various enzymes or toxic metabolites resulting in tissue necrosis or dissolution of cell walls releasing the cell contents that may serve as nutrient source. The fungal physiology was the subject of great interest during the first half of the 20th century. Although several enzymes produced by fungal pathogens were described, their role in pathogenesis and virulence of the pathogen could not be established for most of the enzymes using physiological and biochemical techniques. The discovery of the structure of DNA by Watson and Crick (1953) was hailed as the momentous achievement of the last century in biological sciences. The introduction of recombinant DNA technology later, was considered as an important milestone leading to breath-taking accomplishments in molecular genetics. The effectiveness of molecular cloning and other molecular approaches to address the basic questions regarding plant pathogenesis was demonstrated by different researchers during the past three decades. The dramatic revelation of the process of infection by *Agrobacterium tumefaciens* leading to production of tumors in plants has been the torch bearer lighting the path for the others attempting to understand the intricacies of interactions, at cellular and molecular levels, between fungal, bacterial and viral pathogens and their host plants. Applications of various molecular techniques that have widened information base for studying different phases of disease development in individual plants in vitro and the factors influencing disease incidence on populations of plants and spread in vivo are discussed in two chapters of the volume 2.

1.1 Disease Development in Individual Plants

Molecular genetic tools have allowed the identification and detailed functional analysis of genes involved in the interplay of microbial plant pathogens and their host plants. A pathogen has to overcome a series of barriers during different phases of

P. Narayanasamy, *Molecular Biology in Plant Pathogenesis and Disease Management: Disease Development, Vol. 2*,

pathogenesis. The involvement of conserved signal networks in disease initiation by microbial pathogens with contrasting infection strategies in different host plant species has been indicated (Lee et al. 2003; Talbot 2004). Different genes are involved in signal events which accompany and control all stages of infection and colonization processes.

The fungal pathogenicity genes necessary for disease development may not be essential for the completion of pathogen life cycle. The success of a pathogen in initiating infection depends primarily on its ability to perceive and respond to signals generated by the plant especially in the early stages of infection (recognition) and also at later stages involving different types of cells and tissues. Identification of pathogenicity determinants with certainty has not been possible in several pathogen–host interactions, because of functional redundancy of genes thought to be involved in pathogenesis (Gold et al. 2001; Yoder and Turgeon 2001). The fungal pathogens appear to have broadly two types of mechanisms and the associated genes that specifically counteract and dismantle plant defense components and mechanisms that assist the pathogens to avoid or suppress activated plant defenses. The effects of preformed antimicrobial compounds (phytoanticipins) and compounds that are synthesized postinfectionally (phytoalexins) have to be countered by the invading pathogen. The presence or absence of saponins has been correlated with disease development, as in the interaction of *Gaeumannomyces graminis* with oat and wheat. The mutant strains (*ava1*) are unable to detoxify avenacin and hence they cannot infect oat producing avenacin A, while they were pathogenic on wheat which does not naturally produce saponins (Papadopoulou et al. 1999). The pathogenesis-related (PR) proteins are both preformed and induced antifungal proteins that are suggested to play a role in providing protection by direct toxicity to the invading fungal pathogen and also by release of fungal elicitors that can potentially activate plant defenses (Selitrennikoff 2001).

Successful fungal pathogens seem to have evolved mechanisms that help them avoiding plant detection system and preventing induction of plant defenses or they may develop pathways to suppress the manifestation of these responses. The *Phytophthora GIPs* have been indicated to prevent the release of fungal elicitors and consequently prevent the activation of plant defense systems (Rose et al. 2002). Another group of genes expressed in planta are involved in the establishment and maintenance of infection and they are not directly associated with acquisition of nutrients from the host plant. The genes *CLTA1* in *Colletotrichum* spp. (Dufrense et al. 2000) and *CIH1* in *C. lindemuthianum* (Perfect et al. 2000) have been reported to prevent elicitation of host defenses.

Bacterial pathogens employ diverse strategies to undermine plant defenses and target core components of plant resistance such as hypersensitive response (HR)-based programmed cell death (PCD), cell wall-based defenses, jasmonic acid (JA) signaling and the expression of defense genes (Abramovitch and Martin 2004). The type III secretion system (TTSS) is an essential virulence system employed by many Gram-negative bacterial pathogens to inject the effector proteins into host cells. The TTSS is important to pathogenicity of bacterial pathogens belonging to the genera *Pseudomonas, Xanthomonas, Ralstonia, Erwinia* and *Pantoea.* These pathogens colonize the intercellular spaces (apoplast) of plants and are generally capable of eliciting plant cell death at some stage of disease development. As most of them are host-specific, they exhibit compatibility with the plant hosts. The TTSS pathway is encoded by the *hrp* and *hrc* genes. Some of the effector proteins possess double functions by enhancing virulence in compatible interactions and also by acting as avirulence genes the products of which elicit defense-related responses in incompatible interactions (Alfano and Collmer 2004).

Plant viruses have comparatively small genomes. Nevertheless, they adopt effective strategies to infect the plant hosts as severely as the bacterial and fungal pathogens. In addition, the viral pathogen can also infect their natural vectors (exhibiting propagative relationship) involved in the transmission of viruses from plant to plant. There are no detectable physiological functions such as respiration in viruses. But they are able to bring the host's synthetic machinery under their control by effectively suppressing the defense mechanisms. Plant virus genomes carry genes for synthesis of coat protein, movement protein, replicase and virus-coded nonstructural proteins such as HC-Pro. It is intriguing and equally amazing to note that with such small genomes, the viruses deplete efficiently the resources of infected plants to their advantage to reach high titers and induce various kinds of external and internal symptoms characteristic of virus infection (Narayanasamy and Doraiswamy 2003).

1.2 Disease Development in Populations of Plants

Molecular biological approaches have provided large quantities of genomic data for microbial plant pathogens that are useful for epidemiological analyses. New information about genetic sequences and gene expression has formed the basis for predictions about epidemic features and outcomes and for understanding host resistance and pathogen evolution. Microarrays have opened up the possibility of analyzing expression simultaneously for thousands of genes. These tools can be expected to contribute to plant disease epidemiology by providing necessary information about which resistance or pathogenicity genes are present in individuals or population. In addition, it is possible to identify genes other than those directly involved in resistance and virulence important in epidemics, the role of phenotypic status of hosts and pathogens and the role of the status of the environmental metagenome. Gene expression within individual organisms and in populations of organisms can be assessed, based on studies on models of group dynamics derived from population biology and ecology. By determining pathogen gene frequencies, changes within and among populations as a result of both natural selection of genes and population increases and spread can be assessed for tracking disease outbreaks (Zwankhuizen et al. 1998), for development of predictions about sources of inoculum (Cortesi and Milgroom 2001) and for understanding the evolution of pathogen virulence (Escriu et al. 2003). Improved diagnostic systems and genomic data for all microbial pathogens causing economically important diseases on various crops have to be developed to determine precisely the abundance of pathogen and its strains in a range of environmental settings. The presence of pathogenicity genes and other genes related to toxin production and epidemiological features has to be detected to determine more reliably the genotypes in a population responsible for disease incidence, especially when the infection is still at very low levels, in order to eradicate the pathogen before it becomes well established (Garrett et al. 2006).

The importance and utility of the molecular techniques for gaining a deeper insight into the intricacies of plant–pathogen interactions may not have any lasting value unless it translates into real impacts in crop protection. It is considered that novel methods of managing crop diseases more efficiently may be developed in the near future based on the findings that conservation of signal pathways involved in pathogenesis by a diversity of microbial pathogens may be vulnerable generic targets for intervention in the life cycles of plant pathogens. The effectiveness and applications of molecular techniques for developing suitable disease management systems for various crops grown in different agroecosystems are discussed in six chapters included in volume 3 of this treatise.

References

Abramovitch RB, Martin GB (2004) Strategies used by bacterial pathogens to suppress plant defenses. Curr Opin Plant Biol 7:356–364

Alfano JR, Collmer A (2004) Type III secretion system effector proteins: double agents in bacterial disease and plant defense. Annu Rev Phytopathol 42:385–414

Cortesi P, Milgroom MG (2001) Outcrossing and diversity of vegetative compatibility types in populations of *Eutypa alata* from grapevines. J Plant Pathol 83:79–86

Dufrense M, Perfect S, Pellier A-L, Bailey JA, Langin T (2000) A GAL4-like protein is involved in the switch between biotrophic and necrotrophic phases of the infection process of *Colletotrichum lindemuthianum* on common bean. Plant Cell 12:1579–1589

Escriu F, Fraile A, Garcia-Arenal F (2000) Evolution of virulence in a plant virus. Evolution 57:755–765

Garrett KA, Hulbert SH, Leach JE, Travers SE (2006) Ecological genomics and epidemiology. Eur J Plant Pathol 115:35–51

Gold SE, Garcia-Pedrajas M, Martinez-Espinoza AD (2001) New and used approaches to the study of fungal pathogenicity. Annu Rev Phytopathol 39:337–365

Lee N, D'Souza CA, Kronstad JW (2003) Of smuts, blasts, mildews and blights: cAMP signaling in phytopathogenic fungi. Annu Rev Phytopathol 41:399–427

Narayanasamy P, Doraiswamy S (2003) Plant viruses and viral diseases, New Century Book House, Chennai, India

Papadopoulou K, Melton RE, Leggett M, Daniels MJ, Osbourn AE (1999) Compromised disease resistance in saponin-deficient plants. Proc Acad Natl Sci USA 96:12923–12928

Perfect SE, Pixton KL, O'Connell RJ, Green JR (2000) The distribution and expression of a biotrophy-related gene *CIH 1,* within the genus *Colletotrichum.* Mol Plant Pathol 1:213–221

Rose JKC, Ham K-S, Darvill AG, Albersheim P (2002) Molecular cloning and characterization of glucanase inhibitor proteins: coevolution of a counterdefense mechanism by plant pathogens. Plant Cell 14:1329–1345

Selitrennikoff CP (2001) Antifungal proteins. Appl Environ Microbiol 67:2883–2894

Talbot NJ (2004) Emerging themes in plant–pathogen interactions. In: Tallbot NJ (ed) Annual plant reviews plant–pathogen interactions, vol 11. Blackwell Publishing, Oxford, UK, pp 1–26

Watson JD, Crick FHC (1953) Molecular structure of nucleic acid. Nature 171:737–738

Yoder OC, Turgeon BG (2001) Fungal genomics and pathogenicity. Curr Opin Plant Biol 4:315–321

Zwankhuizen MJ, Govers F, Zadoks JC (1998) Development of potato late blight epidemics: disease foci, disease gradients and infection sources. Phytopathology 88:754–763

Chapter 2
Molecular Biology of Plant Disease Development

Abstract The microbial pathogens vary widely genetically and structurally. Consequently their requirements for their development and the potential for overcoming the barriers formed by host plants also show distinct differences. Various kinds of enzymes elaborated by fungal and bacterial pathogens at different stages of pathogenesis have critical role in pathogen development and survival. The involvement of various pathogen genes and effectors in subduing the defense responses of host plants has been demonstrated in several pathogen–host interactions. The interaction between gene products of pathogen and host plant and consequent increase or decrease in the intensity of disease symptoms have highlighted the importance of molecular biological approaches in providing reliable and conclusive evidences. The fungal pathogens produce a cascade of enzymes encoded by specific genes whose expression at different stages of pathogenesis hold the key for their successful infection. Some of them produce host-specific toxins that are able to induce the symptoms just like the pathogens themselves. The secretion of enzymes and other effector proteins by bacterial pathogens is controlled by type II or type III secretion systems (TTSS). The products of these secretion systems function as pathogenicity factors. The products of TTSS are required for both pathogenicity and elicitation of resistance. The molecular interaction between pathogenicity factors/effector proteins and compounds constitutively produced and those produced in response to infection in various pathosystems have been discussed to provide a comprehensive understanding of the phenomenon of pathogenesis. Viruses and viroids being molecular pathogenic agents have to adopt strategies different from those of cellular pathogens. The usefulness of molecular techniques to probe deeper into the pathogen perception, gene silencing and suppression of RNA silencing, reflecting the constant attempts of host and pathogen to gain the upper hand is elucidated.

Microorganisms existing in the environment may be divided broadly into two groups based on their ability or inability to induce disease(s) in plants. The microorganisms possessing the unique genes for pathogenicity involved in the induction of characteristic symptoms in plants are the pathogens. In contrast, the microorganisms belonging to the second group lacking the pathogenicity genes cannot induce

P. Narayanasamy, *Molecular Biology in Plant Pathogenesis and Disease Management: Disease Development, Vol. 2*,

disease in plants. They are known as saprophytes involved in the degradation of dead organic matter and act as scavengers, making the environment clean and safe for human habitation. The microbial plant pathogens – oomycetes, fungi, bacteria, mycoplasma, viruses and viroids – have specific genes for adhesion to plant surface (or receptors), production of enzymes and toxins, resistance to oxidative stress or chemicals, degradation of plant cell walls, in planta nutrition and breaching host defense strategies. Perception of pathogen signals by plants involves a complex suite of cellular responses that involve cross-talk of numerous signal transduction pathways culminating in disease induction (compatible interaction) or resistance (incompatible interaction).

The ability to discriminate between self and nonself is considered as a key step in the evolution of eukaryotic immune systems. The plants have evolved, dedicated and variable receptor families for recognition of nonself structures / organism-mediated products. They possess a repertoire of receptors that function as a radar system for detecting pathogen-derived nonself molecules. The outcome of the interactions between plants and the microbial pathogens that invade them depends on the effectiveness of the surveillance system of the plant species. Only when there is a match between a strain-specific pathogen effector and its corresponding plant host resistance (*R*) gene product, the plant, in question, may be able to recognize their intruder(s) and to mount an effective resistance response. Such a recognition of the pathogen effector by a plant R-receptor results in rapid death of plant host cells at sites of attempted penetration / invasion, as part of immune response of the plant. When the plant fails to recognize the pathogen effector, infection by the pathogen is initiated. Depending on the ability of the pathogen to overcome the effects of inhibitory compounds produced by the plants to arrest the pathogen development, the future course of disease development is determined (Narayanasamy 2002).

The microbial pathogens have evolved highly specialized tactics to overcome plant defense responses and to promote disease development. Successful pathogens have the ability to avoid or actively suppress plant defense responses, due to the pathogen factors that induce susceptibility in an otherwise resistant or tolerant host plant species. The pathogen factors include host selective toxin (HSTs) (Wolpert et al. 2002) and small molecule suppressors (Shiraish et al. 1997) from fungal pathogens, type III secretion system effectors and toxins from bacterial pathogens (Bender et al. 1999) and suppressors of posttranscriptional gene silencing (PTGS) from viral pathogens (Moissiard and Voinnet 2004). Susceptibility or resistance to a pathogen is seen, if the pathogen expresses virulence or avirulence gene(s). When the avirulence (Avr) protein is recognized by the host plant, hypersensitive reaction (HR)-based programmed cell death (PCD) leads to death of cells around points of pathogen's attempted penetration into the host plant surface. PCD limits the pathogen establishment and spread by killing both the pathogen and the infected host cells. The molecular basis of pathogenicity is discussed to highlight the various strategies adopted by different pathogens to overcome the strategies of host plant for containing the pathogen development and disease progression.

2.1 Fungal Pathogens

Under favorable conditions, pathogenesis or the process of disease development has five distinct phases: (i) attachment of fungal pathogen to plant surface, (ii) germination of spores or pathogenic unit, (iii) penetration / entry into the host tissue, (iv) colonization of host tissue and (v) symptom expression. Biochemical and histological investigations have provided some basic information on different stages of pathogenesis. Studies to understand the molecular basis of pathogenesis are yet to provide a clear picture of the role of pathogenicity genes in the process of disease development.

2.1.1 Attachment of Fungal Pathogens to Plant Surfaces

The fungal and oomycetes pathogens, after their natural dissemination, may land on different organs of the host plant species. Some pathogens may show organ specificity. In such cases, initiation of infection can occur only if the pathogen is able to reach the right tissue or organ. An extreme case of organ specificity is seen in the case of *Colletotrichum acutatum* within the morphologic species. The isolates of *C. acutatum* causing post-bloom fruit drop (PFD) disease can infect only flower petals of most citrus cultivars. On the other hand, isolates of *C. acutatum* causing Key lime anthracnose (KLA) disease can infect only Key lime, inducing anthracnose on twigs, shoots, leaves and flowers (Fagan 1979; Timmer and Brown 2000). The fungal gene required for pathogenicity on Key lime leaves and the gene KLAP1 (*K*ey *L*ime *a*nthracnose *p*athogenicity) encoding a putative Cys2 His2 transcription activator required for pathogenicity of *C. acutatum* have been identified (Chen et al. 2005).

Attachment of spores or other infection units to the appropriate host plant surface is the first step in the initiation of infection, as demonstrated in the case of *Colletotrichum graminicola* (Mercure et al. 1994a, b) and *Phytophthora megasperma* (Ding et al. 1994). The extracellular enzymes secreted by fungal pathogens facilitating the adhesion of spores to plant surface have been implicated. The cutinases and esterases may have a role in the adhesion of infection structures of *Colletotrichum graminicola* (Pascholati et al. 1993) and *Uromyces viciae-fabae* (Deising et al. 1992), whereas the spore tip mucilage of *Magnaporthe grisea* was responsible for the adhesion of conidia to culticle through hydrophobic action (Hamer et al. 1988). The nature of compounds assisting in spore adhesion has been shown to be proteins or glycoproteins in *Colletotrichum musae* (Sela-Buurlage et al. 1991), *Nectria hematococca* (Kwon and Epstein 1993), *Uromyces viciae-fabae* (Clement et al. 1993) and *Colletotrichum graminicola* (Mercure et al. 1994a, b).

The fibrillar "spore coat" present on the conidia of *Colletotrichum lindemuthianum* and its cell wall facilitate the adhesion to aerial plant surfaces for initiation of infection. A monoclonal antibody (UB20) capable of reacting with a glycoprotein on the conidial surface was employed to localize the glycoprotein. This compound was concentrated in the outer surface of the spore coat. The attachment of conidia was inhibited by UB20 IgG to hydrophobic surface in an antigen-specific manner, suggesting a role for specific glycoproteins in the initial attachment of conidia (Hughes et al. 1999). The presence of specific glycoproteins on the conidia, germ tubes, appressoria and primary hyphae of *C. lindemuthianum*, but not on surface of the secondary hyphae was demonstrated by immunofluorescence labeling. This finding indicates that the fungal cell surface may be modified during necrotrophic growth, with none of the glycoproteins associated with early stages of infection-process being produced (Perfect et al. 2001). *Peronospora parasitica* causing downy mildew disease, forms germ tubes and appressoria which produce a fibrillar matrix containing β-1, 3 glucans, confined to germling-substrate interface and another matrix consisting of protein spreading beyond the contact interface as a thin film. As these matrices provide tenacious adherence, they may contribute to the germling attachment to plant host surfaces (Carzaniga et al. 2001).

2.1.2 Germination of Spores and Penetration of Host Plant Surfaces

Oomycete zoospores formed inside the sporangia (as in *Phytophthora* spp.), are biflagellated with anterior flagellum conferring more motile force. Cysts are formed after shedding off the flagella and they germinate to produce germ tubes and appressoria later. The appressoria penetrate the cuticle. The molecular events that control these developmental process are not clearly understood. A protein kinase gene *Pipkz1* induced during cleavage of sporangial contents of *Phytophthora infestans* was cloned to help characterization of signal transduction pathways active in spore biology (Judelson and Roberts 2002). The *Pipkz1* gene induced a protein kinase during zoosporogenesis in *P. infestans*. It interacted with a putative bZIP transcription factor. The gene *Pibzp1* for the transcription factor was a single copy and expressed in all tissues. Transformants of *P. infestans* stably silenced for *Pibzp1* were obtained. The zoospores from silenced strains (mutants) spun in tight circles, instead of the normal pattern of straight swimming punctuated by turns. Encystment of the zoospores was not, however, altered in the silenced strains. The cysts of the mutants formed long germ-tubes and the percentages of germinated cysts producing appressoria was significantly reduced compared with non-silenced controls. Thus silencing of *Pibzp1* affected pathogenicity of the strains of *P. infestans* resulting in failure of infection of potato plants. However, the mutants were able to colonize wounded tissues. The results indicated that *Pibzp1* is a key regulator of several stages of zoospore-mediated infection pathway (Blanco and Judelson 2005).

Just prior to spore germination, three genes of *Phytophthora infestans*, causing potato late blight disease, are activated. These genes forming a cluster, code for novel proteins known as Car-(cyst-germination-specific acidic repeat) proteins. The Car proteins, transiently expressed during spore germination and appressorium production, are localized at the surface of germ tubes. They may have a dual role of protecting germ tubes from desiccation and of adhering the germ tubes to the plant surface (Görnhardt et al. 2000). Pathogenic fungi produce soluble molecules and enzymes that suppress hypersensitive response (HR)-based plant defenses. *P. infestans* produces soluble glucans in its spore germination fluids that may suppress the oxidative burst and HR in potato (Shiraishi et al. 1997). Pathogen infection can induce cells to die in a regulated manner with morphological features similar to programmed cell death (PCD). Both resistant (HR) and susceptible interactions may show PCD. Fungal pathogens can coopt plant PCD pathways to establish infections and subsequent colonization of host tissues (Gilchrist 1998; Richael and Gilchrist 1999; Greenberg and Yao 2004). Arachidonic acid (AA), a polyunsaturated fatty acid that is widely dispersed in the lipids of *P. infestans* is released from germinating spores during plant infection (Ricker and Bostock 1992).

Arachidonic acid induces local cell death and transcription-dependent systemic responses to *P. infestans* (Coquoz et al. 1995). A tomato gene *DEAI*, that is induced by *P. infestans* and its PCD-inducing elicitor, arachidonic acid (AA) were characterized. *DEA1* was expressed in response to PCD-inducing AA within 8 h following treatment of tomato leaflet 16 h prior to the development of visible cell death. Infection by *P. infestans* affected the *DEA1* transcript levels as determined by differential display (DD)-PCR. At 48 h postinocualtion, *DEA1* expression in the water control decreased, while expression in the *P. infestans* treatment remained at higher levels, until the leaf was entirely colonized by the pathogen at 120 h postinoculation, after which period *DEA1* level decreased (Weyman et al. 2006).

In addition to PCD initiated by the host activities, cell death induced by compounds produced by the expression of genes of bacterial and fungal pathogens has been reported in certain pathosystems. A class of cell-death-inducing proteins designated Nep1-like proteins (NLP) induces nonspecific necrosis in a range of dicotyledonous plants. *Phytophthora infestans* produces three NLPs viz., PiNPP1.1, PiNPP1.2 and PiNPP1.3. PiNPP1.1 was found to induce cell death in tomato and *Nicotiana benthamiana*, by delivering through agroinfection with a binary *Potato virus X* vector. Expression analyses revealed that PiNPP1.1 was up-regulated during late stages of infection of tomato by *P. infestans*. The cell death induced by PiNPP1.1 was shown to be dependent on the ubiquitin ligase-associated protein SGT1 and the heat-shock protein SGT1, following virus-induced gene silencing procedure. Furthermore, cell death due to PiNPP1.1 was also dependent on the defense-signaling proteins (COI1, MEK2, NPR1 and TGA2.2) suggesting distinctly different signaling requirement from that of INF, an elicitin, triggering HR in *Nicotiana* spp. However, enhancement of cell death was observed due to combined expression of PiNPP1.1 and INF in *N. benthamiana*, suggesting a synergistic interaction of these proteins (Thirumala-Devi et al. 2006). The pattern of gene expression in soybean and *Phytophthora sojae* during an infection time course was investigated. A gene microarray for host plant and pathogen cDNA transcripts was constructed, large changes in the ratio of host and pathogen RNA occurred between 12 and 24 h after infection, reflecting the rapid growth and proliferation of the pathogen genes expressed during infection and reaching a maximum at 24 h. The pathogen appears to transit from biotrophy to necrotrophy between 12 and 24 h after infection (Moy et al. 2004).

The rice blast pathogen *Magnaporthe grisea* has *MPG1* gene that is expressed at the initiation of pathogenesis. It is conserved in many races and host specific forms of *M. grisea*. *MPG1* codes for a gene product that forms a component of the host surface perception process, in addition to its requirement for efficient appressorial formation and cellular differentiation. In addition, *MPG1* encodes a fungal hydrophobin secreted by the pathogen on the rice leaf surface. Dramatic reduction in virulence following abolition of MPG1 function was observed, indicating its role as virulence gene for pathogenicity (Talbot et al. 1993; Talbot 1998). The *PTH* genes have been shown to have a role in pathogenicity of *M. grisea*. Of the eleven *PTH* genes identified by mutational analysis, *PTH11* was predicted to encode a novel transmembrane protein. The *pth11* mutants were found to be nonpathogenic because of a defect in appressorium differentiation. Probably *pth*11 was unable to synthesize a protein encoded by *PTH11*, involved in host surface recognition (Sweigard et al. 1998; DeZwaan et al. 1999). Mutants of *M. grisea* impaired in pathogenicity were identified by restriction enzyme-mediated DNA integration (REM1) mutagenesis. Many mutants defective in conidiogenesis and appressorial functions were detected. One of them exhibited distinct differences in pathogenicity on susceptible rice cultivars (reduced pathogenicity) and on barley (with no change in pathogenicity) compared with wild parent (Balhadere et al. 1999). The peroxisomal carnitine acetyl transferase (CAT) activity in *M. grisea* has been shown to be essential for the appressorium function, especially for elaboration of primary penetration hyphae. The *PTH2* gene in *M. grisea* encodes for the major CAT activity which registers enhanced expression in response to acetate and lipid. It is regulated by cyclic-AMP (cAMP) response pathway. In addition, a Pth2-GFP fusion colocalizes with a peroxisomal marker protein. The nonpathogenic mutants obtained by targeted deletion of *PTH2*, lacked CAT activity and they could not utilize a range of lipid substrates. A delay in lipid reserve mobilization from the germ tubes into developing infection cells and abnormal chitin distribution in infection structures may be due to impairment of appressorium function in *Δpth2* mutants. These findings indicate a role for Pth2 in the generation of acetyl CoA pools required for appressorium function and rapid elabortion of penetration hyphae during host infection (Bhambra et al. 2006).

The involvement of a cyclic-AMP (cAMP) signaling mechanism in appressorial differentiation was indicated by different studies. A *CPKA* gene encoding a catylic subunit of AMP-dependent protein kinase A may control appressorium formation by regulating cAMP signaling (Xu et al. 1997). It is possible that higher levels of a cAMP seen during conidial germination may be responsible for differentiation of appressoria (Choi et al. 1998). However, even in the presence of a cAMP, the *apf1* mutants,lacking *apf1* gene could not form appressoria, indicating that *apf1* acts independently of cAMP signaling pathway required for differentiation appressoria in *M. grisea* (Siluè et al. 1998). Marked reduction in glycogen and lipid mobilization occurring during conidial germiantion was observed in ΔcpkA mutant that lacks the catalytic subunit of cAMP dependent protein kinase A (PKA) (Thines et al. 2000). In the case of *Colletotrichum lagenarium* causing anthracnose disease in cucurbits, the role of cAMP signaling was studied. The *cpk1* and *cac1* mutants knockout mutants of *CPK1* and the adenylate cylclase gene *CAC1*, were nonpathogenic on cucumber. Both mutants germinated poorly, suggesting the involvement of cAMP signaling in the germination of conidia. The conidia of the mutants germinated but the appressoria were nonfunctional. The results showed that Cmk1 mitogen-activated kinase (MAPK) regulates germination, appressorium formation and infectious growth. The cAMP signaling seems to control multiple steps of fungal infection in cooperative regulation with Cmk1 MAPK in *C. lagenarium* (Yamauchi et al. 2004).

A mitogen-activated protein kinase (MAPK) gene PMK was suggested to regulate appressorium formation and growth of the infectious hyphae in fungal pathogens (Xu and Hamer 1996). In *M. grisea*, *PMK1* has been shown to be essential for pathogenesis. Generation of hydrostatic turgor by accumulating molar concentrations of glycerol was also required and degradation of glycogen occurred rapidly during conidial germination. In a *Δpmk1* mutant lacking MAP kinase required for appressorial differentiation, no mobilization of glycogen and lipid could be noted (Thines et al. 2000). The possibility of the same MAP kinase being essential for infection by *Botrytis cinerea,* causing gray mold diseases in fruits, was assessed. The *Botrytis* MAP kinase required for pathogenesis (BMP), MAP kinase gene was found to be highly homologous to the *M. grisea PMK1*. The *bmp1* mutants obtained by gene replacement were non-pathogeneic on carnation flowers and tomato leaves. Restoration of normal growth rate and pathogenicity of *bmp1* mutants could be obtained by reintroduction of wild-type *BMP1* allele. The *bmp1* mutants germinated on plant surfaces, but could not penetrate and macerate plant tissues, as the wild-type. The results indicated that BMP1 is required for infection of plants by *B. cinerea* and the MAP kinase pathway may be widely conserved in many fungal pathogens for regulating infection process (Zheng et al. 2000).

Mitogen-activated protein kinases (MAPKs) are involved in the transduction of extracellular signals. They function in a cascade of hierarchical components viz., MAPK kinase kinase, MAPK kinase and MAPK. Sequential activation of these enzymes mediates cellular response to specific external stimuli usually including changes in gene expression (Gustin et al. 1998). In fungal pathogens, *Fus3/Kss1* homologs have been shown to be required for appressorium formation and infection (Di Pietro et al. 2001). A mutant of *Botrytis cinerea*, defective in *Fus 3/ Kss1* homolog BMP1 was nonpathogenic and partially defective in germination (Zheng et al. 2000; Doehlemann et al. 2006). The MAPK, orthologous to yeast Slt 2 (Levin 2005) was involved in the maintenance of cell-wall integrity and various aspects of saprophytic and pathogenic development of filamentous fungi (Kojima et al. 2002; Mehrabi et al. 2006). A *B. cinerea* knockout mutant in the *bmp3* gene encoding region was detected and replaced by the hygromycin resistance cassette. Deletion of *bmp3* led to reduced vegetative growth on various media, strongly impaired conidiation and loss of sclerotia formation in *Δbmp3* mutant. The *Δbmp3* mutant did not exhibit any change in susceptibility to cell wall damage induced by glucanase, Caloflour White or Nikkomycin Z, but it was more susceptible to the oxidizing agent paraquat and the phenylpyrrole fungicide fludioxonil. *Δbmp3* showed defect in surface sensing. After penetration, retardation of progress of necrotic lesion induced by *Δbmp3* mutant was observed. All these defects, however, could be restored by genetic complementation of the mutant with the wild-type *bmp3* (Rui and Hahn 2007).

The *PMK1* gene is homologous to the *FUS3* and *KSS1* genes that regulate the transcription factor STE12 in yeast. The MST12 gene of *M. grisea* that is homologous to yeast STE12 was isolated by a PCR-based approach. The *mst12* mutant isolated by gene replacement, were nonpathogenic on rice and barley leaves. In contrast to *pmk1* mutants that did not produce appressoria, *mst12* mutants formed typical dome-shaped and melanized appressoria. However, the appressoria produced by these mutants could not penetrate the onion epidermal cells. Furthermore, *mst12* mutants, when wound inoculated, failed to form spreading lesion and appeared to be defective in infectious growth, suggesting that transcription factor(s) other than MST12 may be present in *M. grisea* (Park et al. 2002) During an incompatible interaction between pepper (chilli) and *Colletotrichum gloeosporioides,* a pepper esterase gene (Pep EST) was highly expressed. The recombinant PepEST protein expressed in *Escherichia coli* was applied after inoculation of compatible unripe pepper fruit. No symptom of the disease was observed, though the recombinant protein did not have fungicidal activity. But it significantly inhibited appressorium formation in a dose-dependent manner. In addition, the recombinant protein inhibited appressorium formation by *M. grisea* also and this effect could be reversed by treatment with cAMP or 1,16-hexadecanediol. The results suggest that the recombinant protein regulates appressorium formation by modulating the cAMP-dependent signaling pathway in this pathogen (Kim et al. 2001).

The MAP kinase cascades have an important role in plant growth and development as well as biotic and abiotic stress responses. In rice, five MAPKs, including three related to the host defense responses have been characterized and 17 members of the rice MAPK gene (*OsMPK*) family were identified. Nine of the 17 *OsMPK* genes were induced, upon inoculation with *M. grisea,* at the mRNA level during early, late or both states of infection. Four of the *M. grisea*-induced *OsMPK* genes were associated with host-cell death in the lesion-mimic rice mutant. The genome-wide expression analysis suggested that about 50% of the rice MAPK genes appeared to be associated with infection by *M. grisea* and host defense response (Reyna and Yang 2006).

In *M. grisea*, MAP kinases PMK1 gene has been shown to control appressorium formation and infectious growth. The yeast Fus 3 and Kss1 MAP kinases homologous to Pmk1, are regulated by the Ste20 PAK kinase for activating the pheromone response and filamentation pathways. Two PAK genes *CHM1* and *MST20* in *M. grisea* were characterized. Reduction in growth of aerial hyphae and conidiation was noted in mutants-disrupted in *MST20.* But growth rate, appressorium formation and plant infection remained unaffected in these mutants. In contrast, in *chm1* mutants all these parameters were adversely affected. CHM1 seems to play a critical role in appressorium formation and penetration, whereas MST20 is dispensable for plant infection. CHM1 and MST20 may have redundant function in *M. grisea,* playing no critical role in activating the PMK1 MAP kinase pathway during early infection stage (Li et al. 2004).

The MAP kinase (MEK) Mst 7 and MEK kinase (MEKK) Mst11 that activate Pmk1 in *M. grisea* were identified. Like the *pmk1* mutant, the *mst11* and *mst7* deletion mutants were found to be defective in appressorium formation and were nonpathogenic indicating that Mst11-Mst7-Pmk1 MAP kinase cascade may be conserved for regulating infection-regulated morphogenesis (Zhao et al. 2005). The later investigation, revealed that formation and growth of the appressoria of *M. grisea* was regulated by the Mst11-Mst 7-Pmk1 MAPKinase cascade. The *MST50* gene that directly interacted with both *Mst11* and *Mst7,* was characterized. The *mst 50* mutant was also defective in appresosrium formation, sensitive to osmotic stresses and were nonpathogenic. Expressing a dominant active MST7 allele complemented the defect of the *mst50* mutant in appressorium formation, but not its pathogenicity. *MST50* functioned as an upstream component of the PMK1 pathway for regulating appressorium formation. MST50 physically interacted with Mst11 via the SAM domain in the yeast two-hybrid assays. Deletion of SAM in Mst50 eliminated its interaction with Mst11 and its activity in appressorium formation. The interaction between Mst50 and Mst7 or Mst11 was detected by coimmunoprecipitation assays in developing appressoria. Mst 50 seemed to bind to different formation components of PMK1 MAP kinase pathway and function as an adaptor protein (Park et al. 2006).

The *CMK1* gene encoding a MAP kinase has been reported to have a critical role in conidial germination, formation of melanized appressoria, penetration of plant surface and subsequent colonization of tissues in *Colletotrichum lagenarium* causing anthracnose disease of cucumber. *CMK1* could complement appressorium formation of the PMK1-MAP kinase mutant of *M. grisea.* Loss of *CMK1* led reduction in conidiation and complete lack of pathogenicity on cucumber, since the conidia did not germinate and form appressoria in contrast to *pmk1* mutants of *M. grisea*. MAP kinase signaling pathways seemed to be important for the formation pathogens (Takano et al. 2000). Mutants of *M. grisea* defective in pathogenicity on leaves or roots of cereals were also detected (Dufrense and Osbourn 2001).

The requirement of melanization of appressoria for mechanical penetration of plant surface by fungal pathogens was indicated by different investigations. Melanin was shown to be a pathogenicity factor for the development of high internal turgor pressure for direct penetration of rice epidermal cells by *Magnaporthe grisea* (Howard et al. 1991). Melanin deficient, nonpathogenic mutants *M. grisea* were transformed with a cosmid clone pMBR1 that carries melanin biosynthesis genes *ALM, BRM1* and *BRM2* of *Alternaria alternata*. The mutants became pathogenic to rice, as the melanin synthesis was restored (Kawamura et al. 1997). Similar restoration of pathogenicity to nonpathogenic albino mutants of *Colletotrichum lagenarium* was also reported earlier (Kubo et al. 1991). However, the expression of three melanin genes *C. lagenarium* in the *cmk1* mutant did not occur and consequently the conidia of the mutant failed to germinate. The results suggest that CMK1 plays a role in gene expression essential for melanization of appressorium in *C. lagenarium* (Takano et al. 2000).

The darkly melanized appressoria formed at the end of the germ tubes from the conidia of *C. lagenarium* penetrate the host plant surface. The *CMK1* gene, a homolog of *Saccharomyces cerevisae FUS3 / KSS1 / MAPK* genes, regulates conidial germination, appressorium formation and invasive growth. The *CST1* gene from *C. lagenarium* the homolog of *Ste 12* of the yeast, was isolated and characterized. The *cst1* Delta strains were nonpathogenic on cucumber leaves, but could form lesion when wound-inoculated. Infectious hyphae did not form from the appressoria of *Cst1* Delta strains, suggesting that *CST1* was required for penetration of host surface by infectious hyphae. Rapid disappearance of lipid droplets during appressorium formation in *Cst1* Delta strains may have a bearing on the failure of penetration of the host surface (Tsuji et al. 2003). In another investigation, the role of other MAPKs in *C. lagenarium* was studied. The role of cAMP signaling was evaluated by functional characterization of the regulatory subunit gene of the CAMP-dependent protein kinase (PKA). By applying PCR based screening, the *RPK1* gene encoding the PKA regulatory subunit was isolated. The *rpk1* mutants had reduced vegetative growth and conidiation and were nonpathogenic on cucumber plants. However, they were able to form lesion when inoculated through wounds. Proper regulation of the PKA activity by the *RPK1* encoded regulatory subunit appears to be needed for growth, conidiation and appressorium function in *C. lagenarium* (Takano et al. 2001). The MAPK gene *MAF1* related to yeast *MPK1* and *M. grisea MPS1* genes, was functionally characterized. The *maf*1 gene replacement mutants showed reduced conidiation and pathogenicity. The germ tubes produced were elongated without appressoria both on plant surface or glass. The results suggest a role for *MAF1* in the early differentiative phase of appressorium formation, whereas CMK1 may be involved in the maturation of appressoria of *C. lagenarium* (Kojima et al. 2002).

In *C. lagenarium*, the gene *ClaPEX6* (PEX6) encoding a protein involved in peroxisome biosynthesis was identified through analysis of nonpathogenic mutants of *C. lagenarium* (Kimura et al. 2001). The requirement of *PEX6* gene for pathogenicity was demonstrated by knockout analysis of this gene. The *pex6* deletion mutant formed small appressoria and severely reduced melanization resulting in failure of generating penetration hyphae. Peroxisomal metabolic functions appeared to be important for appressorium functionality, especially for appressorial melanization. In order to determine the role of peroxisome associated metabolism in pathogenicity, the *ICL1* of *C. lagenarium* was isolated and functionally characterized. *ICL1* encodes isocitrate lyase involved in the glyoxylate cycle in peroxisomes. The *icl1* mutants could not utilize fatty acids and acetate for growth. These mutants formed highly pigmented appressoria, suggesting that glyoxylate cycle was not required for melanin biosynthesis in appressoria. However, *icl1* mutants had severely reduced virulence level and penetration hyphae failed to develop in the host plant, suggesting that *ICL1* was involved in the invasion of host tissues. Partial restoration of virulence of *icl1* mutant occurred on exogenous supply of glucose. The results showed that *PEX6* was required for appressorial melanization and suggested that peroxisomal pathway may have functional roles in appressorial melanization and subsequent host invasion steps and the latter step requires the glyoxylate cycle (Asakura et al. 2006).

Genetic manipulation of the fungal pathogens to establish their identity and analysis of genes required for pathogenicity or virulence have provided information to understand the pathogenic potential of different fungi. The molecular basis of delineating pathogenicity of *Colletotrichum acutatum* was studied by applying restriction enzyme-mediated integration (REM1) mutagenesis. By this approach, non-pathogenic mutants were produced and the fungal gene required for pathogenicity in *C. acutatum* was identified. Two distinct isolates have been shown to induce Key lime anthracnose (KLA) and postbloom fruit drop PFD diseases of citrus. The gene *KLAP1* is essential for pathogenicity of isolates of *C. acutatum* on Key lime leaves, but apparently not required for flower infection, as the *klap1* – null mutants were able to infect flowers of both Key lime and sweet oranges. Targeted gene disruption of KLAP1 produced mutants that were blocked in the penetration stage and were entirely defective in pathogenicity on Key lime leaves, but remained pathogenic to flower petals. *KLAP1* contains a nuclear targeting sequence and multiple DNA-binding sites and has strong similarity to putative transcription activations. Pathogenicity on Key lime could be restored in a *klap*-null mutant by complementation with full-length *KLAP1* gene clone. The results clearly show that *KLAP1* is an important pathogenicity factor in *C. acutatum* (Chen et al. 2005).

Three G protein α subunit genes-*magA, magB* and *magC* – are required for growth, development and pathogenicity of *Magnaporthe grisea*. Deletion of *magC* resulted in reduced conidiation, whereas disrupting *magB* resulted in reduction in mycelial growth, conidiation and appressorium formation (Liu and Dean 1997). Likewise, the involvement of a G α-protein-encoding gene *ctg1* in controlling conidial germination in *Colletotrichum trifolii* was demonstrated. Inactivation of *ctg1* gene led to the failure of germination of conidia (Truesdell et al. 2000). These results indicate that G protein α-subunit genes may be involved in signal transduction pathways that control several functions such as growth, conidiation, conidium adhesion, appressorium formation, mating and pathogenicity.

The role of histidine kinases (HK) in the conidiation of *Botrytis cinerea* has been investigated. The gene encoding the single group III HK, BOS1 was inactivated. The null mutants of *BOS1*, in addition to changes in osmosensitivity and resistance to fungicides, exhibited additional characteristics. BOS1 was found to be essential for normal macroconidiation and full virulence. In standard culture media, the mutants very rarely formed conidiophores and those few conidiophores did not produce conidia. Application of 1 M sorbitol restored the ability to produce the conidiophores and conidia, suggesting that another BOS1-independent cascade may be involved in macroconidiation. Further, the null mutants had significantly reduced virulence and in planta growth of the pathogen was severely reduced. This report appears to be the first in indicating that this class of HK can be a pathogenicity factor in this ascomycete (Viaud et al. 2005). The signal transduction pathways of *Botrytis cinerea* involved host infection were identified using heterologous hybridization and a PCR-based approach to isolate two genes (*bcg1* and *bcg2*) encoding alpha subunits of heterotrimeric GTP-binding proteins. The gene *bcg2* exhibited sequence similarities to the *magC* gene of *M. grisea* and the *gna-2* gene of *Neurospora crassa*. Both *bcg1* and *bcg2* were expressed at very early stages in infected bean leaves, as revealed by RT-PCR assays. The *bcg1*- null mutants did not secrete extracellular proteases and had reduced pathogenicity on bean and tomato. Although the conidial germination and penetration into the host surface were not altered in *bcg-1-null* mutants, the infection process was inhibited after formation of primary lesions. In contrast, the pathogenicity of *bcg2-* null mutants was reduced slightly. Restoration of characteristics of wild type was possible by complementation of *bcg1* mutants with wild-type gene copy (Schulze Gronover et al. 2001).

The plants have an external barrier, cuticle that can prevent the penetration by the fungal pathogens. This hydrophobic host surface has to be sensed by the fungus, initiating the appropriate signaling pathway(s) for its further development. A protein kinase gene from *Colletotrichum trifolii*, causing alfalfa anthracnose disease, was characterized and designated lipid-induced protein kinase (*LIPK*). Purified plant cutin specifically induced *LIPK* and the inhibitors of the kinase prevented appressorium formation and to a lesser extent, spore germination also. Multiple abnormally branched appressoria were formed, when LIPK was overexpressed. Gene replacement of *lipk* resulted in nonpathogenic strains on intact host surface, but they could colonize through wounded tissue. The results showed that LIPK has a central role in triggering infection structure formation and it is specifically induced by the components of plant cuticle. The pathogen is able to sense the nature of host plant surface inducing the relevant protein-kinase-mediated pathway for its successful infection (Dickman et al. 2003).

By using appropriate mutants of *Colletotrichum lindemuthianum*, causing bean anthracnose, blocked at different stages of appressorial development viz., differentiation, maturation and function, signal exchanges between the pathogen and host plant were studied. The strain *H*18 was blocked at the appressorium differential stage. No genuine appressoria were formed in this strain. The strain *H*191 blocked at appressorium maturation stage, exhibited a pigment defect and developed only partial internal turgor pressure. The third strain *H*290, impaired in appressorium function, could not penetrate into bean tissues. The mutant strains *H*18 and *H*191 with a defect in differentiation of appressorium, produced long, branching germ tubes suggesting a relationship between germ tube length and defective appressorium differentiation.The mutant *H*290 produced nonexpanding lesions, whereas *H*191 strain did not form lesions even on wounded leaves. Likewise, mutant strain *H*18 also was unable to colonize plant tissues when inoculated onto wounded plant surfaces (Veneault-Fourrey et al. 2005).

Several genes essential for pathogenicity have been identified using mutants of *Colletotrichum*. CMK1 from *C. lagenarium*, a homolog of the yeast MAPK KSS1 that regulated invasive growth and activated the transcription factor Ste12p- was identified. A STE12-like gene CST1 was later isolated from *C. lagenarium* (Tsuji et al. 2003). The role of STE12-like gene (*CLSTE 12*) encoding transcriptional factor essential for appressorium-mediated host penetration was studied in *C. lindemuthianum*. A spliced variant of *C. lindemuthianum* whose expression was negatively regulated during early stages of pathogenesis, whereas the correctly spliced mRNA expressed up to the penetration step, suggesting distinct roles for these two transcripts. Comparison of the cell surface protein patterns generated from the wild-type strain and the *Clste* 12Δ mutant revealed the distinct presence of a major protein (Clsp1p9) of 16-kDa in the wild type strain but not in the mutant

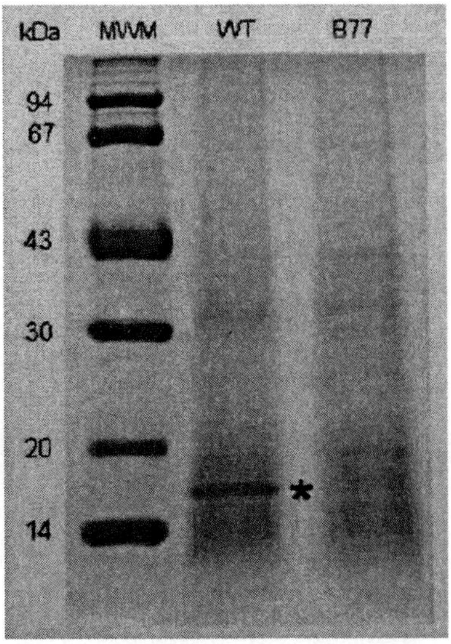

Fig. 2.1 Analysis of cell surface protein patterns obtained from wild-type strain and CLSTE12 mutant B77 of *Colletotrichum lindemuthianum*
Note the presence of a major band (marked with *) only in the wild-type, but not in the mutant B77. (Courtesy of Hoi et al. 2007; Blackwell Publishing, Oxford, UK)

strain (Fig. 2.1). Clsp1p 1p protein belonged to a new family of wall-associated proteins found only in Euascomycetes. The results suggested that the activity of CLSTE 12 could be modulated by a regulated alternative splicing mechanism and this factor may be involved in the production of cell surface proteins and host cell wall degrading enzymes (Hoi et al. 2007).

Calcineurin phosphatase and cyclophilinA are cellular components involved in fungal morphogenesis and virulence. Calcineurin is a highly conserved calcium / calmodulin-regulated phosphatase. The functional enzyme is a heterotrimer composed of a catalytic subunit (calcineurin A, CNA), a regulatory subunit (calcineurin B,CNB) and calmodulin (Dickman and Yarden 1999). Cyclophilins are a conserved family of proteins present in all organisms. The cytosolic cyclophilin A is the cellular primary target of the immunosuppressive drug cyclosporin A (CsA) (Marks 1996). The cyclophilin / Csa complex is a strong inhibitor of calcineurin. The CYP1 cyclophilin was probably involved in the regulation of virulence-related functions including appressorium turgor generation and lipid synthesis in *Magnaporthe grisea.* The gene *CYP1* putatively encodes a mitochondrial and cytosolic form of cyclophilin. The targeted gene replacement showed that *CYP1* acted as a virulent determinant in *M. grisea. cyp1* mutants exhibited reduced virulence and were impaired in associated functions like the penetration peg formation and appressorium turgor generation (Viaud et al. 2002). Among the expressed sequence tags (ESTs) of *Botrytis cinerea* causing gray mold diseases in wide range of crop plant species, 2839 unique genes including putative sequences coding for calcineurin subunits and cyclophilin A have been identified. The *B. cinerea* putative cyclophilin-coding gene was designated *BCP1* and it was inactivated by homologous recombination. The *bcp1Δ* null mutant was able to develop infection structures, but was altered in symptom development on bean and tomato leaves. In contrast, calcineurin inhibition using cyclosporin (CsA) modified hyphal morphology and prevented infection structure formation. Complementation of 2634 bp fragment staining the *BCP1* gene into the *bcp1Δ* mutant restored full pathogenicity. As calcineurin inhibition in *B. cinerea* prevents infection-related morphogenesis, it was considered that calcineurin has a role in the first steps of plant infection (Viaud et al. 2003).

Expression of genes of *Stagonospora nodorum* infecting wheat was studied. *Gna1*, a gene coding for a GI subunit, a key component of signal transduction pathways was characterized. A slight decrease in transcript levels was observed shortly after spore germination, after which levels steadily increased until sporulation. A defect in direct penetration of host plant surface and consequent reduction in pathogenicity of the *Gna1* mutants was revealed. In addition, *Gna1* mutants did not sporulate and secreted one or more brown pigements into the growth medium. There was no such secretion from the wild-type. *Gna1* appears to be the first signal transduction gene to be cloned and characterized from *S. nodorum* (Tan et al. 2004).

For the fungal pathogens like *Sclerotinia sclerotiorum,* sclerotial development is a basic requirement for the progress of pathogenesis. A highly conserved homolog of ERK-type mitogen-activated protein kinases (MAPKs) from *S. sclerotiorum* (*Smk1*) was required for sclerotial development. During sclerogenesis, the *smk1* transcription and MAPK enzyme activity were induced substantially. Sclerotial maturation was impaired by applying inhibitors of MAPK activation. Addition of cAMP inhibited *smk1* transcription, MAPK activation and sclerotial development. The results showed that this pathogen could coordinate environmental signals (such as pH changes) to trigger a signaling pathway mediated by SMK1 to induce sclerotial production and this pathway was negatively regulated by cAMP (Chen et al. 2004). The intracellular cAMP levels significantly influence sclerotial formation in *S. sclerotiorum*, as the cAMP is a key modulator of cAMP-dependent protein kinase A (PKA). The relative PKA activity levels were monitored during sclerotial development. The PKA activity increased during the white-sclerotium stage in the wild-type strain, whereas in non-sclerotium-producing mutants the PKA levels were low. Application of caffeine induced PKA activity and resulted in the formation of sclerotial initial-like aggregates in the mutants in addition to enhancement of PKA activity. The results indicate an important role for cAMP-dependent PKA in the development of sclerotia, the infection structures for the pathogens (Harel et al. 2005).

A basic endoPG isoform was produced early by *S. sclerotiorum* while infecting soybean seedlings. Within one h of PG-cell contact a remarkable level of cell death was recorded. Activation of programmed cell death (PCD), particularly cytochrome C release in cytoplasm and activation of both caspase 9-like and caspase-3 like proteases were detected. Both increase in Ca (^2+) and cell death were annulled following application of PGIP and PG (Zuppini et al. 2005). Calcineurin, a Ser/Thr phosphatase linked to several signal transduction pathways, has a critical role in the regulation of cation homeostasis, morphogenesis, cell-wall integrity and pathogenesis in fungi. In *Sclerotinia sclerotiorum,* calcineurin expression was altered in a phase-specific manner during development of sclerotia. When calcineurin was inhibited by FK506, cyclosporin A or inducible antisense calcineurin expression resulted in impairment of sclerotial development at prematuration phase and increased germination of preformed sclerotia. Reduced pathogenesis on tomato and *Arabidopsis* was due to induction of antisense clacineurin expression in *S. sclerotiorum.* The cell wall β-1,3-glucan content and sensitivity to cell wall-degrading enzymes and glucan synthase inhibitor capsofungin were also significantly affected following inhibition of calcineurin (Harel et al. 2006).

Application of targeted gene inactivation strategy has shown that cell wall degrading enzymes (ten Have et al. 1998; Valette-Collet et al. 2003) and signal transduction components (Zheng et al. 2000; Schulze Gronover et al. 2001; Klimpel et al. 2002) are the virulence factors in *Botrytis cinerea.* This approach, however, involves potential candidate genes. On the other hand, the production of expressed sequence tags (ESTs) has allowed the identification of large sets genes and putative virulence factors in *B. cinerea.* Six endopolygalacturonase genes (*BCPG1-6*) has been characterized in *B. cinerea.* These genes are differentially expressed, when *B. cinerea* is grown in liquid culture on different carbon sources. *BCPG1* and *BCPG2* which encode basic isozymes are constitutively expressed. Galacturonic acid induced the expression of *BCPG4* and *BCPG6.* Low pH of the culture medium was responsible for the induction of expression of *BCPG3* gene. Expression of *BCPG4* induced by galacturonic acid, was repressed by glucose (Wubben et al. 1999, 2000). *BCPG1* is the major transcript with a high basal expression in different media and host plants, while other *BCPG* genes have more complex expression regulation with induction by galacturonic acid or low ambient pH. Furthermore, BCPG1 was required for full virulence (ten Have et al. 2001, 2002). In a later investigation a cDNA library was produced from *B. cinerea* grown in axenic culture. From this library, 6559 ESTs were produced. Only BCPG1 was present in the ESTs and *BCPG1* gene seemed highly expressed (37 ESTs). A signaling pathway depending on calcium and calcineurin was identified in the EST set. This pathway is involved in fungal morphology and virulence. The EST set also provided the *BcPLS1* gene, the homolog to *MgPLS1* that is necessary for appressorium-mediated penetration of *M. grisea* into host plants. *BcPLS1* and *MgPLS1*- null mutants had similar penetration defects (Viaud et al. 2005).

Analysis of ESTs produced from the cDNA libraries of *Magnaprothe grisea*, representing a variety of growth conditions and cell types showed that about 23,000 of the ESTs could be clustered into 3050 contigs, leaving 5127 singleton sequences of the pathogen DNA. Analysis of ESTs provided the criteria for identification of fungal genes involved in pathogenesis. A family of metallothionein present in *M. grisea* was identified (Ebbole et al. 2004). *M. grisea* Guy 11 was transformed with a promoterless enhanced green fluorescent protein (EGPP) construct to isolate hygromycin-resistant transformants. The transforming vector was inserted in a predicted gene designated *MIR1* and resulted in a Mir11-107-EGFP fusion. Quantitative real-time RT-PCR analysis and expression assays of *MIR1-EGFP* fusion constructs indicated that the expression of *MIR1* was highly induced during rice plant infection. *MIR1,* a specific gene of *M. grisea* was found to be highly conserved among the field isolates belonging to *M. grisea* species complex. No deffects in appressorial penetration and rice infection could be seen in the *mir1* mutants.

When overexpressed with RP27 promoter, nuclear localization of the Mir1-EGFP fusion was detected in conidia and vegetative hypha. The results suggested that the reporter genes based on *MIR1* could be employed for monitoring infectious growth in *M. grisea* (Li et al. 2007).

The temporal expression of the antioxidant, the putatively secreted large subunit catalase *CATB* gene was investigated in order to find out whether *M. grisea* could counteract the toxic burst of H(2)O(2) localized beneath the site of attempted penetration. The *CATB* gene was very highly up-regulated (600 fold) in vivo coincident with penetration and moderately up-regulated, in response to exogenous H(2)O(2). The *catB* mutant obtained by targeted gene replacement of *CATB*, showing growth abnormalities was severely less pathogenic than Guy 11 on barley and rice. In addition, exposure to H_2O_2 resulted in further reduction in pathogenicity. Complement of the *catB* mutant with *CATB* gene restored the wild-type phenotype. There appeared to be no evidence suggesting a role for CATB in detoxification of the host derived H(2)O(2) at the site of penetration by *M. grisea*. On the other hand, strengthening of the fungal wall by CATB may help *M. grisea* during forceful entry into the host tissue (Skamnioti et al. 2007).

A number of pathogenicity genes in plant pathogenic fungi has been identified by applying the targeted gene disruption strategies. The fungal pathogens are capable of sensing changes in the environment and respond and adapt to new situations. The mitogen-activated protein kinases (MAPKs) are involved in the transduction of a variety of extracellular signals and in the regulation of growth and development. *Mycosphaerella graminicola* (anamorph – *Septoria tritici*), a nonappressorium-forming wheat pathogen penetrates the leaves through stomata without differentiating into appressorium, an infection structure formed by many fungal pathogens such as *Magnaporthe grisea*. An MAPK-encoding gene in *M. graminicola* strain IPO323 with high homology to the orthologous *Fus3* gene of *Saccharomyces cerevisiae* was identified and it was designated *MgFus3*. Several developmental processes were altered in the *MgFus3* mutants. They showed neither melanization of hyphae nor produced pycnidia as the wild-type strain. Further, the *MgFus3* mutants failed to produce any lesions even after prolonged incubation of up to 34 days, whereas necrotic lesions and pycnidia were formed in leaves at 11 days post-inoculation with wild-type strain. Histological investigations revealed that *MgFus3* mutants did not penetrate stomata, due to impaired recognition by germ tubes which grew along stomatal slits without penetrating them resulting in failure of mesophyll colonization by the mutants (Fig. 2.2). The results indicated that the *MgFus3* mutants were nonpathogenic and *MgFus3* can be considered as a multifunctional pathogenicity factor for *M. graminicola* (Cousin et al. 2006).

The phytopathogenic fungi secrete complex extracellular matrices (ECMs) in the early stage of infection, especially the ones associated with conidia. Visual evidence for the presence of an amorphous material attaching the germ tubes of *Stagonospora nodorum*, causing leaf and glume blotch disease of cereals, to the wheat leaf surface at 3–9 days post-inoculation (Karjalainen and Lounatmaa 1986). Three major groups of ECMs have been recognized: (i) pre-formed, non-secreted conidial ECMs which are closely associated with the outer layer of the conidial

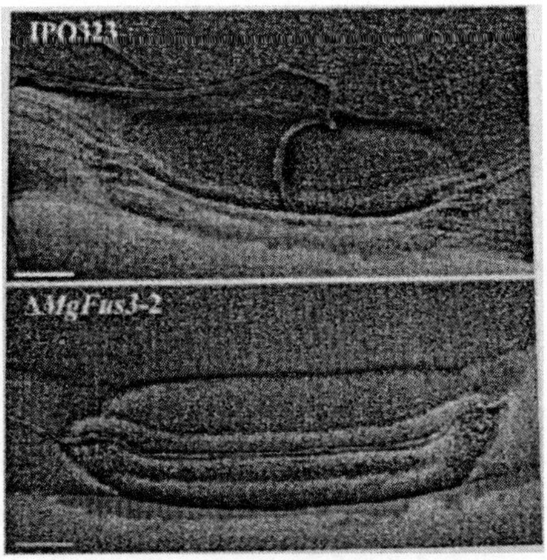

Fig. 2.2 Scanning electron micrographs depicting penetration behavior of *Mycosphaerella graminicola* IPO323 and ΔMgFus3-2 on primary leaves of wheat cv. Obelisk
Note the successful stomatal penetration by wild-type strain (*top*) and penetration failure of the mutant strain (*bottom*). (Courtesy of Cousin et al. 2006; Blackwell Publishing, Oxford, UK)

cell as in *Colletotrichum lindemuthianum* (Hughes et al. 1999); (ii) preformed, nonsecreted conidial ECMs which require no metabolic activity and are released rapidly upon hydration or contact with a surface as in the uredospores of *Uromyces viciae-fabae* (Deising et al. 1992) and (iii) actively secreted ECMs which include all ECMs that require metabolic activity for secretion as in *Peronospora parasitica* (Carzaniga et al. 2001).

Fungal pathogens are known to produce host-specific toxins (HSTs) that facilitate them for colonization of host tissues. The formation of such a HST by *Alternaria brasicola* at very early stage of infection during spore germination has been reported. The AB-toxin was released from the conidia of *A. brassicola* when they germinated on *Brassica* plants, but not on nonhost plants, plastic plates or culture media (Otani et al. 1998). Likewise, *A. panax* was also found to produce a protein HST during conidial germination (Quayyum et al. 2003). The AB-toxin production by *A. brassicola* was studied in detail. The spore germination fluids (SGFs) contained AB-toxin as revealed by leaf necrosis assay. The AB-toxin was released in SGFs incubated for 12 h on host leaves and the toxin production increased with increase in incubation period. The SGFs were concentrated 10-fold and analyzed by sodium dodecyl sulphate (SDS)-polyacrylamide gel electrophoresis (SDS-PAGE) technique. One protein band (38 kDa) was detected in the SGF supplied with active fraction. A few additional bands other than the most abundant, 38-kDa protein were also present. A host derived factor that induced AB-toxin production was found to be released from host leaves after spore germination had commenced. The purified compound was shown to be an oligosaccharide of 1.3 kDa (Oka et al. 2005).

Cutinase produced by fungal pathogens capable of penetrating the plant surface directly, may have a role in the initiation of infection. The cutinase activity of a mutant of *Nectria haematococca* with reduced pathogenicity showed marked reduction in the cutinase activitiy (80%–90%). *Mycosphaerella* sp. (a wound pathogen of papaya fruits) transformed with the cutinase gene of *N. haematococca,* was able to infect unwounded papaya fruits, indicating an increase in the virulence of

Mycosphaerella sp. (Dickman et al. 1989). However, the mutant of *N. haematococca* produced by transformation-mediated gene disruption, deficient in cutinase, was as pathogenic as the wild-type despite the loss of cutinase activity (Stahl and Schäfer 1992). Likewise, the presence of cutinase activity appears to be unimportant for infection of rice by *M. grisea.* The transformants generated by targeted gene disruption were pathogenic on three host species (Sweigard et al. 1992). Expression of cutinase A (*cutA*) in *Botrytis cinerea* during early stages of infection of gerbera flowers and tomato fruits could be detected. The cutinase A-deficient mutants of *B. cinerea* were able to penetrate and induce symptoms on gerbera flowers and tomato fruits, as efficiently as the wild type strains (van Kan et al. 1997). The loss of cutinase activity in the isolates of *Fusarium solani* f.sp. *cucurbita* did not affect their pathogenicity on cucurbit fruits (Crowhurst et al. 1997). In contrast, the requirement of cutinase for cuticular penetration of oilseed rape by *Pyrenopeziza brasicae* causing light leaf spot disease was reported. The single copy cutinase gene *Pbc1* was disrupted by a transformation-mediated approach. The absence of cutinase activity in culture supernatants from the mutant was confirmed by using p-nitrophenyl butyrate as the substrate. In addition, entry into the host by direct penetration by the wild type, but not by the mutant, was visualized by scanning electron microscopy (SEM). However, demonstration of the requirement of cutinase gene for pathogenicity by complementation of the mutant with the *Pbc1* leading to restoration of pathogenicity, will provide strong support to conclusion on the requirement of cutinase for penetration of cuticle (Li et al. 2003). The role of cutinase as a pathogenicity factor may be important in certain pathosystems, but many pathogens may produce different enzymes acting synergisticaly making the presence of cutinase as not a critical factor for pathogenicity.

Several fungal pathogens are known to secrete cell wall-degrading enzymes. In *Cochliobolus carbonum* infecting maize, no change in pathogenicity was noted, following disruption of the activity of a pectin-degrading polygalacturonase (PG) (Scott-Craig et al. 1990). In contrast, the saprophytic growth and host invasion by *Colletotrichum lindemuthianum* require essentially the endopolygalacturonase (endo PG) activity under the control of the genes *CLPG1* and *CLPG2*. By using specific antibodies raised against the protein encoded by *CLPG1*, it was possible to detect this protein in planta in infected plants. This protein was associated in tissues with extensive degradation of host cell wall. Application of RT-PCR technique revealed that *CLPG1* was expressed at the beginning of the necrotrophic stage of infection, indicating a major role for *CLPG1* in secretion of endoPG by *C. lindemuthianum* (Centis et al. 1997). The requirement of the gene *Bcpg1* for the full virulence of *Botrytis cinerea* for infection of tomato leaves was demonstrated by ten Have et al. (1998). The mutants of *B. cinerea* obtained by elimination of the *Bcpg1* that encodes an endoPG, by partial gene replacement could initiate primary infection as the wild type. But appreciable reduction in the secondary infection by mutant was recorded, indicating *Bcpg1* gene is essential for rapid invasion of host tissues. Likewise, the importance of a single gene *Pat1* with major effects for pathogenicity of *Botryotinia fuckeliana* was indicated by the report of Weeds et al. (1999).

The antibodies raised against the epitope present in pectin, polygalacturonic acid, xyloglucan and callose were used to study the modification of cell wall components during penetration of epidermal cell walls by *Uromyces fabae,* causing cowpea rust disease. The density of pectin and xyloglucan epitopes was found to be reduced, suggesting that the pathogen may degrade the plant cell wall at the penetration site by secreting the respective enzymes (Xu and Mendgen 1997). The stage-specific secretion, development and ultrastructure of extracellular matrix (ECM) sheaths by conidia and germ tubes of *Stagonospora nodorum* was investigated, during the first six hour post-inoculation in wheat leaves. A combination of immunological, histochemical and ultrastructural procedures were applied. One antibody (SN-MG11) specifically recognized a conidial surface protein, while the second antibody (SN-CH9) reacted with a carbohydrate epitope present on an antigen in the ECM and the germ tube cell wall. The lectins tested showed that the cell walls of germlings were rich in α-mannose and α, β, N-acetyl glucosamine. Three major phases of ECM released on the leaf surface were recognized: (i) commencing less than 15 min after contact of conidium with the host, composed of a carbohydrate core and protein halo; (ii) coinciding with the emergence and growth of germ tubes, composed of proteins and carbohydrates and (iii) spreading around the conidia after germ tube production, not seen on the artificial surfaces. The release of the protein-based ECM, spreading around ungerminated spores of *S. nodorum* in phase 1 may be triggered by a contact stimulus, since the same pattern was observed on nonhost and host surfaces. Immunogold labeled cross sections of germinated conidia indicated that larger amounts of dense ECM were secreted from the germ tubes compared with that secreted from conidia (Zelinger et al. 2004) [Appendix 1].

Mycosphaerella graminicola infecting wheat penetrates the host indirectly through stomata. The mitogen-activated protein kinase (MAPK) *MgFus3,* a homolog of *Saccharomyces cervisiae Fus3* was identified and it was required for penetration and MAP kinase gene *MgMK1* was shown to be essential for pathogenicity (Mehrabi et al. 2004). The MAP kinase pathways, also known as extracellular signal-regulated kinase pathways are involved in the transduction of a variety of extracellular signals. A second MAP kinase in *M. graminicola* designated MgSlt2, a homolog of the *S. cerevisiae Slt2* gene was isolated and characterized. The expression of *MgSlt2* was studied along with β-tubulin as control, by RT-PCR under in vitro and in planta conditions, resulting in a specific 757-bp amplification product. Both genes showed similar expression patterns. The *MgSlt2* knockout strains in *M. graminicola* isolate IPO323 were generated, while *MgFus3* was found to be essential for penetration. *MgSlt2* knockouts could penetrate the host, equally well as the wild type. The infection hyphae were present in substomatal cavities and intracellular areas of the mesophyll, suggesting that *MgSlt2* was not involved in early recognition events. However, infectious hyphae of the mutant were not able to branchout and colonize the mesophyll. This resulted in highly reduced virulence and prevented pycnidia formation. It was concluded that *MgSlt2* is a new pathogenicity factor in *M. graminicola* possibly required for tissue colonization (Mehrabi et al. 2006).

Fusarium graminearum, causal agent of corn ear rot disease, is important because of the significant yield loss as well as the health hazards due to consumption of grains contaminated with the mycotoxins produced by this pathogen. Trichothecene toxins produced by *Fusarium graminearum* (*F*g complex) act as virulence factors on some hosts. Strain-specific differences in trichothecene metabolite profiles (chemotypes) are not well correlated with the *Fg* complex phylogeny. A 19 kb region of trichothecene gene cluster was sequenced in 39 strains chosen to represent global genetic diversity of species in the *Fg* complex and four related species of *Fusarium*. Phylogenetic analyses of the sequences of this fragment (19-kb) showed that polymorphism within these virulence-associated genes has persisted through multiple speciation events and appears to have been maintained by balancing selection acting on chemotype differences that originated in the ancestor of extant species within the B-trichothecene lineage of *Fusarium* (Ward et al. 2002).

Many molecular investigations focused on trichothecene biosyntehsis (in *F. graminearum*) and population structure (Brown et al. 2004; Cumagan et al. 2004; McCormick et al. 2004). The trichodiene synthetase gene (*TR15*), as a virulence factor was the earliest to be characterized. The *tri5* deletion mutants, though had normal growth and development, exhibited reduced virulence on Wheaton wheat, common winter rye and maize. Deoxynivalenol (DON) (the mycotoxin) production was shown to be important for allowing the pathogen to spread within colonized spikes (Bai et al. 2002). In addition to *TR15*, two MAPK genes, *MGV1* and *GPMK1* were demonstrated to be essential for pathogenicity of *F. graminearum* (Hou et al. 2002; Jencmionka et al. 2003). Furthermore, disruption of the homolog of *Cochliobolus heterostrophus CPS1*, a novel virulence factor, also resulted in reduced virulence in *F. graminearum* (Lu et al. 2003). Restriction enzyme-mediated integration (REM1) mutagenesis demonstrated to be an efficient approach for the identification of pathogenicity genes was applied for generating mutants of *F. graminearum*. Eleven pathogenicity mutants defective in colonizing corn silks and flowering wheat heads were isolated. Genetic analyses indicated that defects in pathogenicity were tagged by the transforming vector in six mutants. The *cbl-1* and *msy1* deletion mutants were methionin auxotrophic and had significantly reduced virulence on corn silks and wheat heads, indicating that methionine synthesis is critical for plant infection by *F. graminearum*. In addition, two genes, the putative b-ZIP transcription factor gene and the transduction β-subunit-like gene disrupted in mutants M7 and M75 respectively appeared to be novel virulence factors in *F. graminearum* (Seong et al. 2005).

The biotrophic fungal pathogens causing powdery mildews, downy mildews and rusts adopt different strategies for infection which may be similar to necrotrophic pathogens in some respects. Production of cellulases by *Erysiphe graminis* f.sp. *hordei* (*Egh*) causing barley powdery mildew, as in the case of *Phytophthora infestans*, was demonstrated during penetration of the plant surface. By applying immunofluorescence (IF) using two MABs specific to cellobiohydrolases, antigen localization at the germ- tube tips of *Egh* was revealed. Cellobiohydrolase I was present at the primary germ- tube tip, whereas cellobiohydrolase II was detected at the appressorial germtube tip. A combination of enzyme activity and mechanical pressure exerted by the appressoria of *Egh* led to the penetration of host surface (Pryce-Jones et al. 1999). The modification of cell wall components during penetration of epidermal cell walls by *Uromyces fabae*, causing cowpea rust disease, was investigated. Antibodies raised against pectin, polygalacturonic acid, xyloglucan and callose were used. The density of pectin and xyloglucan epitopes were reduced, suggesting that this pathogen could degrade the plant cell wall at the penetration point by secreting appropriate enzyme (Xu and Mendgen 1997).

Various fungal genes are expressed in different infection structures during early stages of pathogenesis. Haustoria required for absorption of nutrients by the fungal pathogens, are formed in host cells. As many as 31 different in planta-induced genes (*PIGs*) have been identified in the haustoria of *Uromyces fabae*. Some of the *PIGs* were highly expressed and were present as a single or low copy number genes in the rust pathogen genome (Hahn and Mendgen 1997). By screening a cDNA library constructed from haustorial DNA of *U. fabae,* the *PIG2* gene encoding a protein with high homologies to fungal amino acid exporters was isolated. The putative amino acid exporter protein, formed following expression of *PIG2* mRNA was localized to the plasma membrane of the haustoria, as revealed by the immunofluorescence (IF) microscopic observation. The involvement of haustoria in the uptake of nutrients by rust pathogens was confirmed by the molecular evidence through IF microscope technique (Hahn et al. 1997).

The haustoria of biotrophic fungal pathogens may perform another function, in addition to their involvement in absorption of nutrients from host cells. They may be actively involved in establishing and maintaining the biotrophic relationship. A rust transferred protein 1 from *U. fabae* (*Uf*-RTP1p) was detected in the host parasite interface, the extrahaustorial matrix as well as inside the infected plant cells by IF and electron microscopy. A homolog of *Uf*-RTP1p was identified in *U. striatus* (*Us*-RTP1p). The presence of these two proteins inside infected plant cells was detected by employing four idependently raised polyclonal antibodies. The concentration of *Us*-RTP1p increased depending on the developmental stage of haustoria and it was detected in the host cell nucleus later. Putative nuclear localization signals (NLS) could be observed in the predicted RTP1p sequences. Western blot analysis indicated that *Uf* –RTP1p and *Us*-RTP1p possibly gain entry into the host cell as N-glycosylated proteins (Kemen et al. 2005). The genes encoding haustorially-expressed secreted proteins (HESPs) were identified by screening a flax rust pathogen (*Melampsora lini*) haustorium-specific cDNA library. Among 429 unigenes, 21 HESPs were identified, one of which corresponded to the AvrL 567 gene. Three other HESPs cosegregated with independent AvrM, AvrP4 and AvrP123 loci (Catanzariti et al. 2006).

Blumeria (*Erysiphe*) *graminis* f.sp. *hordei* (*Bgh*) an obligate biotrophic fungal pathogen causes powdery mildew disease in small grain cereals. The pathogen remains ectophytic. The penetration peg formed under the appressorium [10–12 h after inoculation (ai)] breaches the culticle and epidermal cell wall, swells within the host cell (12–15 hai) differentiating into a haustorium with numerous digitate processes for absorbing nutrients from the host cells. Two distinct host responses may arrest fungal development even in susceptible plants. A defensive papilla formed by the plant epidermal cell may prevent penetration. Effective papilla defense leads to increase in the ability of cells adjacent to the attacked cells to form papillae in response to subsequent attacks (Lyngkjaer and Carver 2000) indicating the existence of intercellular communication in a consequence of the response. Papilla-based defences can be suppressed by MLO calmodulin-binding membrane protein (Kim et al. 2002a). The second type of response is, enhanced epidermal cell death occurring more frequently in resistant genotypes of barley than in susceptible ones. Plants with major gene resistance such as with alleles at the *Mla* locus will elicit localized HR (Panstruga and Schulze-Lefert 2002). In *Mla1* barley epidermal HR is seen either prior to or soon after haustorium formation (Zeyen et al. 1995).

Stomata function as important regulators of plant interactions with their environment and their movements control transpiration. The relationship between stomatal behavior and powdery mildew development in susceptible barley has been studied. The disrupted stomatal behavior and subsequent leaf water conductance(g_1) were assessed following inoculation of *Bgh* on barley genotypes Pallas and Risø-S (susceptible), PO1 (with *Mla1* conditioning a HR) and P22 and Risø-R (with *mla5* conditioning papilla-based penetration resistance). In susceptible lines stomata closed in darkness after infection, but opening in light was persistently adversely affected, as revealed in cryofixed leaf segments by using scanning electron microscope images. In the inoculated Pallas leaves, none of the stomata was fully open, a few were partially open (16%) and the majority (85%) were closed. In contrast, in healthy leaves virtually all stomata were fully open (96%) (Fig. 2.3). On the other hand, stomata recovered nearly complete function by about 30 hai. In PO1, stomata became locked open and were unable to close in darkness, soon after epidermal cells died due to HR (Prats et al. 2006).

Studying *Blumeria graminis* f.sp. *hordei* (*Bgh*) infection process has been found to be difficult, since appressoria and haustoria are embedded in the barley epidermal cell layer. By applying the differential display technique, the problem could be overcome and the fungal genes involved in the early stages of infection could be identified. Of the 33 cDNA fragments isolated, nine were shown to represent differentially expressed *Bgh* genes by RT-PCR technique. The cDNAs that were specifically up-regulated during the formation of primary appressoria and haustoria were designated *Barley-induced genes* (BIGs). The cDNA and genomic clones of three novel genes, *BIG1 BIG2* and *BIG4* were characterized. They were predicted to encode secreted proteins and to be specific for the *Bgh*-barley interaction (Grell et al. 2005).

A novel PCR-based technique was developed for the amplification of multiple copies of rDNA sequences in a single conidium of the powdery mildew pathogen. A number of transcripts in a single conidium of *Bgh* infecting the coleoptile epidermis were the potential targets for in situ cDNA synthesis (Matsuda et al. 1997). Later micro-needle technique for the removal of cellular contents from target single cells, was used. The isolated conidia were subjected to PCR amplification of the 5.8S rDNA and its adjacent ITS sequences followed by nested PCR to attain sensitivity high enough to amplify target nucleotide sequences. It was possible to amplify transcripts expressed in single conidia. Conidia at pre- and post-germination were

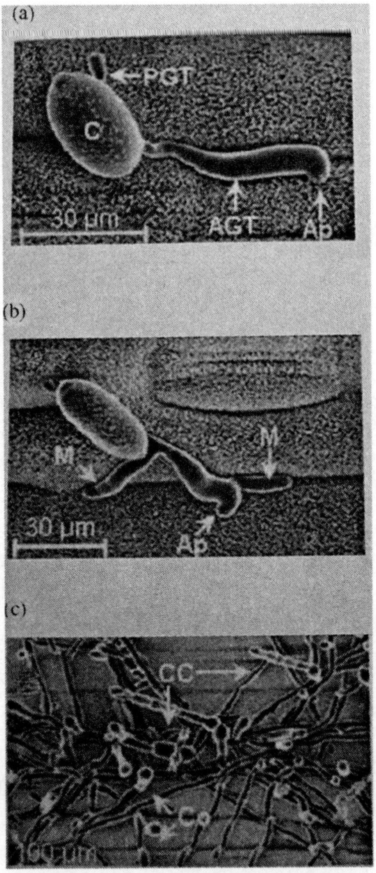

Fig. 2.3 Cryo-SEM, transmitted light and fluorescence (blue light elicitation) micrographs of key stages in the development of *Blumeria graminis* var. *hordei* (*Bgh*) and barley cell responses
(a) Appressorium (Ap) formation- emergence of primary germtube (PGT) and appressorial germtube (AGT) from *Bgh* conidium at 5 hai; **(b)** Mycelial and haustorial growth- development of colonies at 30 hai with ectophytic mycelium (M) from the appressorium; **(c)** Formation of extensive mycelium producing conidiophores (Co) and chains of conidia (CC) at 96 hai. (Courtesy of Prats et al. 2006; Oxford University Press, Oxford, UK)

tested. The β-tubulin homolog *TUB2-ol* was expressed at pre- and post-germination stages and the cutinase homolog *CUTI-ol* was only expressed post-germination. Combination of observation under digital microscope and two-step PCR amplification has the potential for use in the investigation of individual propagules on the plant surface (Matsuda et al. 2005).

The transcript profiles of *B. graminis* f.sp. *hordei* (*Bgh*) were analyzed by high density cDNA microarrays (2027 unigenes), throughout asexual life cycle of- and development of infection by *Bgh*. RNA was obtained from four stages preceding penetration and four stages after penetration of the host cells. Presence of global switch in the expression between the pre-and post-penetrative stages was discernible. Accumulation of RNA-encoding protein biosynthesis genes was observed in the late stages. Other functional clusters such as virulence-related genes and sterol metabolism genes were found to be upregulated in pre-and penetration stages respectively. The expression of *cap20* was related to the abundance of a group of RNAs and they may be pathogenicity factors in *Bgh*, as in the case of *Colletotrichum gloeosporioides* requiring *cap20* for its virulence (Both et al. 2005).

Some fungal pathogens belonging to Oomycetes have a biotrophic mode of nutrition that requires access to living host plant cells at an early stage in the development of infection as in most *Phytophthora* spp. All in Peronosporaceae (downy mildews) and Albuginaceae (white rusts) are obligately biotrophic in nature and cannot be cultured in cell-free media easily. Oomycetes share with many bacterial and fungal pathogens some characteristics in the phenomenon of pathogenesis. Oomycetes accomplish parasitic colonization of plants by modulating host cell defenses through an array of disease effector proteins. The effectors may be defined as molecules that manipulate host cell structure and function, thereby facilitating infection (virulence factors / toxins) and / or triggering defense responses (avirulence factors / elicitors). The dual functions of effectors have been elucidated more clearly in the case of plant-bacterial pathosysems (Volume 2, Section 2.2). Two classes of effectors targeting distinct sites in the host plant have been recognized. The apoplastic effectors secreted by the pathogen into the plant extracellular space, interact with extracellular targets and surface receptors. The cytoplasmal effectors are translocated inside the plant cell, possibly through specialized structures like infection vesicles and haustoria that invaginate inside the living host cells (Kamoun 2006).

The effectors, in most cases, have been identified by their ability to activate defense responses and innate immunity in the host plant. The effector proteins have to be secreted by the oomycete so that they can interact with the cellular targets at the intercellular interface between the plant and the pathogen or inside the host cell. Most of the secreted proteins are exported through the general secretory pathway via short, N-terminal amino-acid sequences designated signal peptides. The PexFinder (*P*hytophthora *ex*tracellular protein) algorithm was applied to predict the secreted proteins in the culture filtrates of *P. infestans*. The prediction correlated with the proteomic identification with a high degree of accuracy (Torto et al. 2003).

The oomycetes such as *Phytophthora* have evolved a mechanism, like bacterial pathogens, to escape the enzymatic activity of pathogenesis-related (PR) proteins such as glucanases, chitinases and proteases. These pathogens contain little chitin in their cell wall and hence they are not affected by plant chitinases. In addition, they have evolved active counter defense mechanisms by secreting inhibitory proteins that target host glucanases and proteases (Kamoun 2003). *P. sojae* secreted the glucanase inhibitors GIP1 and GIP2 that could inhibit the soybean endoβ-1,3-glucanase EGase A (Rose et al. 2002). The serine protease inhibitor EPI1, a protein secreted by *P. infestans* was shown to protect several secreted proteins of this pathogen (Kamoun 2006). Another class of secreted protease inhibitors of *P. infestans* are EPIC1 and EPIC2. These two inhibitors target an apoplastic papain-like cysteine protease of tomato. Elicitin such as INF1 to INF7 form a family of structurally related proteins capable of inducing HR in test plants. All elicitin genes encode putative extracellular proteins that share a 98-amino acid elicitin domain with a core of six conserved cysteines. Strains of *P. infestans* deficient in the elicitin INF1 produced lesions in *Nicotiana benthamiana* leaves suggesting that INF1 was involved in conditioning avirulence in this plant species (Kamoun et al. 1998). The studies on the biological functions of elicitins in *Phytophthora* demonstrated that class I elicitins can bind sterols such as ergosterol and function as sterol-carrier proteins. As *Phytophthora* spp. cannot synthesize sterols, elicitins appear to have biological function of essential importance to these pathogens (Mikes et al. 1997, 1998). The secreted proteins include avirulence (AVR) proteins that activate the defense response of the plant species. There appears to be a dynamic evolutionary battle between host plant and pathogen to achieve and evade detection respectively resulting in a molecular "arms race" (Birch et al. 2003).

The pathogens infecting roots adopt different strategies for successful infection leading to disease development, as the undergound plant parts show variations in structure and functions from that of aerial plant parts. *Fusarium oxysporum* f.sp. *lycopersici* causing wilt disease in tomato produces endopolygalacturonase encoded by the gene *pg1*. Expression of *pg1* in roots and lower stems of infected plants was detected by applying RT-PCR assay. The *pg1* gene was cloned and sequenced. The transformed isolates of *F. oxysporum* f.sp. *melonis* deficient in *PG1* were as pathogenic as the wild type, indicating the endoPG is not essential for infection of plants (Pietro and Roncero 1998). The presence of cell wall components, cellulose, xylan, pectin and lignin in the roots of healthy wheat plants and those infected by *Gauemannomyces graminis* var. *tritici* was detected by employing enzyme-gold and immunogold labeling techniques. The labeling densities for cellulose, xylan and pectin in the cell walls of infected roots were significantly reduced compared to those of healthy tissues. Degradation of the cell wall components used as substrates for the activities of the cell wall-degrading enzymes was indicated by labeling intensities. Quantification of labeling densities of lignins showed only slight increase in the cell walls of infected roots, whereas greater deposition of lignin was noted in the cell walls of healthy roots indicating the successful initiation of infection by the root infecting pathogen (Kang et al. 2000).

Studies on the changes in the chitinases present in the plants during pathogenesis can be expected to throw light on the complex interaction between fungal pathogens and plants, since the fungal cell walls contain chitin. The melon cultivars Galia (susceptible) and Bredor (resistant) were examined at different periods of inoculation with *F. oxysporum* f.sp. *melonis*. By employing an antibody against a class III chitinase, the presence of this enzyme was traced by both Western blotting analysis and immunolocalization. The Western blotting detected constitutive expression of chitinase III in stem base tissues of the susceptible cultivar, Galia. Three bands (26, 30 and 35 kDa) were detected prior to infection. Upon infection two bands (30 and 35 kDa) became faint and the third one also was markedly reduced. Immunolocalization experiments showed that chitinase could be seen in the intercellular spaces surrounding a few isolated cells or cell groups dispersed through cortical region in both susceptible and resistant cultivars. Chitinase become undetectable in susceptible cultivar after eleven days after infection, whereas chitinase could be detected till 18 days after infection in resistant Bredor melon (Table 2.1) (Baldé et al. 2006).

The knowledge about the signal exchange between plant roots and pathogenic fungi is only fragmentary. Both seed and root exudates may stimulate the germination of spores (Nelson 1991, 2004). Infection by microbial pathogens is generally initiated just behind root apex in the zone of elongation. *Nectria haematococca*

Table 2.1 Detection of chitinase in the intercellular spaces of stem sections of susceptible (Galia) and resistant (Bredor) melon cultivars at different periods after inoculation with *Fusarium oxysporum* f.sp. *melonis*

Cultivar	Treatments	Number of days after inoculation						
		4	7	9	11	14	16	18
Susceptible (Galia)	Infected	++	−	+	++	−	−	*
	Control	++	−	−	−	−	−	−
Resistant (Bredor)	Infected	++	+	+	++	+	++	+++
	Control	++	−	−	−	−	−	−

(+) Root cortical tissue with a few labeled isolated cells
(++) Root cortical tissue with a few labeled cells or groups of cells
(+ + +) Several cells with abundant signal in the central cylinder
(−) none of the cells showing label
(*) dead plants
Source: Baldé et al. (2006)

(*Fusarium solani* f.sp. *pisi*) causing foot and root rot of pea, has been a model for studying some aspects of root infection. Flavanoid-induced spore germination was partially inhibited by H89, a cAMP-dependent inhibitior of protein kinase A (Ruan et al. 1995). A mutant of *N. haematococca* exhibited reduced virulence due to significant reduction (80%–90%) in its cutinase activity. However, loss of cutinase activity in the cutinase-deficient mutants of a highly virulent strain of *N. haematococca* generated by transformation-mediated gene disruption, did not reduce the pathogenicity (Stahl and Schäfer 1992).

The root tips are considered to be protected from fungal infection by the formation of root border cells (Gunawardena and Hawes 2002; Woo et al. 2004). In pea-*N. haematococca* pathosystem differential spore germination and pathogenicity gene expression or defense gene expression in relation to localized root infection were investigated. Infection in susceptible pea was initiated predominatly in the zone of elongation, whereas the remainder of the root was found to be resistant. The failure to infect root tip was not due to a failure to induce spore germination at this site, or suppression of pathogenicity genes in the fungus or increased expression of plant defense genes. On the other hand, exudates from the root tip induced rapid spore germination by a pathway that is independent of nutrient-induced germination. Then, a factor produced during infection and death of border cells at the root apex seemed to selectively suppress fungal growth and prevent sporulation. Host-specific mantle formation in response to border cells appears to represent an unrecognized form of host-parasite relationship common to diverse species (Gunawardena et al. 2005).

2.1.3 Colonization of Host Tissues

The fungal pathogens colonize the host tissues, after penetration, by producing different kinds of enzymes or toxins or by preventing development of resistance of the host tissues by adopting different strategies depending on the pathosystems. The successful colonization results in increase in the resistance / susceptibility levels of the host plant species to the pathogen concerned. A sensitive method of quantifying fungal and plant DNA in *Arabidopsis thaliana* leaves infected by *Alternaria brassicola* and *Botrytis cinerea* was developed based on real-time quantitative PCR protocol. This procedure allowed to monitor disease progression by quantifying fungal abundance reliably from the very beginning of the infection time course. The relative abundance of the cutinase A and ASK genes was quantified over a wide dynamic range (6.25 pg-16.7 ng DNA), providing excellent sensitivity to the real-time PCR technique which is very fast and economical. Furthermore, this protocol allowed to score the outcome of both compatible and incompatible interactions between *Arabidopsis* and the fungal pathogens which led to dramatic differences in plant fungal colonization (Gachon and Saindrenan 2004).

2.1.3.1 Enzymes Produced by Fungal Pathogens

The enzymes elaborated by fungal pathogens play important role in different stages of pathogenesis. Different types of primary cell wall constituents are present in higher plants. Pectin forms a major structural component of the middle lamella of primary cell walls and it consists of rhamnogalacturonan and homogalacturonan. The enzymes produced by the fungal pathogens degrade the pectin by depolymerization (as in the case of polygalacturonases, pectin and pectate lyases) or altering its structure (as done by pectin acetylesterases and pectin methylesterases) (Kars and van Kan 2004). In general optimal activity of fungal pectin methylesterases (PMEs) occurs between pH4 and 6, whereas plant and bacterial PMEs peak between pH6 and 8 (Alebeek et al. 2003). Molecular studies of PMEs have been useful for characterization of several related genes in the pathogens as well as in plants.

The growth of the fungal pathogen *Botrytis cinerea* in planta has been shown to be accompanied by degradation of pectin and endoPG genes (*Bcpg*) are expressed during invasion of different plant tissues. The importance of endoPGs for virulence of *B. cinerea* was indicated by ten Have et al. (1998) and Kars et al. (2005a). A purified glycoprotein from *B. cinerea* (strain T4), identified as endoPG1 (T4BcPG1) has been found to be both a virulence and avirulence factor that activates defense reactions in grapevine cv. Gamay (Poinssot et al. 2003). Observations on the phenotype of the *Bcpme1* mutant suggested that PME action preceding pectin deploymerization was essential for its virulence (Valette-Collet et al. 2003). The expression in planta of *B. cinerea* genes *Bcpme1* and *Bcpme2* was analyzed in a time-course experiment. The transcripts of both genes were detected 24 hpi and increased during the course of infection process in a manner similar to *Bcpg1* and the constitutively expressed BcactA which was considered as a measure of fungal biomass (ten Have et al. 1998; Benito et al. 1998). *B. cinerea* produces four exo-PGs. The PABs produced against one of the exoPGs (60 kDa) recognized two exoPGs with molecular mass of 66 and 70 kDa. Expression of exo-PGs in cucumber leaves inoculated with conidia of *B. cinerea* was studied using immuno-histochemical analysis. The concentration of exo-PGs increased progressively at 9 h after inoculation, indicating that exo-PGs may be required at early stage of infection and also for tissue maceration in infected plants (Rha et al. 2001). A functional analysis of the two *Bcpme* genes in a *B.cinerea* strain (BO5.10) was carried out using a gene-replacement approach. Replacement of both *Bcpme1* and *Bcpme2* genes did not alter the virulence of *B. cinerea*, when tested on tomato and grapevine (Kars et al. 2005b). In another investigation, the phytotoxic metabolite botrydial, another virulence factor of *B. cinerea* was reported to act in a strain-specific manner (Siewers et al. 2005). It is possible that *B. cinerea*, a versatile pathogen, may possess multiple infection mechanisms with functional redundancy.

The role of endoPGs in the pathogenicity of *Alternaria citri* (causing black rot disease) and *A. alternata* (inducing brown spot disease), morphologically indistinguishable, was evaluated. The genes of these pathogens encoding endoPG were mutated by gene targeting. The biochemical properties of the PGs and the nucleotide sequences of the genes were highly similar. However, the phenotypes of the mutants exhibited distinct differences. An endoPG mutant of *A. citri* had reduced capacity in inducing black rot symptoms on citrus as well as in the maceration of potato tissue and it could not colonize citrus peel segments. On the other hand, an endoPG mutant of *A. alternata* remained unaltered in its pathogenicity compared with the wild-type. The study revealed different roles of endoPGs depending on the type of symptoms induced rather than the taxonomy of the fungal pathogen(s) (Isshiki et al. 2001).

The G (alpha) subunit BCG1 has been shown to have an important role during infection of host plants by *B. cinerea.* The mutants (*Δbcg1*) could conidiate, penetrate host leaves and induce small primary lesions. However, further development of symptoms was arrested in contrast to the uninhibited disease development in infection by wild-type. The fungal genes, whose expression on the host plant was specifically affected in *bcg1* mutants, were identified by employing suppression subtractive hybridization (SSH). The genes (22) that differentially expressed during infection by the wild-type, were predicted to encode proteases, enzymes involved in secondary metabolism and others encoding cell wall-degrading enzymes. These genes were not expressed in the mutants in vivo, but some were expressed under certain in vitro conditions as in wild-type. The results suggest that BCG1 is involved in at least one additional signaling cascade in addition to the cAMP-dependent pathway (Schulze-Gronover et al. 2004). Colonization of plant tissues by *B. cinerea* is accompanied by rapid killing of plant cells. All pathogenic strains of *B. cinerea* are known to secrete a non-specific phytotoxin the sesquiterpene botrydial. The functional characterization of *bcbot1*, encoding a P450 monooxygenase provided the evidence that it is involved in botrydial pathway. The botrydial biosynthetic gene *bcbot1* was expressed in planta. Its expression in vitro and in planta was controlled by an α-subunit of a heterotrimeric GTP-binding protein BCG1. Deletion of *bcbot1* resulted in reduced virulence only in strain T4 among three strains tested, indicating that the effect on virulence of *B. cinerea* was strain-dependent (Siewers et al. 2005).

The role of endoPG in colonization of citrus tissues by *Alternaria citri*, as well as the spatial and temporal regulation of endoPG gene expression was investigated. The colonization of an *A. citri* mutant expressing a gene encoding green fluorescent protein (GFP) from the jellyfish *Aequorea victoria* in citrus tissues was monitored

(Chalfie et al. 1994). The gene encoding GFP has been successfully expressed and used as a vital marker for growth of fungal pathogens. The plasmid pTEFEGFP carrying a GFP gene was introduced into the wild-type (G1) and its endoPG-disrupted mutant (GM4). The transformants G1 and GM4 showed strong green fluorescence in germinating spores on cellulose membranes. The spores, germ-tubes, appressoria and infection hyphae of G1 and GM4 transformants emited green fluorescence. The hyphae of only G1 transformant vertically penetrated the orange peel. However, the hyphae of both G1 and GM4 could spread efficiently in the juice sac area of citrus fruit. Another transformant of *A citri* EPG7 carrying a GFP gene under control of the endoPG gene promoter of *A. citri* was induced by pectin in the peel during the infection stage, but repressed completely in the juice sac area (Isshiki et al. 2003). The relationship between PG activity and pathogenicity of *Geotrichum candidum*, causing sour rot disease of citrus was investigated. Expression of the *S31pg1* gene of citrus race was detected in the mycelium of race 31 grown in liquid cultures of citrus and also in inoculated lemon peel. But no transcript of *S63pg1* gene of noncitrus race could be detected, both in culture and inoculated lemon peel, indicating an important role for PG in development of sour rot symptoms (Nakamura et al. 2001).

The interaction between fungal endoPGs and polygalacturonase-inhibiting proteins (PGIPs) found in plant cell walls has been studied in some pathosystems. The involvement of EPGs and PGIPs in the production of specific signals (oligosaccharins) during fungal pathogenesis has been suggested (Kemp et al. 2004b). *Sclerotinia sclerotiorum*, a necrotrophic pathogen, secretes oxalic acid and endoPGa and endoPGb, being acidic and basic in nature. The activity of PGb 3 was stimulated by oxalic acid (oxalates) and PGa activity was maintained at pH 3.6 also. PGa escaped inhibition of soybean PGIP, possibly at this pH the PGIP present in the plant tissue was inactive. RT-PCR assay revaled that during infection of soybean by *S. sclerotiorum,* the expression of the putative *pga* gene was delayed in comparison to the basic one. The different temporal expressions of PGa and PGb and their differential responses to pH, oxalate and PGIP appear to be consistent with possible maximization of the fungal PG activity in the soybean tissue (Favaron et al. 2004).

The role of endoPGs in pathogenesis has been studied in some detail. But the involvement of fungal enzymes that degrade cellulose, one of the major cell wall components in most plants has not been established clearly. Cloning and characterization of the *cel5A* gene from *B. cinerea* was done and the effect of disruption of *cel5A* was also studied. The gene *cel5* appears to be single copy gene as revealed by Southern blotting of the genomic DNA of *B. cinerea*, after digestion with *Bam* HI, *Xba* I or *Sal* I and probing with the *Cel5A* gene. The levels of *cel5A* transcripts and extracellular β-1,4-glucanase activity were regulated by the carbon source. Carboxy methylcellulose (CMC) induced the formation of *cel5A* transcripts and enzyme activity, whereas glucose repressed both function. The *cel5A* mRNA was detected during infection of tomato leaves by *B. cinerea*. Disruption of *cel5A* by transformation with the gene replacement vector did not lead to any change in virulence of the mutants on tomato leaves and Gerbera petals. No significant difference between mutants and wild type either in lesion size or the rate of disease development could be recorded. Likewise, no distinct difference in the extracellular β-1,4-glucanase activity and growth rate of the mutants in comparison with wild type was seen (Espino et al. 2005).

The involvement of other enzymes, in addition to pectinases and cellulases, in the degradation of host plant cell walls during pathogenesis, has been reported. By applying immunogold labeling technique, the presence of homogalacturonan (pectin) in cell walls of rye ovaries was detected. Deposition of labels increased at rye-*Claviceps purpurea* (ergot disease) interface. Two xylanases genes *cpxyl1* and *cpxyl2* were isolated from a genomic library of *C. purpurea*, using the genes of *Cochliobolus carbonum* (*xyl1*) and *Magnaporthe grisea* (*xyn33*) as heterologous probes. Evidence for the activities of PG, PME, xylanases and cellulases was obtained in this pathosystem. The genes *cpxyl1* and *cpxyl2* were found to be single copy genes and were expressed both in vitro and in planta during all stages of infection (Giesbert et al. 1998; Tenberge 1999). The gene *xyn11A* coding for an endo β-1,4-xylanase belonging to family 11 of glycosyl hydrolases was expressed in planta infected by *Botrytis cinerea.* Disruption of *xyn11A* reduced the xylanase activity moderately and growth rate of *B. cinerea*. However, deletion of the gene significantly affected virulence reflected by delay in appearance of secondary lesions and reduction in lesion size by more than 70%. Complementation of the gene restored virulence in the mutant to a level equal to that of wild-type (Brito et al. 2006).

Extracellular hydrolytic enzymes encoded by multiple fungal genes are required for degrading the physical barriers present in plants for successful infection and subsequent colonization of tissues. Two inducible pectate lyase (PL) genes (*pel*) were identified in *Nectria haematococca* infecting peas. While *pelA* is induced by pectin, *pelD* is induced in planta. No reduction in virulence of *N. haematococca* was discernible, following disruption of either *pelA* or *pelD* genes. But drastic reduction in virulence of *N. haematococca* occurred, if both *pelA* and *pelD* were disrupted indicating PL is a virulence factor in *N. haemotococca* (Rogers et al. 2000). In order to assess the contribution of PL to enhancing pathogenicity of *Colletotrichum gloeosporioides*, infecting avocado, *pel* gene from the isolate Cg-14 was expressed in *C. magna* isolate (L-2.5) that could induce minor symptoms in watermelon seedlings and avocado fruits. Western blot analysis with antibodies against Cg-14 PL detected a single PL secreted by the transformed *C. magna* isolate. This isolate containing the 4.1 kb genomic clone of *C. gloeosporioides* had higher maceration capacitites on avocado fruits compared to wild-type *C. magna.* In addition, watermelon seedlings were more severely infected, when inoculated with the transformed *C. magna* isolate. This study clearly shows that PL is a pathogenicity factor required for penetration and colonization of host tissues (Yakoby et al. 2000). In a later investigation *pelB* gene of *C. gloeosporioides* was disrupted via homologous recombination. The mutants did not show any reduction in growth on glucose. But they did not secrete pectate lyase B (PLB) and had lower PLB and PG activities compared with wild-type. When PLB mutants were inoculated onto avocado fruits, the decay diameter was reduced by 36%–45%. PLB seems to be an important virulence factor and probably it may have a role in induction of host defense (Yakoby et al. 2001). The influence of pH levels in plant tissues on colonization by fungal pathogens has been examined. *C. acutatum* secretes seven isozymes of PG. The genes *capg1* and *capg2* encoding two of these isozymes were differentiately regulated with respect to pH and carbon sources. The expression pattern a *capg1* was found to be positively correlated with total PG activity (Mc Evoy et al. 2003). Accumulation of ammonia and consequent tissue alkalinization predisposes avocado fruits to infection by *C. gloeosporioides*. The transcript level of *pelB* increased in parallel as a function of pH. The ambient pH and the source of N (organic or inorganic) were indicated to be independent regulatory factors for processes linked to PL secretion and virulence of *C. gloeosporioides* (Prusky et al. 2003).

The role of enzymes other than cell wall-degrading enzymes to pathogen origin, in disease development has been elucidated. Evidence suggesting that ornithine decarboxylase (ODC) is essential for the virulence of *Stagonospora* (*Septoria*) *nodorum* infecting wheat was provided by Bailey et al. (2000). Following the targeted gene replacement procedure, a knockout strain lacking the ODC allele was obtained. As polyamines from host plants have to be made available for the development of *S. nodorum,* the mutants were not able to obtain the supply of ODC due to the disruption of the gene concerned. The MAGB encoding a G alpha subunit from *Magnaporthe grisea* was identified. Disruption of MAGB resulted in defects in a number of cellular responses including appressorium formation, conidiation, sexual development, mycelial growth and surface sensing. In the site-directed mutagenesis experiment, conversion of glycine 42 to arginine was predicted to abolish GTPase activity which in turn may constitutively activate G protein signaling in *magB*. The mutation resulted in reduced virulence of mutants, in addition to several growth abnormalities (Fang and Dean 2000).

The pea-*Mycosphaerella pinodes* pathosystem can be considered a good model system to investigate how pathogenicity factors contribute to the breakdown of plant defense responses in establishing basic compatability. *M. pinodes* secretes two well-characterized fungal signal molecules, a glycoprotein elicitor (approx. 70-kDa) and a polypeptide suppressor (<5-kDa) in the spore germination fluids (Shiraishi et al. 1978; Oku et al. 1980) The suppressors Gal NAc – *O* – Ser-Ser-Gly (supprescin A) and Gal-Gal NAc-*O*-Ser-Ser-Gly-Asp-Glu-Thr (supprescinB) play an important role as a pathogenicity factor by counteracting elicitor-induced defense responses (Shiraishi et al. 1992, 1997). Molecular studies indicated that the suppressors inhibited the defense signaling induced by *M. pinodes* elicitors, including activation of cell wall and plasma membrane, ATPases and protein kinases (Shiraishi et al. 1997). In contrast, the suppressor either inhibited or delayed, in the presence of elicitors, the activation of defense genes and superoxide generation (Yamada et al. 1989; Kato et al. 1995). Later suppressor induced S-64 was identified as a 12-oxophytodienoic acid reductase (OPR) (Ishiga et al. 2002). The identity of specific signaling steps involved in suppression-mediated breakdown of elicitor-induced response has to be clearly established. The involvement of a MAPK pathway in *M. pinodes* elicitor-induced defense response in pea and the interruption of the elicitor-induced MAPK cascade(s) in pea by the suppressor have been demonstrated. A polypeptide fungal suppressor from *M. pinodes* inhibited the polysaccharide elicitor (secreted by *M. pinodes*) – induced activation of only p44 kinase, a myelin basic protein (MBP)-dependent kinase, but not p48 kinase. The defense-inducing signaling molecules chitosan and salicylic acid (SA) activated both p44 and p48 kianses, but methyl jasmonate (MeJA) did not alter activity of the kianses. The results suggest that *M. pinodes* suppressor may act through inhibition of a MAPK (p44) facilitating the establishment of basic compatability during infection of pea (Uppalapati et al. 2004).

The microbial plant pathogens have to counteract the effects of preformed protective compounds and antimicrobial substances produced in response to initiation of infection. Some fungal pathogens have been shown to be able to secrete enzymes that may convert the antimicrobial compounds into nontoxic ones. *Gloeocerospora sorghi* produces cyanide hydratase that can convert cyanide released by sorghum tissues into nontoxic formamide (Wang et al. 1992). However, cyanide hydratase was shown to be not essential for pathogenicity of *G. sorghi*, since the mutant generated by specific gene disruption, was as pathogenic as the wild-type, despite its sensitivity to cyanide (Wang and Van Etten 1992). The inability of *Gaeumannomyces graminis* var. *tritici* to infect oats is associated with the absence of a gene that can produce an enzyme to metabolize avenacin present in oats. In contrast, *G. graminis* var. *avenae*, because of the presence of avenacinase gene, could successfully infect oats. Disruption of avenacinase gene reduced the pathogenicity of the mutants on oats (Schäfer 1994). Further, the ability of fungi to colonize roots of oats seemed to depend on the level of tolerance to avenacin. The fungi capable of tolerating avenacin could colonize oat roots, whereas the fungi sensitive to avenacin were eliminated from the rhizosphere of oats (Carter et al. 1999).

Developing fruits contain antimicrobial compounds capable of inhibiting the development of pathogens. A cDNA encoding PGIP from mature apple fruit was characterized after cloning. The transcript levels of the PGIP varied depending on fruit maturity levels. Areas showing decay and tissues adjacent to inoculated sites of *Penicillium expansum* and *Botrytis cinerea* (causing blue and gray mold diseases) had very high levels of PGIP transcripts, suggesting different roles for PGIP depending on stages of fruit development (Yao et al. 1999). Polygalacturonases (PGs) produced by the fungal pathogens may be inhibited by the PG-inhibiting proteins (PGIPs) produced by the host plants in response to biotic and abiotic stresses. The BcPG1, an important pathogenic factor of *Botrytis cinerea* may be inhibited by the PGIPs encoded by two genes *AtPGIP1* and *AtPGIP2*. Both genes are induced by pathogen infection through different signaling pathways. Expression of an antisense *AtPGIP1* gene in *Arabidopsis thaliana* plants resulted in reduced AtPGIP1 inhibitory activity and enhancement of susceptibility to *B. cinerea* infection. The results indicated that PGIP contributes to basal resistance to *B. cinerea* that has to be breached for establishing successful infection (Ferrari et al. 2006).

Tomato plants contain α-tomatines, a steroidal glycoalkaloid saponin which has been implicated as a preformed antimicrobial compound effective against many potential fungal pathogens (Martin-Hernandez et al. 2000). The tomato pathogens such as *Botrytis cienrea* (Quidde et al. 1998), *Septoria lycopersici* (Woods et al. 2004) and *Fusarium oxysporum* f.sp. *lycopersici* (Ito et al. 2002) have been shown to degrade α-tomatinase by the activities of extracellular enzymes produced by them. A strong correlation between tolerance to α-tomatine, the ability to degrade α-tomatine and pathogenicity of pathogens infecting tomato was reported by Sandrock and Van Etten (1998). In contrast, *Phytophthora infestans* and *Pythium aphanidermatum* can infect tomato plants despite their inability to degrade α-tomatine. Further, *Cladosporium fulvum* sensitive to α-tomatine can successfully induce disease in tomato. However, the transformants of *C. fulvum* expressing a heterologous expression of cDNA encoding tomatinase from *Septoria lycopersici* showed increase in sporulation in susceptible tomato lines and the symptoms caused were more severe in resistant lines compared to wild-type (Melton et al. 1998).

F. oxysporum f.sp. *lycopersici* (FOL) and *F. oxysporum* f.sp. *radicis lycopersici* (FORL) have been reported to produce tomatinases with same mechanisms involving hydrolysis of α-tomatine (Lairini et al. 1996). But the gene encoding tomatinase seems to be different between these two formae speciales (Ito et al. 2004b). The tomatinase gene *FoToml* capable of degrading α-tomatine to less toxic metabolites, is present in FOL and also in some strains of *F. oxysporum* belonging to formae speciales nonpathogenic to tomato. Four *FoFoml*- positive strains (2, 3, 11 and 17) of *F. oxysporum* nonpathogenic to tomato colonized not only the roots and basal stems, but also upper stems of tomato plants. The strain #3 transformed with β-glucuronidase (GUS) reporter gene colonized vascular vessels of all tomato cultivars tested. The *FoToml* gene was expressed in the roots and stems of plants, as revealed by RT-PCR assays, suggesting that *FoToml* was important for the four *FoToml*-positive strains to survive in tomato plants. Prior inoculation of tomato plants with either #2 or #3 strain resulted in suppression of vascular wilt of tomato caused by FOL, due to activation of defense genes as indicated by accumulation of transcripts of acidic chitinase gene (*Chi3*) (Ito et al. 2005).

In addition to enzymatic degradation of preformed antimicrobial compounds present in the host plants, the production of host-specific toxins (HSTs), considered as virulence or pathogenicity factors, assist in the pathogenesis of fungi. The tomato pathogens *Alternaria alternata* tomato pathotype *A. alternata* f.sp. *lycopersici* (causing Alternaria stem canker) and *Corynespora cassiicola* (causing Corynespora target spot) produce AAL-toxin (AL-toxin) and CCT-toxin (CC-toxin) respectively (Bottini et al. 1981; Otani et al. 1998). Spore germination of *A. alternata* pathogenic and nonpathogenic to tomato and of *C. cassiicola* pathogenic to tomato was not affected by α-tomatine (0.1 mM). But spore germination of *C. cassiicola* nonpathogenic to tomato was significantly inhibited. Germinating spores of *A. alternata* and *C. cassiicola* resistant to α-tomatine detoxified it by degrading it to a less polar product (Table 2.2) as confirmed by the results of thin-layer chromatography (TLC) (Fig. 2.4). *A. alternata* tomato pathotype and *C. cassicola* pathogenic to tomato are able to detoxify α-tomatine and the detoxification is essential for host tissue colonization by pathogens that also, produce host-specific toxins (Oka et al. 2006).

2.1.3.2 Toxins Produced by Fungal Pathogens

Fungal pathogens are known to produce host-specific and non-specific toxins which aid in establishment of infection and disease development to varying extents. *Cochlibolus victoriae* (*Helminthosporium victoriae*) causing the victoria blight

Table 2.2 Percentages of spore germination as a measure of detoxification of α-tomatine using *Alternaria brassicola*

Isolate / host	Inhibition (%)*
Alternaria alternata	
AS – 27 / tomato-pathogen	14.6
0–94 / tomato non-pathogen	0.0
Corynespora cassiicola	
LC 93009 / tomato	0.0
LC 93020 / tomato	11.7
A. brassicola / cabbage	
0–264	100

*Spore suspension (10^5 spores/ml) in 0.1 mM tomatine (in 1% methanol) was incubated for 24 h and the cell-free solution was assayed for the presence of α-tomatine using *A. brassicola* spores as test organism. Source: Oka et al. (2006)

disease of oats produces a host-selective toxin victorin which is the primary pathogenicity determinant of this pathogen. Isolates of *C. victoriae* that produce victorin are pathogenic, while those that do not produce victorin are nonpathogenic (Meehan and Murphy 1946; Luke and Wheeler 1955). Oat cultivars showed similar levels of resistance to both toxin and pathogen, indicating a genetic similarity of resistances to the pathogen and its toxin. The response of host plants to victorin is governed by a single dominant gene (*Vb*) and paralleles the response to the pathogen itself (Scheffer 1976). It is of interest to note that victorin can function

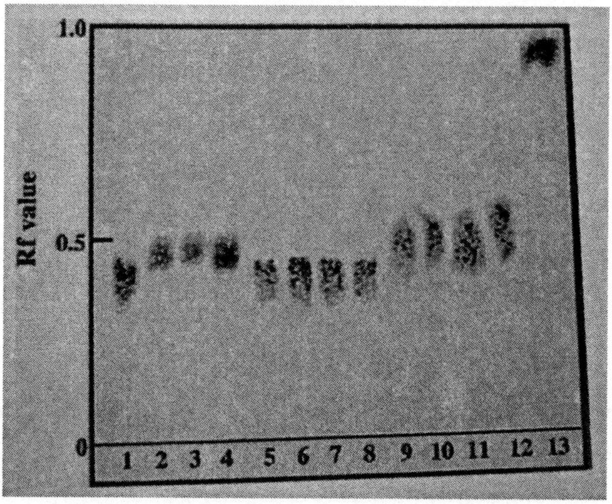

Fig. 2.4 α-tomatine detoxified products by *Alternaria alternata* and *Corynespora cassiicola* isolates analyzed by thin-layer chromatography (TLC) technique, after incubation of fungal isolates with α-tomatine for 24 h
Lane 1: α-tomatine (standard); Lane 2–8: Isolates of *C. casiicola*; Lane 9–12: Isolates of *A. alternata*; Lane 13: Tomatidine (standard). (Courtesy of Oka et al. 2006; The Phytopathological Society of Japan and Springer-Verlag, Tokyo)

not only as toxin, but also as a specific elicitor of phytoalexin accumulation and stress-responding genes in victorin-sensitive oatlines when applied at a low concentration (Mayama et al. 1986; Tada et al. 2000). Further, victorin has been reported to induce programmed cell death (PCD) in compatible *Vb* oat lines with characteristic features of animal apoptosis, such as mitochondrial permeability transition (Curtis and Wolpert 2002), chromatin condensation (Yao et al. 2001) and nuclear DNA laddering-fragmentation of chromatin in multiples of 180-bp nucleosomal units, a process known as DNA laddering as well as chloroplast dysfunction (Navarre and Wolpert 1999). During apoptotic cell death induced by victorin, rRNA and mRNA were specifically degraded in oat cells. Northern analysis showed that rRNA species from cytosol, mitochondria and chloroplasts were all degraded via specific degradation intermediates during victorin- induced apoptotic cell death. Specific cleavage of mRNA of house keeping genes such as *actin* and *ubiquitin* during victorin-induced cell death was observed (Hoat et al. 2006). A single locus *TOX2* in *Cochliobolus carbonum* infecting maize governs the production of the host-specific HC-toxin (Scheffer and Yoder 1972). The strains of *C. carbonum* that could not produce HC-toxin were not pathogenic to maize. The maize plants that could inactivate HC-toxin were not susceptible to HC toxin-producing strains (Schäfer 1994). The HC-toxin producing Tox strains($TOX2^{+}$) alone contained the *ToxC* gene which encoded HC-toxin synthetase and had three functional copies located in the same chromosome. Targeted gene disruption of all three functional copies resulted in generation of mutants that did not produce HC-toxins. These mutants were nonpathogenic to maize, indicating that *ToxC* is essential for pathogenicity and virulence (Ahn and Walton 1997). Later, another gene *TOXF* of all *TOX2* locus was characterized. *TOXF* was present as 2 or 3 copies in all HC-toxin producing $TOX2^{+}$ strains and loss of this gene by disruption led to inability to produce HC-toxin and consequent failure of infection by mutants (Cheng et al. 1999). Progenies of crosses between HC toxin producing and toxin non-producing strains, had lost one or more copies of the *TOX2* genes. However, most of the strains retained their virulence (Pitkin et al. 2000).

Cochliobolus heterosporus race T, causing the destructive southern corn blight disease on maize lines with Texas male sterile (tms) cytoplasm, produces a host-specific toxin (HST), T-toxin (a family of C(35) to C(49) polyketides). Production of T-toxin is under the control of two unlinked loci, *TOX1A* and *TOX1B* carried on 1.2 Mb of DNA not present in race O, a mildly virulent race which does not produce T-toxin. This toxin is not produced by any another *Cochliobolus* spp. or other related fungi. T-toxin production depends on *PKS1* a polyketide synthase (PKS)-encoding gene at *Tox1A* and *DEC1*, a decarboxylase-encoding gene at *Tox1B*. The requirement of *PKS2*, a second PKS-encoding gene that maps at *Tox1A* for T-toxin biosynthesis and high virulence to maize was demonstrated by ligation specificity-based expression analysis display (LEAD), a comparative AFLP/gel fractionation / capillary sequencing procedure (Baker et al. 2006). Nonribosomal peptide synthetases (NRPSs) are multifunctional proteins required for biosynthesis of small peptides, independently of the ribosomal protein synthesis machinery. The genome of the maize pathogen *C. heterostrophus* contains 12 NRPS-encoding genes (*NPSs*). *NPS6* has been shown to be involved in virulence of *C. heterostrophus* to maize.

Deletion of *NPS6* resulted in concomitant reduction in virulence and enhancement of sensitivity to H_2O_2 (Lee et al. 2005). In a later investigation, deletion of *NPS6* orthologs in *Cochliobolus miyabeanus* (brown spot of rice), *Fusarium graminearum* (Fusarium head blight of wheat), *Alternaria brassicola* (leafspot of *Arabidopsis*) led to reduced virulence and hypersensitivity to H_2O_2. The *Δnps6* mutant of *C. heterostrophus*, when complemented with the *NPS6* ortholog from *Neurospora crassa* regained the original level of virulence to maize as the wild type and tolerance to H_2O_2, indicating functional conservation in pathogen and saprobes belonging to Ascomycetes. Exogenous application of iron increased the virulence of *Δnps6* strains of *C. heterostrophus, C. miyabeanus, F. graminearum* and *A. brassicola* to respective host plant species. Further, NPS6 was in control of the biosynthesis of extracellular siderophores by *C. heterostrophus, F. graminearum* and *A. brassicola.* It was hypothesized that the extracellular siderophores may be involved in the supply of essential nutrient (Fe) to the pathogen in planta but may not function as phytotoxin (Oide et al. 2006).

The AK toxin, a host specific toxin is elaborated by*Alternaria alternata*, causing black spot disease of Japanese pear. The genes involved in AK toxin biosynthesis were tagged by employing restriction enzyme-mediated integration (REMI) mutagenesis technique. Of the two genes identified, *AK1* encodes a member of the class of carboxyl-activating enzymes, whereas the other gene AK2 governs the production of a protein of unknown function. The requirement of these genes for AK-toxin production and pathogenicity was demonstrated by transformation-mediated gene experiments (Tanaka et al. 1999). The presence of two additional genes, downstream of AK2 involved in toxin biosynthesis was detected later. The DNA gel blot analysis with pulsed-field gel electrophoresis (PFGE) indicated that all the four genes and their homologs were located in the same chromosome of *A. alternata* (Tanaka and Tsuge 2000). *Rhizoctonia solani* causing rice sheath blight disease produces a HST, RS toxin. The RS toxin-specific antibodies (IgG) detected pathogen antigens in rice leaf sheaths inoculated with *R. solani* and also in culture filtrates in ELISA tests. A positive correlation between disease intensity and levels of antigen in the infected tissues was observed (Shanmugam et al. 2002).

Another fungal pathogen *Pyrenophora tritici-repentis* infecting wheat produces a host-selective (specific)-toxin (HST) encoded by *ToxB* gene. Six of the estimated nine copies in multiple ToxB open reading frames (ORFs) from race 5 were cloned and analyzed. The identical 261-bp ORFs present in all six copies of *ToxB* encode the same form of Ptr ToxB. The *ToxB* loci reside on two chromosomes. A related gene *toxb* present in nonpathogenic race 4 showed high similarity (86%) to *ToxB* and it appeared to be a single-copy gene. Such similarity between pathogenic and nonpathogenic isolates (races) in a gene(s) encoding HSTs has not been noted in any other fungal pathogens producing HST (Martinez et al. 2004).

The tremendous success of green fluorescent protein (GFP) from the jelly fish *Aequorea victoria* when used as an in vivo reporter has revolutionized the understanding of fungal biology and fungal pathogens-host interaction (Lorang et al. 2001; Jensen and Schulz 2004; Czymmek et al. 2005). Further advances in fluorescent protein technology made it possible to expand the color palette of fluorescent transformation vectors available for use in the studies on filamentous fungi. The vector pCT74 was employed to use the promoter from *Pyrenophora tritici-repentis* necrosis-inducing HST gene *ToxA* for driving expression of the synthetic GFP gene, *sGFP* (Chiu et al. 1996; Ciuffetti et al. 1999). This vector conferred bright fluorescence to plant pathogens (Lorang et al. 2001). The development of cyan fluorescent protein (CFP) (Heim and Tsien 1996) and yellow fluorescent protein (YFP) (Ormö et al. 1996) help in simultaneous visualization of two distinct GFP variants (Ellenberg et al. 1998; Stuurman et al. 2000) and this approach provided an ideal pair for fluorescent energy transfer (FRET) analysis (Tsien 1998; Dye et al. 2005). The fluorescent proteins EYFP, ECFP and mRFP1 (variants of GFP) in addition to, sGFP were included in *ToxA* promoter- driven from expression vectors. The nine copies of *ToxB*, a chlorosis-inducing HST gene from *P. tritici-repentis* have the potential to offer multiple promoters for heterologous expression of proteins in fungal pathogens. The toxA promoter successfully drove expression of all three fluorescent proteins EYFP, ECFP and mRFP1 in the cytoplasm of both *P. tritici-repentis* and unrelated *Verticillium dahliae* causing wilt disease, as revealed by fluorescence and confocal microscopy. Like the *ToxA* promoter, *ToxB* promoter also could drive strong expression of sGFP in conidia, conidiophores and hyphae of *V. dahliae*. The results indicate the possibility of using the *ToxA* and *ToxB* promoters to drive expression of flurorescent fusions to determine protein distribution or label cell components or organelles (Andrie et al. 2005)

Tox A was shown to be a protein and product of a single gene (Ciuffeti et al. 1998, 1999). Transformation of a nonpathogenic isolate of *P. tritici-repentis* with *ToxA* gene rendered it pathogenic on *ToxA*-sensitive wheat lines. The *ToxA* gene encoded a preprotein that contained a single sequence to target the protein and a pro-sequence (N-domain) that was necessary for proper folding that might be removed prior to secretion (Tuori et al. 2000). Introduction of the mature toxin into the apoplastic space of a sensitive plant produced necrotic response similar to the disease symptoms induced by ToxA-producing isolates, indicating the ability of Tox A protein to induce cell death in the absence of the pathogen. Unlike most HSTs that are products of enzymatic pathways, at least two toxins produced by *P. tritici-repentis* were found to be proteins and products of single genes. Sensitivity to these toxins in the host was conferred by a single gene for each toxin. The toxin-sensitive and-insensitive wheat cultivars were treated with ToxA followed by proteinase K (PK) to study the site of action of PtrToxA (ToxA). ToxA was resistant to protease but only in sensitive leaves, suggesting internalization of ToxA into toxin-sensitive cells. Protection of ToxA from PK, the intracellular detection of ToxA by immunolocalization and the direct visualization of a functional GFP-ToxA fusion protein in cytoplasm and in association with chloroplasts showed that ToxA was internalized into toxin-sensitive cells. Intracellular expression of ToxA by biolistic bombardment into both toxin-sensitive and -insensitive cells resulted in cell death suggesting that the ToxA internal site of action was present in both cell types. The results showed that ToxA protein was capable of crossing the plant plasma membrane from the apoplastic space to the interior of the plant cell in the absence of the pathogen (Manning and Ciuffetti 2005).

Fusarium graminearum, incitant of scab or Fusarium head blight (FHB) disease in barley and wheat, infects the individual spikelets as soon as the spike erupts from the flag leaf sheath, near the time of pollination (Briggs 1978). The primary site of initial infection could not be determined by microscopical techniques. Hence, a strain of *F. graminearum* was transformed with *gfp* to study the patterns of infection in barley and *Arabidopsis* model systems. After inoculation with the transformant, infected tissues were examined for fluorescence under the blue light with a dissecting microscope equipped with a 470 nm excitation filter and 535 nm emission filter. The observations showed that the most rapid route of infection involved the brush hairs at the extruded seed tip. Colonization followed a pattern of rapid basipetal growth along the pericarp epithelium, accompanied by slower growth inward through the pericarp and testa. In the case of *Arabidopsis thaliana*, the transformant readily infected the leaves, producing abundant spores within distant leaves. This technique has the potential for tracing the growth of *F. graminearum* without any misinterpretation due to contaminating fungi and also for determining the pathogen development in genotypes with different levels of resistance (Skadsen and Hohn 2004). The cis-active elements in *Ustilago maydis* mig 2 promoters conferring high-level of activity during the growth of the pathogen in maize were identified. A 350-bp mig 2–5 promoter fragment contained all elements sufficient to confer differential promoter activity. This region was fused to the GFP reporter gene. The core promoter elements required for high-level promoter activity from elements conferring inducible expression in plant was subjected to mutational analysis. The detection of all six *mig2* genes did not reduce the ability to induce tumor formation in infected maize plants (Farsing et al. 2005).

The pathotype of *Alternaria alternata* infecting tangerines (*Citrus reticulata*) causes Alternaria brown spot and the pathotype infecting rough lemon (*C. jambhiri*) induces Alternaria leafspot. These pathotypes produce chemically distinct host selective toxins (ACT-and ACR-toxins) and they can be distinguished based on the host range (Kohmoto et al. 1991, 1993; Akimitsu et al. 2003). The chemical structures of ACT-toxin and ACT-toxin-I produced by the tangerine pathotype have been defined (Gardner et al. 1985; Kohmoto et al. 1993). The ACTT genes considered to be involved in the biosynthesis of ACT-Toxin were identified by Masunaka et al. (2000). The strain BC3-5-1-OS2A from a leaf spot on rough lemon was pathogenic on both cv. Iyokan tangor and rough lemon and also produced both ACT-toxin and ACR-toxin. It contained both genomic regions, one of which was present only in ACT-toxin producers and the other was found only in ACR-toxin producers. The pulse field gel electrophoresis (PFGE) analysis of seven rough lemon pathotype isolates and nonpathogenic *A. alternata* isolates showed that all rough lemon pathotype isolates contained unique chromosomes of 1.2–1.5 Mb that were absent from all nonpathogenic isolates tested. The presence of this chromosome was correlated with production of ACR toxin and pathogenicity to rough lemon. The probe based on the 408-bp PCR product following random priming of isolated 1.5 Mb chromosome of rough lemon pathotype isolate HC1, hybridized to BC3-5-1-OS2A. This indicated the presence of the chromosome required for the production of ACR-toxin. The quantity of ACT-toxin and ACR-toxin produced by BC3-5-1-OS2A in culture filtrate appeared similar to that produced by each pathotype independently (Masunaka et al. 2005) [Appendix 2].

The variants (7) of *Alternaria alternata* produce different host selective toxins (HSTs) that cause necrosis in different host plant species and they are designated "pathotypes" of *A. alternata* (Kohmoto et al. 1995). The tomato pathotype produces a HST named as AAL-toxin (Grogan et al. 1975), while *Corynespora cassiicola* infecting tomato produces a HST called as CCT-toxin (Otani et al. 1998). Detoxification of α-tomatine mediated by HSTs produced by these tomato pathogens has been investigated. Spores of *A. alternata* and *C. cassiicola* nonpathogenic to tomato, germinated, following inoculation of tomato leaves and formed appressoria. However, they did not proceed to form infection hyphae required for penetration of host tissues. When the HST CCT-toxin produced by *C. cassiicola* was added to nonpathogenic spores, colonization within leaves was observed only in *A. alternata,* but not in *C. cassiicola*. In contrast, when the spores of *C. cassiicola* nonpathogenic to tomato were suspended in spore germination fluid of nonpathogenic *A. alternata* with α-tomatine activity, this isolate was induced to colonize leaves in the presence of CCT-toxin. The results indicated that detoxification of α-tomatine has to precede, before host colonization can be promoted by activity of HSTs (Oka et al. 2006).

Alternaria alternata, strawberry pathotype causes Alternaria black spot and produces a HST known as AF toxin (Maekawa et al. 1984; Nishimura et al. 1984). A cosmid clone pcAFT-1 from NAF8 strain of the strawberry pathotype was shown to contain AFT genes involved in AF-toxin biosynthesis (Hatta et al. 2002). Four *AFT* genes *AFT1-1, AFT3-1, AFTR-1* and *AFTS-1* were detected in 20-kb region of pcAFT-1 (Hatta et al. 2002; Ito et al. 2004a). Targeted mutation of *AFT1*-1 and *AFT3-1* abolished the ability to produce AF toxins by the mutants and consequently they were nonpathogenic to strawberry plants. Nevertheless, they could grow and sporulate normally in culture indicating the dispensability of the 1.05-Mb chromosome containing all genes involved in toxin biosynthesis for saprophytic growth. The AF-toxin biosynthetic genes appear to be housed on a conditionally dispensible (CD) chromosome (Hatta et al. 2002). In a further study, the complete sequence (∼30 kb) of pcAFT-1 was worked out and two additional transposon-like sequences, TLS-S2 and TLS-S3 downstream of AFT3-1 were found, in addition to TLS-S1. The structure and genomic distribution of five TLS found in cosmid clones pcAFT-1 and pcAFT-2 were investigated. The distribution of TLS-S2 copies in the strawberry pathotype strains was determined by DNA gel blot analysis. Nine strains from six other pathotypes and a nonpathogenic strain of *A. alternata* were also analyzed to determine the distribution of the TLS-S2 copies. Strains from the Japanese pear and tangerine pathotypes showed the presence of the TLS-S2 copies, whereas it could not be detected in other pathotypes or in the nonpathogenic strain (Fig. 2.5) (Hatta et al. 2006).

Some fungal pathogens have been reported to produce nonspecific phytotoxic compounds that have different actions on host plant tissues. *Cercospora kikuchii* infecting soybean elaborates a polyketide designated cercosporin. Light-enhanced accumulation of cDNA clones of mRNA transcripts of *C. kikuchii* has been

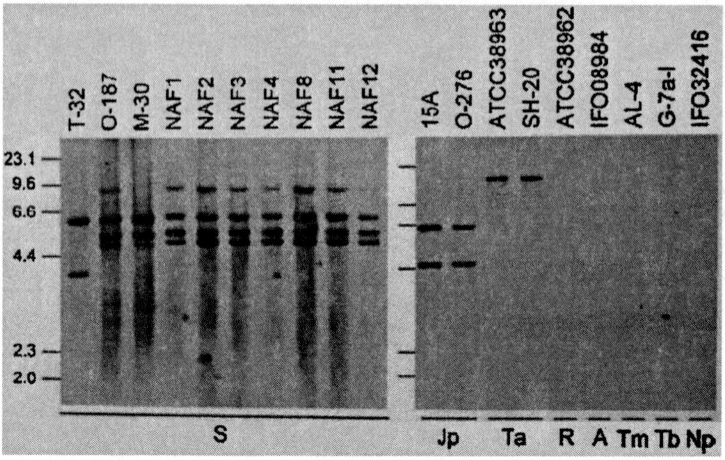

Fig. 2.5 Distribution of TLS-S2 copies in pathotypes and non-pathogenic strains of *Alternaria alternata* infecting Japanese pear.
Total DNA was digested with *Eco*RV, electrophoresed in 0.8% agarose gels and probed with TLS-S2-1 fragment from pS2. *Left*: Pathotypes of strains: S- strawberry; *Right*: Jp, Japanese pear; Ta, tangerine; R, rough lemon; A, apple; Tm, tomato; Tb, tobacco; Np, nonpathogen. (Courtesy of Hatta et al. 2006; The Phytopathological Society of Japan and Springer-Verlag, Tokyo)

demonstrated. Targeted disruption of the genomic copy of the mRNA transcription of one of the cDNA clones-designated CFP (cercosporin facilitator protein) led to drastic reduction in production of cercosporin and consequently to reduced virulence on soybean. The results suggested that *CFP* encodes a cercosporin transporter that contributes to resistance to cercosporin by actively exporting cercosporin, thus maintaining low cellular concentrations of the toxin (Callahan et al. 1999). Many *Cercospora* spp. produce cercosporin, a light-activated non-host-selective toxin. Cercosporin has been shown to play a pivotal role in pathogenicity of *C. nicotianae* and lesion formation (Choquer et al. 2005, 2007). A polyketide synthase gene (*CTB1*) was functionally identified and shown to play a key role in cercosporin biosynthesis by *Cercospora nicotianae* causing frog eye spot disease of tobacco. *CTB1* encoded a polypeptide with a deduced length of 2196 amino acids containing ketosynthase (KS), an acetyl transferase (AT), a thioesterase/claisen cyclase (TE/CYC) and two acyl-carrier protein (ACP) domains, and high levels of similarity to many fungal type I polyketide synthase. Targeted disruption of *CTB1* led to the loss of both *CTB1* transcript and cercosporin biosynthesis in *C. nicotianae*. The *ctb1*-null mutants induced a fewer necrotic lesions on inoculated tobacco leaves in comparison with wild-type. Pathogenicity was, however, restored to null mutants by complementation with a full-length *CTB1* clone in addition to production of cercosporin at levels comparable to wild-type isolate. The results conclusively showed that cercosporin is synthesized via polyketide pathway and that cercosporin is an important virulence factor in *C. nicotianae* (Choquer et al. 2005). In another investigation, another gene *CTB3* immediately adjacent to the *CTB1* was also shown to be essential for the biosynthesis of cercosporin (Dekkers et al. 2007), suggesting the possible involvement of a cluster genes in cercosporin biosynthesis. This situation was demonstrated to be true by Chen et al. (2007) who described a core gene cluster comprising of eight genes (*CTB1–CTB8*) associated with cercosporin production. Expression of these genes was co-ordinately induced under cercosporin-producing conditions and was regulated by the Zn(II) Cys_6 transcriptional activator, *CTB8*. Disruption of the *CTB2* gene encoding a methyl-transferase or the *CTB8* gene yield in mutants that were entirely defective in cercosporin production and expression of other genes in *CTB* cluster was also inhibited (Chen et al. 2007).

Verticillium dahliae, causative agent of cotton Verticillium wilt, produces a glycoprotein in culture filtrates (CF) with a role in production of wilting symptoms (Chu et al. 1999). The crude toxin, VD-toxin (present in CF) induced damage to the plasma membrane and cell wall in the susceptible callus cells as revealed by transmission electronmicrographs. Invaginated plasmalemma, distinctive plasmolysis and extensive development of membrane-bound vesicles and ultimately agglutinated cytoplasm and disintegration of plasma membrane were adverse effects of VD toxin induced at different periods after treatment (Zhen and Li 2004).

2.1.3.3 Mycotoxins Produced by Fungal Pathogens

Various species of *Aspergillus, Fusarium, Penicillium* and *Alternaria* infect seeds resulting in poor seed quality and also contamination of seeds with mycotoxins that are responsible for serious ailments in humans and animals consuming contaminated grains, foods and feeds. Different immunological methods have been applied for the detection of mycotoxins required for effective disposal of the contaminated materials. Production of mycotoxins depends on the fungal species, crop genotype, storage conditions and the chemicals applied for seed treatment. (Narayanasamy 2005).

Aspergillus flavus and *A. parasiticus* produce aflatoxins in several crops. The lipase gene expression has been shown to have a bearing on aflatoxin production. The gene *lipA* encodes a lipase involved in the breakdown of lipids from aflatoxin-producing *A. flavus* and *A. parasiticus* and also nonaflatoxin isolates of *A. flavus*. The expression of lipase gene under substrate-induced conditions correlated well with production of aflatoxin (Yu et al. 2003a). The *pksA* gene in the strain AF36 required for aflatoxin biosynthesis exhibited a nucleotide polymorphism near the beginning of the coding sequence. The enzyme production and accumulation of aflatoxin were inhibited by a nucleotide change that introduced a premature stop-codon into the coding sequence (Ehrlich and Cotty 2004). The expression of *aflR* and *aflJ* that regulates the activation of an aflatoxin gene cluster depends on *VeA* in *A. parasiticus* infecting peanut kernels. Sclerotial formation and production of aflatoxin intermediates both on culture medium and peanut seeds were annulled following deletion of *VeA* from *A. parasiticus* DNA (Calvo et al. 2004).

Aflatoxins are polyketides with characteristic dihydro (B1 and G1) – or tetrahydro (B2 and G2)–bisfuran rings. The B aflatoxins emit a blue fluorescence, whereas the G aflatoxins produce greenish-blue fluorescence when viewed under UV-light due to structural differences in the A-ring (Yabe et al. 1999; Bhatnagar et al. 2003). The genes (>25) involved in the biosynthetic pathway for the production of aflatoxins from malonyl coenzyme are clustered. Their expression is generally regulated by a single Zn2 Cys6-type transcription factor, AflR which is encoded by one of the genes in the cluster (Yu et al. 1997; Yabe et al. 1999; Bhatnagar et al. 2003). The requirement of P450 monooxygenase for the formation of aflatoxin G1, in addition to Ord A was reported by Yabe et al. (1999). The sequence comparison of the aflatoxin biosynthesis pathway gene cluster upstream from the polyketide synthase gene *PksA*, showed that *A. flavus* isolates do not contain portions of genes (*cypA* and *norB*) predicted to encode respectively, a cytochrome p450 monooxygenase and an aryl alcohol dehydrogenase. Insertional disruption of *cypA* in *A. parasiticus* yielded transformants that lack the ability to produce G aflatoxins, but not B aflatoxins. The substrate for CypA may be an intermediate formed by oxidative cleavage of the A ring of O-methylsterigmatocystin by OrdA, the P450 monooxygenase required for formation of aflatoxins B1 and B2 (Ehrlich et al. 2004).

Fusarium moniliforme (*Gibberella fujikuroi*) and other species of *Fusarium* produce fumonisins contaminating maize. *FUM5*, the putative polyketide synthase gene involved in fumonisin has been characterized (Proctor et al. 1999, 2004). Direct competitive ELISA was applied to detect and quantify total fumonisin (B1, B2 and B3) in corn infected by *F. proliferatum* (Pascale et al. 1999; Bird et al. 2002). Presence of fumonisin produced by *F. verticillioides* could be detected even in single corn kernel by using reflectance and transmittance visible and near-infrared spectroscopy. The color and chemical constituents of infected kernels form the basis for the detection of fumonisin in single kernels, in a nondestructive manner (Dowell et al. 2002). Disruption of *FUM1* encoding a polyketide synthase required for fumonisin biosynthesis, did not result in loss of pathogenicity on maize, as the maize inoculated with mutant produced ear rot disease in maize as the wild strain of *F. verticillioides* (Desjardins et al. 2002). Contamination of maize kernels with deoxynivalenol (DON), a trichothecene, increased in proportion to the incidence of *F. graminearum* (Ngoko et al. 2001). Disruption of the trichodiene synthase gene (*Tri5*) of *F. graminearum* led to loss of the ability to produce DON. In addition, the spread of the mutant from the inoculated wheat spikelets to other spikelets was inhibited, indicating an important role for DON in the disease progress in infected wheat ears (Bai et al. 2002). Non-trichothecene (DON)-producing isolates of *F. graminearum* obtained by gene disruption, were less efficient in colonizing wheat ears compared with wild-type isolates (Eudes et al. 2001).

In a later investigation, the route of infection of wheat spikes by *F. graminearum* and the role of trichodiene synthase gene in its pathogenicity were studied, using a constitutively GFP-expressing wild-type strain and its knockout mutant, preventing trichothecene synthesis. Trichothocenes were not found to be a virulence factor during infection through the fruit coat. However, the pathogen was blocked in the absence of trichothecenes by the development of heavy cell wall thickening in the rachis node of Nandu wheat, a defense inhibited by the mycotoxin. On the other hand, in barley, hyphae were inhibited at the rachis node and rachilla, limiting infection of adjacenet florets through the phloem and along the surface of the rachis (Jansen et al. 2005). The effects of type B trichothecene, deoxynivalenol (DON) considered to be a virulence factor allowing infection of plants by their trichothecene-producing *Fusarium* spp. were assessed in susceptible *Arabidopsis thaliana* plants. The trichothecene-induced lesions exhibited dead cells, callose deposition, generation of hydrogen peroxide and accumulation of salicylic acid. Further, infiltration of trichothecenes induced rapid and prolonged activation of two mitogen-activated protein kinases. In contrast, type A trichothecenes trigger cell death by activation of an elicitor-like signaling pathway in *Arabidopsis*. The results suggest that DON may inhibit translation in host cells without induction of the elicitor-like signaling pathway (Nishiuchi et al. 2006).

Fusarium graminearum secretes several extracellular enzymes considered to facilitate plant infection. The role of lipases during infection of plants by fungal pathogens such as *Magnaporthe grisea* (Thines et al. 2000) and *Botrytis cinerea* (Comménil et al. 1998) has been investigated. Lesion formation by *B. cinerea* on deatched tomato leaves was suppressed by addition of polyclonal antibodies against extracellular lipase of *B. cinerea* to the conidial supension. The extracellular lipolytic activity of *F. graminearum* was strongly induced in culture by wheat germ oil. Based on this activity, a gene *FGL1* encoding a secreted lipase was isolated, cloned and characterized. Expression analysis revealed that *FGL1* was induced by lipid containing substrates and repressed by glucose *FGL1* transcription could be detected at one day post-infection of wheat spikes and the function of the *FGL1* gene product was verified by specifically demonstrating lipase activity after expression in a heterologous host. When wild-type conidia were supplemented with ebelactone B, a known liapse inhibitor, disease severity was substantially reduced. Reduction in extracellular lipolytic activity in culture and virulence to both wheat and maize were observed in mutants generated by transformation-mediated disruption of FGL1 of *F. graminearum* (Voigt et al. 2005).

Polyketides are a class of secondary metabolites produced by many microorganisms, including the mycotoxigenic fungi that produce aflatoxin, zearalenone and aurofusarin. With the availability of the genomic sequences, the genes encoding polyketide synthases (PKSs) have been selectively cloned. Two major groups of PKSs, the reducing PKSs and the nonreducing PKSs have been recognized (Bingle et al. 1999; Nicholson et al. 2001). The genes responsible for production of zearalenone by *Gibberella zeae* (*F. graminearum*) were identified. In addition, a complete set of 15 PKS genes were identified and characterized genetically. The functions of the products of these PKS genes in the life cycle of *G. zeae* were determined by genetically disrupting each of these PKS genes and characterizing the mutant isolates. Five of the 15 PKS genes were found to be involved in the biosynthesis of the mycotoxins zearalenone, aurofusarin and fusarin and the black perithecial pigement. Diverse patterns of expression during grain colonization, plant colonization, sexual development and mycelial growth were observed. This investigation appears to be the first comprehensive analysis of all predicted PKS genes from a single filamentous fungus (Gaffoor et al. 2005).

The red pigementation of *F. graminearum* and related species, has been shown to be due to the deposition of aurofusarin in the cell walls. Aurofusarin deficient mutants were generated by random and targeted mutagenesis in *F. graminearum* and *F. pseudograminearum* to understand the role of aurofusarin (a polyketide) in fungal physiology and pathogenicity. A gene cluster-including the *F. graminearum PKS12* gene was found to be responsible for the biosynthesis of aurofusarin. Targeted mutagenesis was performed to confirm that *PKS12* encodes the precursor for aurofusarin. All mutants were fully pathogenic on wheat and barley. Analysis of aurofusarin deficient mutants, employing HPLC technique confirmed the absence of aurofusarin and an increse in the level of the mycotoxin zearalenone production by the mutants (Malz et al. 2005). Further study was taken up to elucidate mechanism of production of elevated levels of zearalenone (ZON) by some aurofusarin mutants. An analysis of transcripts from polyketide synthase genes identified in *F. graminearum* was carried out for these mutants. *PKS4* was the only gene with an enoyl reductase domain that had a higher level of transcription in the aurofusarin mutants than in the wild-type. The central part of the *PKS4* gene was replaced with a *hygB* resistance gene, using an *Agrobacterium tumefaciens*-mediated transformation protocol. The *PKS4* replacement did not produce ZON, confirming that *PKS4* encodes an enzyme required for the production of ZON in *F. graminearum*. Barley root infection studies indicated that there was no change in the pathogenicity of *PKS4* mutant (Lysøe et al. 2006).

In a similar study, a gene cluster containing *PKS12*, the polyketide synthase gene responsible for the production of the pigement aurofusarin was analyzed by gene replacement employing *Agrobacterium tumefaciens*-mediated transformation to determine the biosynthesis pathway of aurofusarin. Replacement of *aurR1* with *hygB* revealed that it encoded a positively acting transcription factor that was needed for the full expression of *PKS12, aurJ, aurF, gip1* and FGO2329.1 which belong to the gene cluster (Frandsen et al. 2006). Earlier study by Kim et al. (2005b) showed that disruption of the laccase encoding gene *gip1* in the cluster resulted in the loss of aurofusarin production, indicating that an as yet uncharacterized yellow pigement was also present in the wild-type strain of *F. graminearum*. Deletion of AurR1 and PKS12 abrogated the ability to produce both aurofusarin and rubrofusarin. A five-step enzyme catalysed pathway for biosynthesis of aurofusarin with rubrofusarin, as an intermediate is considered to operate based on bio-and chemo-informatics combined with chemical analysis of replacement mutants (Frandsen et al. 2006).

2.1.3.4 Production of Ethylene by Fungal Pathogens

Production of ethylene by host plants and the fungal pathogens has different kinds of effects on them. Ethylene production by infected plants is an early resistance response leading to activation of plant defense pathways. But senescence is acclerated by ethylene and its biosynthesis in infected tissues is enhanced. On the other hand, plant pathogens are also able to produce ethylene affecting both the plants and interacting pathogens (Boller 1991; Chagué et al. 2006). An increase in the rate of ethylene production in the surface layers of sweet potato was observed, in response to invasion by *Ceratocystis fimbriata* causing black rot disease (Okumura et al. 1999; Yoshioka et al. 2001). Ethylene produced by the host specifically at ripening terminates the appressorial dormancy, since it acts as a signal for the induction of appressorium. The activities of PG as well as abscission enzymes and cellulase in oranges were enhanced by treatment with ethylene. This in turn increased infection by *Diplodia natalensis* causing stem-end rot disease, indicating a possible role for these enzymes in pathogenesis (Brown and Burns 1998).

Ethylene has been shown to be involed in the regulation of genes such as chitinase, endochitinase, β-1,3-glucanases and thaumatin-like proteins (TLPs) which are actively expressed, as revealed by the accumulation of transcripts in ripening banana fruit (Medina-Suarez et al. 1997; Clendennen and May 1997). *Colletotrichum acutatum* infects citrus petals and induces premature fruit drop and the formation of persistent calyces. Differential gene expression, in addition to the accumulation of hormones and other growth regulators were investigated. Ethylene evolution showed 3-fold increase, whereas IAA accumulation was as high as 140 times. Northern blot analyses revealed that the genes encoding ACC oxidase or ACC synthase and 12-oxo-phytodienoic acid (12-oxo-PDA) reductase were highly expressed in affected flowers. The gene encoding auxin-related proteins also were upregulated (Lahey et al. 2004). The effects of ethylene on *Botrytis cinerea*, causing grey mold disease, were determined. Exposure of *B. cinerea* to ethylene inhibited its mycelial growth and induced transcriptional changes in the large number of genes. Expression of a putative pathogenicity fungal gene *bcspl1* was increased at 24 h after inoculation in ethylene-producing *Nicotiana benthamiana* plants but only at 48 h after inoculation in ethylene non-producing plants. The results show that a G-protein signaling pathway partially mediate the response of *B. cinerea* to ethylene and ethylene-induced host resistance might involve effects of plant ethylene on both plant and the pathogen (Chagué et al. 2006).

2.1.3.5 Induction of Pathogenesis-Related (PR) Proteins

Pathogenesis-related (PR-) proteins defined as proteins coded for by the host plant, but induced specifically in response to pathogen attack or related situations (Van Loon and Van Strien 1999) have been detected in several pathosystems. The PR-proteins are divided into 14 different families and classified on the basis of similarities in molecular weight, amino acid sequences and their enzymatic and biological activities (Van Loon 1999). The rate of production and distribution of PR-proteins in response to infection by pathogens may vary depending on the resistance / susceptibility levels of cultivars / genotypes. PR-proteins are considered to play important role in plant defense. The major acidic PR proteins (PR-1–PR-5) are localized in the intercellular space of leaf tissues where it is possible for them to come in contact with invading fungal and bacterial pathogens. Barley leaves were inoculated with *Drechslera teres, Puccinia hordei* or treated with *D. teres* toxin. The PR-1a and b, PR-5, β-glucanases and chitinases were found to be the same as with after infection with *B. graminis*, as revealed by immunodetection on Western blots (Reiss and Bryngelsson 1996). Detection and immunolocalization of PR proteins in the seminal root tissues of barley and wheat infected with *Bipolaris sorokiniana* was studied by using respective antisera by Western blot analysis and immunological techniques. Accumulation of PR-1 and PR-5 proteins in the roots of both wheat and barley, as revealed by Western blot analysis, was evident. Root inoculation with *B. sorokiniana* induced PR-protein accumulation in the leaves of barley, but not in wheat as revealed by ELISA format. Immunohistological labeling with antiserum against PR-3a protein showed a strong signal in the cortex tissue of uninoculated plants indicating the constitutive expression of this PR protein. Accumulation of PR proteins in leaves of barley, but not in wheat following inoculation of roots indicated systemic induction of PR genes (Liljeroth et al. 2001).

The accumulation and localization of PR-1, the most abundant PR protein in infected barley leaves, in response to infection by the hemibiotrophic pathogen *B. sorokiniana* were studied using immunohisto-and cytochemical methods (Schäfer et al. 2004). In a later study subcellular localization of a basic PR-1 in barley was investigated. The hyphae of *B. sorokiniana* invades primarily the intercellular space of mesophyll tissue. PR-1 accumulated on the outer electron dense layer of the cell wall of the primary hyphae. In contrast, no PR-1 labeling was detected on the secondary hyphae. PR-1 was also recovered in newly accumulated electron-dense material in the intercellular space. Subcellularly, PR-1 was detected on the outer cell wall layer and cytoplasm of primary hyphae intercellular electron dense material in the junction between host cells and host cell wall appositions (Santen et al. 2005).

Two novel proteins (newly formed PR-17 family) were detected following inoculation with *Blumeria* (*Erysiphe*) *graminis* f.sp. *hordei* in barley. The antisera raised against these proteins HvPR-17a and HvPR-17b proteins, their accumulation in the mesophyll apoplast as well as in leaf epidermis (the only tissue invaded by this pathogen with haustoria) was recorded (Christensen et al. 2002). Accumulation of PR-proteins in the intercellular spaces of leaves of rose shoots inoculated with *Diplocarpon rosae* (incitant of black spot disease) was observed. The serologic relationship of rose PR-proteins with tobacco PR-proteins was indicated by Western blot analysis (Suo and Leung 2002).

2.1.3.6 Expression of Fungal Pathogen Genes

The in planta expressed genes form a group that enclosed genes that are involved in establishment and maintenance of infection but they are not directly associated with acquisition of nutrients from the host tissues. The elements that control in planta gene expression and genes that affect disease development after the initial contact has been established. Various sets of fungal genes operate during pathogenesis. Following penetration and invasion of the first few cells, new genes are required for further disease progression. In necrotrophic fungal pathogens, activation of toxin and cell wall-degrading enzyme (CWDE) are induced. Specific transcription factors such as *Cochliobolus carbonum ccSNF1* that controls in planta expression of CWDE are involved in the regulation of such genes. On ther other hand, in the case of hemibiotrophic and biotrophic fungal pathogens, genes are required for regulation of formation of infection structures such as infection vesicles and haustoria, prevention of elicitation of host defenses (Tudzynski and Sharon 2003). The *CLTA1*, a GAL4-like gene belonging to fungal zinc cluster family of transcriptional activators was isolated from a nonpathogenic mutant of *Colletotrichum lindemuthianum*. The mutant was able to induce small hypersensitive-like necroses and further progress was blocked in transition from biotrophic to necrotrophic phase (Dufrense et al. 2000). The *CLTA1* was hypothesized to be a specific transition factor that activates biotrophic-specific genes in *C. lindemuthianum*-bean pathosystem. The biotrophy-related gene *C1H1* may have such a function (Perfect et al. 2000). The Cih1p glycoprotein was found to be embedded in the extracellular matrix that separates the fungal cell wall from the host plasma membrane (Mendgen and Hahn 2002). The expression of Cih1p which was uniquely present at the interface of the extracellular matrix, was switched off at the onset of necrotrophic development (Perfect et al. 2000).

Potato cv. Russet Burbank (RB), and cv. Kennebec (KB), susceptible and tolerant to *Phytophthora infestans* were used to determine the differential expression of *pal1* and *hmgr2* which encode PAL (phenylalamine-ammonia lyase) and 3-hydroxy, 3-methylglutaryl CoA reductase HMGR, the key enzymes in the phenyl propanoid and terpenoid pathways respectively. The potato cultivars KB and RB were inoculated with two genotypes US-1 and US-8 of *P. infestans*. The accumulation of *pal1* transcripts was weaker in response to US-8 as compared to US-1 and occurred earlier in KB than in RB. The stronger expression of *pal1* in response to US-1 as compared to US-8 was considered to be due to defense gene suppression by the latter. The hmgr2 transcripts, on the other hand, did not accumulate rapidly in RB as compared to KB inoculated with either US-1 or US-8 (Wang et al. 2004a).

Expression of pathogen and host genes during the susceptible interaction between postharvest pathogens and produce has been studied. In the *Penicillium expansum*–apple fruit (cv. Golden Delicious) pathosystem, Northern blot analysis showed that 18 genes were highly expressed. Among the differentially expressed genes, one gene of pathogen origin encoding an unknown protein and two apple genes were identified. The apple genes were homologous to a β-glucosidase and a phosphatase 2C. These genes were exclusively expressed during the infection process. Several up-regulated pathogen genes appeared to mediate adaptive responses to the environment (Sánchez-Torres and González-Candelas 2003). Thioredoxin (TRX) proteins are ubiquitous small proteins existing as four major forms of TRXs named m, f, h and o. TRXh located in cytosol and mitochondria in plant cells may act as messenger proteins in plants. They have been detected as one of the major proteins in the phloem sap and have the capacity to mediate their own cell-to-cell transport through plasmodesmata. The cTRXh cDNA from grapefruit peel tissue was isolated and its expression was induced by infection with *Penicillium digitatum*. The cTRXh had a total length of 785 bp with an open reading frame (ORF) of 369 bp and encoded a predicted polypeptide of 123 amino acids with a molecular mass of 13.4 kDa. Expression analysis experiments showed that cTRXh mRNA levels increased in grapefruit peel tissue upon infection by *P. digitatum* (Hasdai et al. 2004).

In the recent years, the complexity of gene expression during interaction between pathogens, and host plant species is analyzed by using cDNA microarrays developed from expressed sequence tags (ESTs). The ESTs are partial sequences of cDNA clones in an expressed cDNA library and they can be used to identify all of the unique sequences (genes) in order to study their functions. The identified unique cDNA sequences can be used to fabricate a cDNA microarray for functional study. Microarrays have been used to analyze a sample for the presence of gene variations or mutation (genotyping) or for patterns of gene expression (Aharoni and Vorst 2001).

Expressed sequence tags (ESTs) (over 28,000) were produced from cDNA libraries representing a variety of growth conditions and cell types of *Magnaporthe grisea* strains. Both pathogenic strains and nonpathogenic strain bearing a mutation in the PMK1 mitogen-activated protein kinase (MAPK) were examined. Majority of the 23,000 ESTs, were clustered (3050 contigs) leaving 5127 singleton sequences. The estimate of 8177 unique sequences indicated that over half of the genes of *M. grisea* were represented in the ESTs. Growth and cell type-specific patterns of gene expression were identified. In addition, criteria for identification of fungal genes involved in pathogenesis were established in this investigation for further analysis (Ebbole et al. 2004). The MAPK cascades have significant role in plant growth and development including biotic and abiotic stress responses. In *Arabidopsis* and rice, 20 and 5 MAPKs have been identified. In the rice MAPK gene (OsMPK) family, 17 members were identified. Following inoculation with *M. grisea*, 9 of 17 OsMPK genes were induced at the mRNA level during either early, late or both stages of infection. Four of the pathogen-induced OsMPK genes were associated with host-cell death in the lesion-mimic rice mutant and eight of them were differentially induced in response to defense singal molecules such as jasmonic acid, salicylic acid, abscisic acid and ethylene. About half of the rice MAPK genes appear to be associated with infection by *M. grisea* and host defense responses (Reyna and Yang 2006).

Large-scale cDNA and genomic sequencing was taken up to overview the gene content of *Phytophthora infestans*. Expressed sequence tags (ESTs) (75,757) from 20 cDNA libraries represented a broad range of growth conditions, stress responses and developmental stages. From *P. infestans*–potato and tomato interactions, 963 pathogen ESTs were identified. The expressed gene content of mycelial and spore-related stages showed major differences as inferred from cluster analysis of ESTs. Conserved elements of pathogenicity such as class III pectate lyases were recognized based on comparisons with databases of fungal pathogenicity genes. Furthermore, 37 genes encoding components of flagella were also identified (Randall et al. 2005). In another study, the molecular basis of the compatible interaction of *P. infestans* with potato was investigated. cDNA microarrays were applied to identify the genes that were differentially expressed in the host plant. In the infected plants 643 of the 7680 cDNA clones (12.9%) represented in the array were differentially expressed, as compared with mock-inoculated control plants. These genes were classified into 8 groups. Among the three genes derived from two down-regulated clusters, one gene plastidic carbonic an- hydrase (CA) exhibited a very different expression pattern in compatible vs incompatible interactions. Virus-induced gene silencing experiment indicated that suppression of CA increases susceptibility to *P. infestans* (Restrepo et al. 2005).

The patterns of gene expression in soybean and *Phytophthora sojae* during an infection time course were determined, by constructing a 4896-gene microarray of host and pathogen cDNA transcripts. rRNAs from soybean and *P. sojae* were used to workout the ratio of host and pathogen RNA present in the mixed samples. This ratio showed significant changes between 12 and 24 h after infection indicating the rapid growth and multiplication of pathogen within host tissues. The genes encoding enzymes of phytoalexin biosynthesis and defense and PR-proteins were strongly up-regulated during infection as shown by the microarray analysis. The number of pathogen genes expressed during infection reached a maximum at 24 h. The possibility of using a single microarray to simultaneously probe gene expression in two interacting organisms was indicated by this study (Moy et al. 2004).

The genes of *Phytophthora sojae* causing a serious disease of soybeans and newly emerged pathogen *P. ramorum* inducing sudden oak death have been fully sequenced. The *P. sojae* genome contains 499 single nucleotide polymorphisms (SNPs), whereas $\sim 13,643$ SNPs have been identified in the genome of *P. ramorum.* Since the two pathogens have entirely different host ranges, some of their genes involved in the interaction with their respective hosts might have rapidly diverged between the two species as a result of strong selection for effective pathogenesis. The nature of secreted proteins may primarily determine the host–pathogen interactions. The predicted secretomes for *P. sojae* and *P. ramorum* are 1464 and 1188 proteins respectively. It seems that the secretomes are evolving more rapidly than their proteomes. A comparison of the genomes of these two *Phytophthora* spp. reveals a rapid expansion and diversification of many protein families associated with pathogenesis such as hydrolases, ABC transporters, protein toxins, proteinase inhibitors and in especially, a superfamily of 700 proteins with similarity to known oomycete avirulence genes (Tyler et al. 2006).

Plant pathogens, along with their hosts, have coevolved diverse defense and counter-defense strategies resulting in successful infection or development of resistance to the diseases caused by them. *Phytophthora sojae* has evolved a counter defense mechanism to nullify the effects of β-1,3-glucanases produced by soybean plants. *P. sojae* secretes glucanase activity (Rose et al. 2002). Many species of *Phytophthora* produce a diverse family of Kazal-like extracellular Ser protease inhibitors with about 35 members (Tian et al. 2004). Two Kazal-like inhibitors EPI1 and EPI10 of 14 inhibitors produced by *P. infestans,* could bind and inhibit the PR-P69B subtilisin-like Ser protease of the host plant tomato, indicating their counter defense activity (Tian et al. 2004, 2005). Further studies on unigenes of *P. infestans* with predicted signal peptides showed that a novel family of putative protease inhibitors with cystin-like domains (EPIC1–EPIC4) was also produced as a counter-defense strategy. The genes *epiC1* and *epiC2* were up-regulated during infection of tomato, suggesting a role for these gene products during infection. The biochemical functional analysis revealed that EPIC2B interacted with and inhibited a novel papain-like extracellular Cys protease designated Phytophthora Inhibited

Protease 1 (PIP1). The results suggested an interplay between host proteases of diverse catalytic families and pathogen inhibitors as a general defense-counterdefense process operating in plant–pathogen interactions (Tian et al. 2007).

In the cotton-*Fusarium oxysporum* f.sp. *vasinfectum* pathosystem, different gene expression profile changes in cotton root and hypocotyls tissues were recorded following microarray analysis of large-scale temporal and tissue-specific plant gene expression changes during a compatible interaction. Enhanced expression of defense-related genes was observed in hypocotyl tissues, whereas few changes in the expression levels of defense-related genes could be seen in the root tissues. In infected roots, more plant genes were repressed than induced especially at the earlier stages of infection. Potential new defense responses were identified such as the biosynthesis of lignans, in addition to many known cotton defense responses including induction of PR-genes and gossypol biosynthesis genes. The involvement of genes coding for phytohormones ethylene and auxin in pathogenesis was implicated by gene expression results (Dowd et al. 2004). *Fusarium oxysporum* causing wilt diseases of different crops interacts with roots or hypocotyls of seedlings and adult plants. It produces necrosis and ethylene-inducing 124-kDa peptide (Nep1) (Bailey et al. 1997). The presence of complex multiple copies of *NEP1* orthologs in five species of *Phytophthora* was reported by Bae et al. (2005). In a later investigation, the responses of *Arabidopsis* to Nep1 were assessed. Nep1 from *F. oxysproum*, following application on *Arabidopsis* inhibited both root and cotyledon growth and triggered cell death resulting in necrotic spots. Nep1 was localized to the cell wall and cytosol as revealed by immunolocalization observations. The contents of most water-soluble metabolites were reduced at 6 h after Nep1 treatment, indicating loss of cellular membrane integrity. The results of quantitative PCR confirmed that of microarray analysis and showed that the induction of genes localized in the chloroplast, mitochondria and plasma membrane and genes responsive to calcium / calmodulin complexes, ethylene, jasmonate, ethylene biosynthesis and cell death was the resultant of Nep1 treatment. Nep 1 can facilitate cell death as a component of diseases induced by necrotrophic pathogens of plants (Bae et al. 2006) (Fig. 2.6).

The expression profiles in two peanut genotypes susceptible or resistant to late leaf spot disease (*Cercosporidium personatum*) were determined by using cDNA microarray technique. Microarrays containing 384 unigenes selected from two EST cDNA libraries challenged by abiotic and biotic stresses were obtained. A total of 112 spots representing 56 genes in several functional categories were detected as up-regulated genes. Seventeen genes, each matching gene with known function in Gen- Bank were selected for validation of their expression levels using real-time PCR. These genes were more intensively expressed in resistant genotype as a result of response to inoculation with *C. personatum* than in susceptible genotype (Luo et al. 2005).

The ability of RNA species to induce RNA silencing in *Magnaporthe grisea* (rice blast disease) was assessed by introducing plasmid constructs expressing sense, antisense and hairpin RNAs into an enhanced green fluorescence protein (eGFP) expressing transformant. The fluorescence of eGFP in the transformant was silenced much more efficiently by hairpin RNA of eGFP than by other RNA species. The

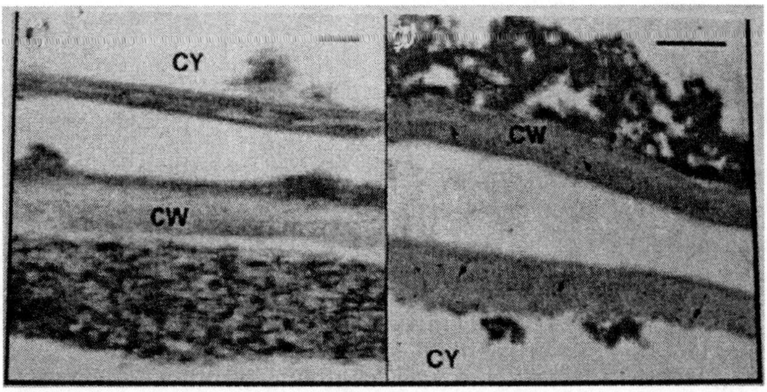

Fig. 2.6 Immunolocalization of Nep 1 in plants grown from sterilized *Arabidopsis* seeds mixed with Nep 1
Left: Absence of gold particles in control plants; *Right*: Presence of gold particles over the plant cell wall (CW)/membrane and cytosol (CY). (Courtesy of Bae et al. 2006; The American Society of Plant Biologists, Rockville, MD, USA)

accumulation of eGFP mRNA was substantially reduced in the silenced transformants, but no methylation of the promoter or coding region was involved in it. Furthermore, the presence of small interfering RNAs (siRNAs) was detected only in the silenced transformants. It is of interest to note that the siRNAs consisted of RNA molecules with at least 3 different sizes ranging from 19 to 23 nucleotides and all of them contained both sense and antisense strands of eGFPgene. The results indicate the potential of using RNA silencing operating in *M. grisea*, as a tool for genome-wide gene analysis (Kadotani et al. 2003).

Ustilago maydis induces tumors (galls) in maize in place of earheads/grains. Teliospore production and germination are important steps in its life cycle. Information on the specific molecular events that are involved in the process of teliospore germination is scanty. By performing construction and hybridization of microarrays containing a set of 3918 non-redundant cDNAs, the genes that were differentially regulated during germination, were identified. During teliospore germination, different sets of genes were upregulated or down-regulated relative to their expression levels prior to induction of germination. The genes identified were involved in the early stages of meiosis, initiation of transcription and translation, protein turnover and protein assembly, as well as signal transduction, metabolism and cell defense (Zahiri et al. 2005).

2.1.4 Symptom Expression

After successful penetration of the host tissue / surface by the fungal pathogens, enzymatic degradation of pectin polymers in plant cell walls is considered as an important aspect of parasitism. The degradation process requires the concerted action of a number of extracellular enzymes such as pectin methyl esterase (PME), pectin and pectate lyase and endo-and exo-polygalacturonases produced by the pathogens. Different endo-PG isoforms of *Botrytis cinerea* are encoded by a gene family of at least six genes *Bcpg1–Bcpg6* (ten Have et al. 1998; Wubben et al. 1999). The genes *Bcpg1* and *Bcpg2* are constitutively expressed, while *Bcpg4* and *Bcpg6* are expressed following induction of expression by galacturonic acid (Wubben et al. 2000). The growth of *B. cinerea* in planta was shown to be accompanied by the degradation of pectin and endo-PG (*Bcpg*) genes were expressed during infection of plant tissues. The role of pectin methyl esterase (PME) in inducing lesions in grapevine and tomato leaves appeared to be not significant. All four single *ΔBcpme* mutants (ΔBcpme 1-1B, ΔBcpme-1-11A, ΔBcpme-2-1B and ΔBcpme 2-2B), produced lesions equal to those by wild-type strain in size (Kars et al. 2005a, b).

The primary infectious propagules of most Oomycetes, including *Phytophthora infestans* causing potato late blight disease, are asexual sporangia. They produce motile zoospores, whose primary biological role is to help dispersal of the pathogen to a suitable infection site, mediated in part by their attraction towards chemical signals from the host plant. A protein kinase gene, *Pipak1* induced during early zoosporogenesis plays key roles in several stages of the life cycle of *P. infestans* related to plant infection. Transformants of *P. infestans* stably silenced for *Pibzp1* were generated. None of the five silenced transformants could initiate successful infection as revealed by the absence of any spreading lesion or sporulation (Fig. 2.7), indicating the requirement of this gene for infection process (Blanco and Judelson 2005).

Colletotrichum actutatum causes post-bloom fruit drop (PFD), Key lime anthracnose (KLA) and postharvest anthracnose diseases. Pathologically the PFD isolates infect inducing water-soaked lesions on flowers and young fruits drop and reproduce only on flowers, while the KLA isolates are mostly necrotrophic and infect only Key lime inducing anthracnose on twigs, shoots, leaves and flowers. The genes encoding ethylene and jasmonic acid biosynthetic pathways and IAA regulation were highly up-regulated in PFD-affected petals (Li et al. 2003). These genes markedly enhanced the levels of hormones and growth regulators (Lahey et al. 2004). The detached leaf assay showed that the wild-type isolate was able to induce typical brownish necrosis in Key lime leaves, whereas the *klap-1*–disrupted isolates (Δklap 45 and Δklap48) could not produce any necrotic lesions. These transformants, however, did cause water-soaked orange brown lesions similar to the wild-type on flower petals. After complementation with KLAP1, 16 of the 30 transformants tested, formed necrotic lesions like the wild-type, confirming the role of *KLAP1* in pathogenicity on Key lime (Chen et al. 2005).

Alternaria citri infects several citrus cultivars causing Alternaria black rot, a postharvest fruit disease. The involvement of the pathogen enzyme endo-polygalacturonase (endo-PG) in tissue maceration and consequent rotting has been demonstrated. The endo-PGs were purified and their genes from two different *Alternaria* spp. were cloned: *Acpg1* from *A. citri* and *Aapg1* from *A. alternata* (rough lemon pathotype). Although the sequences of these two genes and the biochemical characteristics of the enzymes they encode, were highly similar, the phenotypes of the mutants obtained by disrupting these genes in the respective

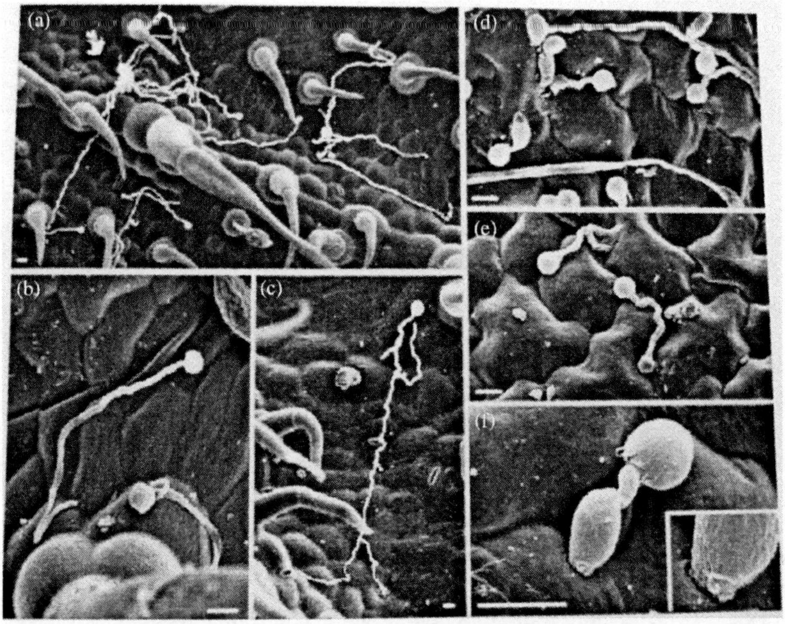

Fig. 2.7 Scanning electron micrographs depicting effect of Pibzp1 silencing on development of *Phytophthora infestans*
(a)–(c) The silenced strain with relatively long germtubes without appressoria; **(d)–(f)** The wild-type strain with short germtubes and appressoria facilitate efficient penetration of leaflet surface; Bar: 10 μm. (Courtesy of Blanco and Judelson 2005; Blackwell Publishing Ltd, Oxford, UK)

pathogens were entirely different. The mutant of *A. citri* was reduced in its ability to cause black rot symptoms. However, no change in the pathogenicity of the mutant of *A. alternata* was noted, indicating different roles of endoPG depending on the species of pathogen tested (Isshiki et al. 2001). The endo-PG minus mutant of *A. citri* induced significantly less black rot in citrus fruit. In a further study, a random mutation analysis of pathogenicity was performed using a restriction enzyme-mediated integration (REM1) procedure. Among 1694 transformants of *A. citri*, pathogenicity was abrogated in three isolates as revealed by the citrus peel assay. One of these three mutants was a histidine auxotroph. The gene *AcIGPD* that encodes imidazole glycerol phosphate dehydratase, the sixth enzyme in the histidine biosynthetic pathway, was identified and cloned. The mutant with disrupted *AcIGPD* could induce less intensive symptom of black rot disease, indicating the involvement of this gene in symptom production by *A. citri* (Katoh et al. 2006).

Alternaria citri, has the gene *AcCreA* encoding a catabolite repression element. A mutant ACOEC2 overexpressing *AcCreA* exhibited normal growth on pectin medium and on segments of citrus fruit peel or juice sac area. In addition, production of endo-PG, an essential virulence factor did not show variation compared with wild-type strain. However, the mutant strain showed repression of endo-PG gene *Acpg1* and consequent loss of endoPG production, when glucose was added to the medium. The wild-type strain affected the central axis of the citrus fruit primarily, without induction of rotting of the juice sac area. In contrast AcOEC2 caused severe black rot symptoms in both the central axis and juice sac areas. The results indicated that virulence and type of symptoms induced by *A. citri* may depend on AcCreA-mediated catabolite repression and that inability of the wild-type strain of *A. citri* to infect the juice sac area might be due to carbon catabolite repression by sugars in the juice of the juice sac area (Katoh et al. 2007).

Botrytis cinerea causing gray mold disease of several fruit crops, produces six endo-PGs (BcPG1–BcPG6). The PGs and their mutated forms of BcPG1 and BcPG2 were expressed transiently in leaves of *Nicotiana benthamiana* using agroinfiltration. Expression of BcPG2 induced the most severe symptoms including wilting and necrosis, whereas the expression of BcPG1, BcPG4 and BcPG5 and the mutants of BcPG1, BcPG4 and BcPG5 and the mutant BcPG1-D203A could cause some symptoms. On the other hand, no symptom was induced by BcPG3, BcPG6 and mutant BcPG2-D192A. BcPG2 had been found to be required for virulence of *B. cinerea*. Coexpression of *Bcpg2* and the *Vvpgip1* gene from *Vitis vinefera* in *N. benthamiana* led to marked reduction in the symptom intensity that could be induced by BcPG2 in planta. There was no in vitro interaction between VvPGIP1 and BcPG2. The results suggested that in planta environment could have favored interaction between VvPGIP1 and BcPG2 resulting in reduction in symptom severity, although such an interaction between these proteins could not be demonstrated in vitro (Joubert et al. 2007).

Fungal pathogens are known to elaborate host-specific (selective) toxins that can induce characteristic symptoms as the pathogen itself in compatible interactions. In these pathosystems, the symptom expression depends on the conditions favoring toxin production by the pathogens. In *Cochliobolus heterosporus* race T, production of T-toxin active against Texas male sterile (tms) lines of maize, is goverened by a single locus (*TOX1*). *C. carbonum* also infecting maize causes leaf spot and ear mold disease. A single locus *TOX2* is responsible for the synthesis of HC-Toxin (Scheffer and Yoder 1972). Failure to produce HC toxin by the pathogen or inactivation of HC-toxin by host plants results in the absence of symptoms. The gene *TOXC* is present only in strains of *C. carbonum* that can produce HC-toxins. In both pathogens pivotal role of toxins in symptom development has been demonstrated clearly (Schäfer 1994).

Programmed cell death (PCD) is a pathological process of cell necrosis as a result of normal tissue homeostasis and stress responses (Wertz and Hanley 1996). Different physiological mechanisms operate in multicellular organisms to remove infected, mutated or damaged cells from healthy tissues (Vaux and Korsmeyer 1999). Activation of PCD is encountered in incompatible interaction to inhibit proliferation of pathogen and this response of the host is designated as hypersensitive response (HR). However, some necrotrophic pathogens also trigger PCD host cells mostly by the action of secreted toxins. The necrotrophic pathogens appear to establish necrotrophic infection by exploiting PCD machinery of the host (Dickman et al. 2001). *Cochliobolus victoriae* produces a host-selective toxin, victorin that induces specific symptoms in oat plants. The response of plants to victorin is governed by a single dominant gene (Vb) and parallels the response of the plant to pathogen itself. All cells of victorin-treated oat leaves died by 12 h after treatment. However, accumulation of pathogenesis-related (PR) proteins (PR-1 and PR-10) was still increasing, possibly because of activity of the defense mechanisms (Yao et al. 2001). Further, rRNA and mRNA were specifically degraded in oat cells during apoptic cell death induced by victorin. The rRNA species from the cytosol, mitochondria and chloroplasts were degraded via specific degradation intermediates due to the action of victorin (Hoat et al. 2006).

Alternaria alternata exists as seven recognized pathotypes (pathogenic variants) which produce host-specific toxins (HSTs). The pathotypes can be differentiated by their host range which is determined mainly by the HST produced by the pathotype concerned (Kohmoto et al. 1993). *A. alternata* infecting tomato produces a HST, AAL-toxin. Mutants deficient on AAL-toxin production were non-pathogenic, indicating the requirement of the HST for pathogenesis (Akamatsu et al. 1999). AAL-toxin an aminopentol ester related structurally to sphingosine and sphingamine induces apoptosis in toxin-sensitive tomato cultivars and animals (Brandwegt et al. 2000; Wolpert et al. 2002). The pathotype *A. alternata* causing Alternaria brown spot of tangerines produces ACT-toxin and the pathotype causing Alternaria leaf spot of rough lemon produces ACR-toxin. But they produce similar symptoms on leaves and young fruits of their respective hosts. A new strain (BC3-5-1-OS2A) isolated from rough lemon in Florida produced both ACT-and ACR-toxins. This strain was shown to carry genomic regions controlling production of both toxins and they were present on two distinct small chromosomes. The dual host specificity and toxin production by BC3-5-1-OS2A makes it unique and it appears to be the only isolate pathogenic to two different citrus hosts (Masunaka et al. 2005). *Pyrenophora tritici-repentis* produces a host selective toxin that induces necrosis of host tissues. This HST is governed by *TOXA* gene (Ciuffetti et al. 1999). The *TOXB* gene with nine copies induces chlorosis (Martinez et al. 2004; Andrie et al. 2005).

Verticillium dahliae has wide host range causing gradual wilting, senescence, sometimes defoliation and ultimate death of infected plants. The physiological and biochemical changes induced by *V. dahliae* include reductions in water potential, stomatal conductance, rate of carbon assimilation and accumulation of proline, soluble sugars and ABA (Goicoechea et al. 2000; Sadras et al. 2000). The protein-lipopolysaccharide complexes and small peptides (considered as pathogenicity factors) released from *V. dahliae,* induce leaf chlorosis and necrosis and inhibition of root growth (Meyer et al. 1994). In the *A. thaliana* and *V. dahliae* interaction, the symptoms include chlorosis, stunting anthocyanin accumulation, induction of early flowering and death of infected plants. A single dominant locus, ***V***erticillium ***d***ahliae-***t***olerance (VET1) was identified and it may function as a negative regulator of the transition to flowering. *VET1* was mapped on chromosome IV. The results indicated that the ability of *V. dahliae* to induce disease symptom may be connected to the genetic control of development and life span in *Arabidopsis* (Veronese et al. 2003).

2.2 Bacterial Pathogens

2.2.1 Initiation of Infection

Several phytopathogenic bacteria multiply or survive on aerial plant parts without inducing any visible symptoms. This asymptomatic phase, among other consequences, allows bacterial populations to reach sizes permitting development of diseases under favorable conditions. The phenomemon of pathogenesis has been understood to some extent largely due to the advancements made in molecular genetics of bacterial pathogens. Moelcular genetics of *Agrobacterium tumefaciens, Clavibacter* spp., *Erwinia* spp., *Pseudomonas* spp. and *Xanthomonas* spp. have been studied more intensively throwing light on the role of bacterial genes involved in pathogenesis.

Prior to the initiation of infection by pathogenic bacterial species, stable biofilm population size has to be established. Epiphytic multiplication is influenced by environmental factors and / or host physiology. Localization in protected sites and clustering in structures such as biofilms leading to enhanced resistance to stress are considered to be adaptive traits of epiphytic bacteria. The most common sites of bacterial colonization have been shown to be stomates, the base of trichomes and depressions along veins. The biofilms, in these environments provide protection to the microbial pathogens from external environmental conditions such as desiccation and a favourable mileu for multiplication. *Xanthomonas axonopodis* pv. *phaseoli* (*Xap*) could survive on leaf surfaces and endophytically (Weller and Saettler 1980). The occurrence of *Xap* population as biofilms on bean leaves was examined during three field experiments on plots established with naturally contaminated bean seeds. The biofilms population sizes remained stable throughout the growing season (around 10^5 CFU/g of fresh weight) while solitary population sizes were more abundant. Enterobacterial repetitive intergenic consensus (ERIC) fingerprinting technique was applied for genomic characterization of solitary and aggregated strain of *Xap*. All 40 solitary and aggregated strains tested were found to have the same genomic profile based on ERIC-PCR analysis (Jacques et al. 2005).

In the recent years evidence is accumulating to indicate that an extensive range of microorganisms have the ability to perceive and respond to the presence of neighboring populations. The term "quorum-sensing" is used to describe this kind of density-dependent phenomenon. This system is accomplished by the extracellular accumulation of small, self-generated chemical signaling moieties that induce a concerted effort on behalf of a population to produce the desired phenotypic effect (De Kievit and Iglweski 2000). The biological phenomenon of microbial communication was first discovered in the maize bacteria such as *Vibrio fischeri* and *V. harveyi* which produce luminescence when they reach high cell densities, but remain non-luminescent at low cell density (Fuqua et al. 1994). Microbially-derived signaling molecules can be divided into two main classes (i) amino acids and short peptide derivatives, commonly utilized by Gram-positive bacteria (Lazazzera and Grossman 1988; Shapiro 1998) and (ii) fatty acid derivatives known as homoserine lactones (HSLs) frequently utilized by Gram negative bacteria (Dunny and Winans 1999; Whitehead et al. 2001). The increase in the luminescence in *V. fischeri* was attributed to the transcriptional regulation of the enzyme luciferase, which in turn corresponded to a threshold density of cells. The bacterial density is assessed based on the release of the chemical signaling molecules or autoinducers. The autoinducer then establishes a communication between the cells that gets reflected in the expression of the luciferase gene (*lux*) (Hastings and Greenberg 1999).

The *lux* gene organization, regulation, function and molecular characterization of the luminescence system of *V. fischeri* MJ1 were investigated by cloning a 9 kb fragment of its DNA that encodes all functions required for autoinducible luminescence in the heterologous host *Escherichia coli.* The bioluminescence gene cluster of *V. fischeri* contains eight *lux* genes (*lux A-E, lux G, lux I* and *luxR*) which are arranged in two bidirectionally transcribed operons separated by about 218-bp (Engerbrecht and Silverman 1987). The products of both *luxI* and *luxR* genes function as regulators of bioluminescence. The *luxI* gene is the only *V. fischeri* gene required for synthesis of the autoinducer, 3-oxo-hexanoylhomserine lactone (3-oxo-C6-HSL) or OOHL in *E. coli* (Engerbrecht et al. 1983; Engerbrecht and Silverman 1987). The *lux A* and *luxB* genes encode subunits of the heterodimeric luciferase enzyme which catalyses the oxidation of an aldehyde. Emission of blue-green light, with a maximum intensity at 490 nm, accompanying the oxidation reaction results in this reaction referred to as bioluminescence. The two bacterial groups viz., Gram-negative and Gram-positive differ in the nature of the signal molecule (Lederberg 1992). The Gram-negative bacteria seem to have the family of N-acyl homoserine lactones (AHLs) as the universal signal factor. The human opportunistic pathogens such as *Pseudomonas aeruginosa* and plant pathogens such as *Erwinia* spp. as well as the symbiotic *Rhizobium* spp. are producers of AHLs as the universal signal factor. An increasing number of Gram-negative bacteria have been reported to have genes similar to either *luxR* or *luxI.* Gram-negative bacteria use luxR-type proteins for autoinduction, whereas Gram-positive bacteria utilize two-component adaptive response proteins for the detection of autoinducers (Pierson and Pierson 1996; Chang and Stewart 1998).

Xanthomonas spp. using non-acyl HSL-based quorum-sensing system is able to gain a competitive advantage over their neighboring bacteria by avoiding such interference and crosstalk. Interspecies communication through the use of auto- inducers has been inferred as a possible mechanism by which the pathogenicity of certain virulent bacterial species such as *Burkholderia cepacia* may be enhanced (McKenny et al. 1995). Evidences indicate a great number of bacteria employ quorum-sensing for regulation of various phenotypes as a part of their pathogenic or symbiotic lifestyles.

Behaviors of several bacterial species are regulated in a cell density-dependent manner of quorum sensing (QS). $N-$ acyl homoserine lactones (AHLs) function as mediation of cell density-dependent communication among Gram-negative bacteria, whereas 2-heptyl-3-hydroxy-4 (1H) – quinolone and autoinducer-2 (4,5-dihydroxy-2,3-pentanedione) are converted into a variety of extracellular molecules (Sperandio et al. 2003). The quinolone signaling molecules appear to have a role in intergenus communication. Swarming motilities have been shown to be regulated by QS in many different bacterial phenotypes. QS is considered to directly regulate the genes for flagellum biosynthesis (Daniels et al. 2004). Flagellum gene regulation may be of two distinct types based on the nature of master regulators. The master transcriptional regulator, FlhDC is used by the bacteria with peritrichous (lateral) flagella for activation of the expression of flagellum genes (Aldrige and Hughes 2002). On the other hand, bacteria with polar flagellum system employ the σ54 dependent on NtrC family of transcriptional activators for flagellum gene expression (Arora et al. 1997).

Burkholderia glumae causing seedling and grain rot in rice produces the phytotoxin toxoflavin which functions as a major virulence factor. The expression of genes involved in the biosynthesis and transport of toxoflavin was shown to be regulated by N-octanoyl homoserine lactone (C8-HSL) (Kim et al. 2004b). The mechanism of flagellum biosynthesis under the control of QS in *B. glumae* was investigated. A major deviation from the current concept for the regulation of bacterial polar flagellum was revealed in *B. glumae.* The C8-HSL-deficient mutant of *B. glumae* was aflagellate and hence unable to swim and swarm at 37°C as the wild-type. A functional role of FlhDC as a regulator of the polarly flagellated *B. glumae* flagella was examined. The QsmR, an IclR-type transcriptional regulator, was identified in the hierarchical gene regulation of *B. glumae* flagella. Since QsmR regulates *flhDC* expression, it represents a novel flagellar regulon controlled by QS. TofR, a cognate C8-HSL receptor, activated *qsmR* expression by binding directly to the *qsmR* promoter region. The *flhDC* homologs were directly activated by FlhDC. Although the nonmotile *qsmR*, *fliA* and *flhDC* mutants produced toxoflavin, the pathogenicity was abolished in these mutants. Variation in the typical regulatory mechanisms of flagellum genes was detected in *B. glumae* which contained FlhDC that is generally present in peritrichous flagellum systems. This study clearly demonstrated that the functional flagella of *B. glumae* have a critical role in pathogenicity suggesting that either QS or flagellum formation could constitute a good target for the control of rice grain rot (Kim et al. 2007).

The bacteria-to-bacteria communication is associated with a quorum-sensing (QS) process based on the production of N-acylhomoserine lactones (HSL). In order to understand the role of HSL throughout plant infection by *Erwinia carotovora* subsp. *atroseptica* (*Eca*), the HSL produced by a specific *Eca* strain virulent on potato was identified. A derivative of this strain that expressed an HSL lactonase gene and produced low amounts of HSL was generated. The results of the experiments showed that HSL production and QS regulate only those traits involved in the second stage of infection, tissue maceration and hypersensitive response in nonhost tobacco plants (Smadja et al. 2004). Among the secreted proteins of *Eca* a low-abundance protein was detected and identified by two dimensional gel electrophoresis and mass spectrophotometry. This protein was a homolog of a *Xanthomonas campestris* avirulence protein with unknown function. This protein designated Svx was shown to be secreted by the type II (out) secretion apparatus which is also responsible for the secretion of the major known virulence factors, Pel C and CelV. Transcription of the *svx* gene was under N-acyl-homoserine lactone (AHL)-mediated quorum sensing control. The mutant of *svx* generated by transposon insertion showed decreased virulence in potato plant assays, revealing a definite role for Svx in pathogenicity of *Eca*. Svx, previously unidentified virulence determinant is secreted by the out machinery and is regulated by quorum sensing, two systems employed by many other virulence factors. The type II secretory system is shown to be a conduit for virulence factors other than the main pectinases and cellulase in *Eca* (Corbett et al. 2005).

The quorum-sensing signal molecule N-(3-oxohexanoyl)-1-homoserine lactone (OHHL) produced by *Erwinia carotovora* subsp. *carotovora* (*Ecc*) has been shown to control seven genes of *Ecc*. The mutants defective in proteins that could play a role in the interaction between *Ecc* and its plant hosts were enriched using TrophoA as a mutagen and Nip (Ecc) and its counterpart in *E. carotovora* subsp. *atroseptica* (*Eca*) were identified. Nip (Ecc) induced necrosis in tobacco, while Nip (Eca) affected potato stem rot. Both proteins affected virulence on potato tubers. They have been shown to be members of a growing family of proteins related to Nep1 from *Fusarium oxysporum* which induces necrotic responses in a variety of dicotyledonous plants. In *Ecc*, nip was repressed weakly by the Lux R family regulator, EccR and it may be regulated by the negative global regulator RsmA (Pemberton et al. 2005).

The *N*-acyl- homoserine lactone (AHL) chemical signals produced by Gram-negative bacteria have been shown to be essential for synthesis of exoenzyme virulence factors in *Erwinia carotovora* (Whitehead et al. 2002). A later investigation provided the first evidence for autoinduction in *E. amylovora* and a role for AHL-type signal. AHL in *E. amylovora* appears to contribute to the expression of virulence factors and symptom development which is similar to the role AHLs play in *E. carotovora*, *E. chrysanthemi* and *Pantoea stewartii* (Whitehead et al. 2002). Two major plant virulence traits, namely production of extracellular polysaccharides (amylovoran and levan) and tolerance to free oxygen radicals, were controlled in a bacterial cell-density-dependent manner. In the stationary-phase cultures of *E. amylovora* two standard autoinducer biosensors NTL4 (of *Agrobacterium tumefaciens*) and BB886 (of *Vibrio harveyi*) detected AHL. The putative AHL synthase gene *eamI*, partially sequenced, showed homology with autoinducer genes from other bacterial genes like *carI, esaI, expI, hslI, yenI* and *luxI*. *E. amylovora* also contained *eamR* a convergently transcribed gene with homology to *luxR* AHL activator genes in other pathogens such as *E. carotovora.* When *Bacillus* spp. strain A24 acyl-homoserine lactonase gene *aiiA* was expressed in *E. amylovora*, induction of AHL biosensors was entirely abolished, in addition to impairment of extracellular polysaccharide production and tolerance to H_2O_2 and reduction in virulence on apple leaves (Molina et al. 2005).

In a later investigation, the protein ExpR, a luxR homolog was found to be an AHL receptor and that it activated transcription of *rsmA,* the gene encoding an RNA binding protein which is a global negative regulator of exoproteins and secondary metabolites. The transcriptional activity of ExpR was neutralized by AHL. The genomes of most strains of *E. carotovora* subspecies tested contained two copies of the *expR* gene: *expR1* and *expR2*. Both variants could activate *rsmA* transcription, with significant differences in the pattern of their AHL interactions, the *rsmA* sequences to which they bound, and their relative efficiencies of activation of *rsmA* transcription. The results suggest that ExpR2 activity may be responsible for regulating exoprotein production primarily by modulating the levels of an RNA binding protein (Cui et al. 2006).

In *Pseudomonas syringae* pv. *syringae*, the AHL synthase, AhlI and the regulator AhlR are required by the AHL-mediated quorum-sensing system and it is also regulated by AefR. The contribution of quorum sensing to the expression of a variety of traits expected to be involved in epiphytic fitness and virulence of *P. syringae* were examined. An *aefR* (^-) mutant and an *ahlI* (^-) (double mutant, deficient in AHL production) exhibited impairment of alginate production, in addition to increased susceptibility to H_2O_2 compared with the wild-type strain. These mutants were hypermotile in culture. In addition, they invaded the leaves at a faster rate resulting in higher levels of incidence of brown spot lesions on bean leaves, after a 48 h post incubation. It is of interest to note that an aef R (^-) mutant showed motility and virulence to the maximum extent. AHL-deficient mutant strains, like the wild-type strain, induced water-soaked lesions on pods. In contrast, AHL-deficient mutants lost the ability for tissue maceration of pods, which occurs at a later stage of infection. Furthermore, significant reduction in the incidence of disease and in planta growth of *P. syringae* pv. *tabaci* was observed in transgenic tobacco plants that produced AHL compared with wild-type plants. The results confirmed that quorum sensing in *P. syringae* regulates traits that contribute to epiphytic fitness as well as to distinct stages of disease development during infection of plants (Quiñones et al. 2005).

The *lux* genes of *Photorhabdus luminescens* have been shown to be useful in monitoring the effects of antibodies on the metabolic activity of human pathogens such as *E. coli* and *P. aeruginosa* isolates (Salisbury et al. 1999; Marques et al. 2005). Binoluminescence is a measure of metabolic activity rather than cell viability and multiplication. The genes of *P. luminescens* were employed to monitor the population dynamics of *P. syringae* pv. *phaseolicola* (*Psp*) in bean plants. The plasmid pGLITE containing the lux CDABE genes was introduced into *Psp* race seven strain 1449B. The growth of *Psp* in the leaf intercellular spaces was assessed by determining the viable counts and the bioluminescence of leaf disc homogenates. Direct measurement of bioluminescence from leaf disc homogenates was consistently positively correlated to the bacterial growth as determined by viable bacterial cell counting. The bioluminescence method provides real-time measurement in planta, in addition to avoiding the use of labor intensive and time-consuming traditional enumeration techniques. Furthermore, it is possible to assess the effects of plants on bacterial cells (Paynter et al. 2006).

The bacterial populations attached to a surface are referred to as biofilms and the bacteria in biofilms are considered to be resistant to adverse conditions such as desiccation and extreme temperatures. It may also function as an important virulence factor. *Pantoea stewartii* subsp. *stewartii* (*Psst*) produces stewartan exo/capsular polysaccharide (EPS) in a cell-density-dependent manner governed by EsaI / EsaR quorum-sensing (QS) system. Analysis of biofilm development and colonization of host tissues by the WT (wild type) and QS regulatory mutant strains was performed. The stewartan EPS synthesis governed by EsaI / Esa R QS system, was required for effective adhesion of bacterial cells and development of spatially defined 3D biofilms. A nonvirulent mutant lacking the *esaI* gene strongly adhered to surfaces in addition to the development of densely packed less structurally defined biofilms in vitro. Treatment with exogenous AHL counteracted the adhesion phenotype. QS mutants lacking EsaR repressor attached poorly to surfaces forming amorphous biofilms heavily enmeshed in excess EPS. The two QS mutant strains were found largely localized at the site of infection, in contrast to efficient dissemination of WT strain within the xylem primarily in a basipetal direction. Specific colonization of annular rings and spiral secondary wall thickening of protoxylem by WT strain was revealed by epifluorescence microscopic imaging of infected tissues. In contrast, indiscriminate growth of mutant strains to fill the xylem lumen was observed (Koutsoudis et al. 2006).

Flagella of bacterial pathogens are considered to be helpful to seek favorable environments or to escape from adverse conditions. The flagellum has a long helical filament composed of a single protein known as flagellin encoded by *fliC* (Aizawa 1996). MotA and MotB are two integeral membrane proteins required for flagellar rotation (Blair and Berg 1991; Chun and Parikson 1988). Flagellar motility is an important virulence factor during infection and they may also induce host defense responses (Young et al. 1999; Shimizu et al. 2003). The flagellin of *Pseudomonas avenae* was shown to induce resistant responses in cultured cells of the incompatible but not in compatible rice cultivars (Che et al. 2000). The flagellin defective *ΔflicC* of *Pseudomonas syringae* pv. *tabaci* did not induce HR in its nonhost tomato plants, while the wild-strains induced such a response on tomato (Shimizu et al. 2003). In order to understand the functions of flagella for pathogenesis of *Erwinia carotovora* pv. *carovora* (*Ecc*) nonpolar in-frame deletion mutants of *fliC* and *motA* homologs were constructed. Both *ΔfliC* and the *ΔmotA* mutants were found to be nonmotile on semisolid medium. The *motA* mutant was peritrichously flagellated like the wild-type strain as revealed by electron microscope observations. On the other hand, the *ΔfliC* mutant was aflagellated and defective in flagellin production. Both mutants were less effective in eliciting soft rot symptoms on Chinese cabbage without any adverse effect on the production of major extracellular enzymes. However, they could induce HR on tobacco leaves. The results suggest motility, but not flagellar component (s) is the critical factor in pathogenesis of *Ecc* (Hossain et al. 2005).

In a further study the role of flagella of *E. carotovora* subsp. *carotovora* (*Ecc*) in biofilm formation was critically examined. A nonflagellate non-motile mutant (*ΔfliC*) and a flagellate nonmotile (*ΔmotA*) mutant of *Ecc* were tested. Both mutants were equally reduced in their ability to form biofilm on the surface of polyvinylchloride (PVC) microtiter plates. Functional flagella rather than nonfunctional flagella are involved in biofilm formation, indicating that flagella mediated motility plays an important role in biofilm formation by *Ecc*. The biofilm forming ability of bacteria may help the bacterial pathogens in host colonization and flagellar motility may be indirectly involved in pathogenicity (Hossain and Tsuyumu 2006).

Many soilborne plant pathogenic bacteria including *Ralstonia solanacearum* have swimming motility which makes an important quantitative contribution to bacterial wilt virulence in early stages of host invasion and colonization. The virulence of nonmotile *fliC* mutants to tomato was significantly reduced (Trans-Kersten et al. 2001). The bacteria may use a complex behaviour known as taxis to sense specific chemicals or environmental conditions and move toward attractants and away from repellants (Blair 1995). The chemotaxis behaviour of *R. solanacearum* strain K60, which was attracted to various chemicals was studied. Qualitative and quantitative chemotaxis assays showed that *R. solanacearum* was attracted to diverse amino acid and organic acids and especially to root exudates from the host plant tomato (Fig. 2.8) (Yao and Allen 2006).

Two site-directed *R. solanacearum* mutants lacking either CheA (cytoplasmic histidine autokinase) or CheW (coupling protein) which are core chemotaxis signal transduction proteins, were completely nonchemotactic, but retained normal swimming motility. Both non-chemotactic mutants were less virulent as nonmotile mutant, demonstrating that directed motility, but not simply random motion, is essential for full virulence. The demonstration that nontactic strains could be as virulent as the wild-type strain, when they were introduced directly into the stem through a cut petiole, indicated that taxis is an important factor in the early stages for successful invasion of host tissues The wild-type strain out-competed the nontactic mutants by 100 folds when coinoculated. The results revealed that chemotaxis is an essential trait required for virulence and pathogenic fitness in *R. solanacearum* (Yao and Allen 2006).

2.2.2 Colonization of Host Tissues

The bacterial pathogens do not produce any specialized structures like fungal pathogens for initation of pathogenesis. They cannot grow through compact tissues of the host plants, but enter the host plant through natural openings, stomata, hydathodes and lenticels and through wounds or injured plant surfaces. Production

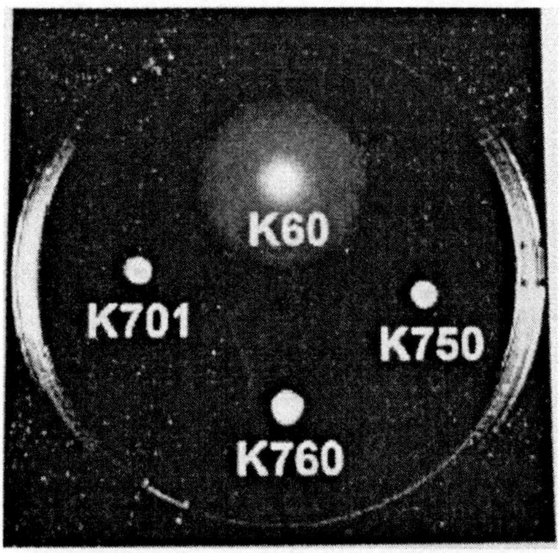

Fig. 2.8 Chemotaxis response of *Ralstonia solanacearum* strains Note typical bacterial taxis halos formed after 2 days of incubation at 28°C in a 1% tryptophan semi-solid agar plate in the wild-type strain K60. The strains K750 (*cheA* mutant) and K760 (*chew* mutant) were motile but nonchemotactic; the strain K701 (*flic* mutant) was nonmotile. (Courtesy of Yao and Allen 2006; The American Society of Microbiology, Washington, USA)

of macromolecules-like enzymes is regulated by four major secretion pathways. Some exo-enzymes and toxins are secreted through the type-I pathway which is *sec*-independent. The proteases produced by *Erwinia chrysanthemi,* causing soft rot is secreted directly into the extracellular environment. This one-step secretion is governed by the type-I secretory apparatus consisting of three accessory proteins (Prt D, E and F) (Pugsley 1993). The Phillippine race 6 (PR6) strains of *Xanthomonas oryzae* pv. *oryzae* (*Xoo*) are not able to infect rice lines expressing resistance gene *Xa21*. The *raxR* and *raxH* genes encoding a response regulator and a histidine protein kinase of two-component regulatory system respectively were identified. Null mutants of PR6 strain PX 099 were defective in either *raxR* or *raxH* or both. These mutants caused significantly longer lesions and reached much higher population levels compared with the wild-type strain in *Xa21*-rice leaves. The mutants (both *raxR* and *raxH*) were complemented to wild-type strain levels of Avr Xa21 activity by introducing expression vectors carrying *raxR* and *raxH* respectively. RaxR expression vectors carrying *raxR* and *raxH* respectively. RaxR expression was confirmed by Western blot and raxR/gfp promoter reporter analyses (Burdman et al. 2004).

The *sec* dependent two step, type II controls the secretion of pectinases and cellulases and it is also known as the general secretory pathway (GSP). Some of these enzymes shown to be virulence factors were secreted through GSP by *E. chrysanthemi* and *E. carotovora* subsp. *carotovora.* The pectate lyase, polygalacturonase and cellulase are exported to the periplasm, but are unable to move through the outer membrane. In the second step, the periplasmic form of these enzymes moves through the outer membrane and then into the extracellular environment. Secretion of major pectate lyases, polygalacturonases and cellulases is controlled by the *out* gene cluster and the number of *out* genes may vary with *Erwinia* spp. (Lindberg and Collmer 1990; Reeves et al. 1993).

The purified pectate lyases from *Erwinia chrysanthemi* named as Pel A, PelB, PelC and PelE are important virulence factors involved in the degradation plant cell walls (Preston et al. 1992). The promoter region of *PelE* gene encoding a major pectate lyase was induced by plant extracts and it was contained in the bacterial regulator. Hyperinduction of this enzyme is regulated by another bacterial gene designated *pir* (Normura et al. 1998; Noueiry et al. 1999). The gene *exuT*, governing the galacturonic acid (Gal UA) uptake has been cloned and sequenced. The *exuT* mutants showed reduction in Gal UA utilization, as well as delay in maceration of potato tuber tissues (Haseloff et al. 1998). *E. chrysanthemi* mutants containing transcriptional fusions of one of the minor pectate lyase genes (*pelI, PelL* and *pelZ*) with reporter gene encoding β-glucuronidase activity were examined for their macerating ability. Strains defective in *pelI* and *pelL* genes exhibited reduced virulence on potato tubers, demonstrating an important role of the enzymes in soft rot disease. In contrast, inactivation of the *pelZ* gene slightly influenced the macerating ability of the mutant. These genes were expressed at higher levels in planta than in culture, indicating the characteristic feature of plant-inducible Pels (Jafra et al. 1999).

The cell wall-degrading enzyme (CWDEs) from the principal virulence determinants of *E. carotovora* subsp. *carotovora* (*Ecc*) and their synthesis is coordinately regulated by a complex network. The ExpS and ExpA proteins encoded by *expS* and *expA* genes have been shown to be members of two-component sensor kinase and response regulator families respectively (Eriksson et al. 1998). The interaction between the proteins may control the virulence gene expression component regulatory system controlling endoPG production and virulence of *Ecc* was demonstrated. The genes coding for the main virulence determinants of *Ecc*, the CWDEs have been shown to be under the coordinate control of global regulator systems including both positive and negative factors. Reduced virulence and impaired production of one of these enzymes, an endoPG (PehA) were shown to be due to mutations in the pehR locus. In contrast, these pehR strains produced essentially wild-type levels of other extracellular enzymes including pectate lyases and cellulases. The pehR locus was later characterized and the DNA sequence was composed of two genes viz. *pehR* and *pehS,* present in an operon. Mutations in either *pehR* or *pehS* caused a Peh-negative phenotype and resulted in reduced virulence on tobacco seedlings. Both genes are required for transcriptional activation of the endoPG gene *pehA* as well as restoration of virulence, as revealed by complementation experiments (Flego et al. 2000).

The principal virulence determinants in *Erwinia carotovora* subsp. *carotovora* (*Ecc*) are controlled by a complex regulatory network. One of the regulators *ExpA,* a GacA homolog, determines transcriptional activation of the extracellular enzyme genes of *Ecc*. A key role of the RsmA-rsmB regulatory system during pathogenesis has been shown to be the integration of signals from the *ExpA* (GacA) and *KdgR* global regulators of extracellular enzyme production in *Ecc* (Hyytiäinen et al. 2001). Likewise, many regulatory factors are involved in the production of pectinase. The main regulatory proteins KdgR, CRP, Pir and PecS mainly from *E. chrysanthemi* have been characterized. By applying Southern blotting, homologs of KdgR and Crp, but not Pir and Pecs were detected in *Ecc*, by introducing plasmids containing the regulatory genes from *E. chrysanthemi*. Pel production was controlled, as predicted from their roles in *E. chrysanthemi* except for PecS, suggesting that KdgR and CRP homologs of *Ecc* may regulate Pel and Peh production, as in *E. chrysanthemi* (Matsumoto et al. 2003b). A new regulatory factor involved in the virulence of *Ecc* was identified in a later investigation. The (Delta) *cytR* mutant was able to produce only low levels of polygalacturonase (Peh) and had no mobility. The virulence of the mutant was much reduced in potato and chinese cabbage in comparison to the wild-type. The CytR homolog of *Ecc* appears to control Peh production and synthesis of flagella in addition to having an important role in its pathogenicity (Matsumoto et al. 2003a).

The PhoP-PhoQ system is a vital factor in pathogenicity of several bacterial pathogens and its presence has been detected in *Erwinia chrysanthemi* EC 16. The involvement of PhoP-PhoQ system in the bacterial resistance to different environmental factors such as acid stress and antimicrobial peptides. The mutants in *phoP-phoQ* operon showed reduction in virulence in several hosts, survival at acid pH in plant tissues, ability to alkalinize plant tissues and production of pectolytic enzymes (Llama-Palacios et al. 2003). PhoQ is a sensor histidine kinase that autophosphorylates in response to environmental conditions and PhoP is a transcriptional regulator that controls the expression of genes that are essential for virulence (Fields et al. 1989). The mutants defective in *phoP* and *phoQ* genes of *E. chrysanthemi* strain EC 3937 were characterized. The growth of the mutants at acid pH was not altered. However, the *phoQ* mutant exhibited reduced ability to survive at acid pH and the susceptibility to antimicrobial peptide thionin was enhanced. In addition, the virulence of phoQ was adversely affected by both low and high concentrations of magnesium, while only low magnesium concentrations reduced the virulence of *phoP* mutants. The Pel activity of both mutants was reduced drastically, in addition to marked reduction in their ability to change apoplastic pH to optimal levels, compared with wild-type strain. The PhoP-PhoQ system appears to play a role in the virulence of *E. chrysanthemi* in planta, although its contribution to bacterial growth at acid pH is not significant (Llama-Palacios et al. 2005).

In a later study, the effect of a mutation in *phoQ*, the gene encoding the sensor kinase PhoQ of the PhoPQ two component regulatory system, on the global transcriptional profile of *E. chrysanthemi*, using cDNA microarrays, was assessed. A mutation in *phoQ* affected transcription of at least 40 genes even in the absence of inducing conditions. Several genes involved in iron metabolism were expressed at higher level in the mutant. The *acs* operon involved in achromobactin biosynthesis and transport was also induced. This siderophore is essential for full virulence of *E. chrysanthemi.* The results indicate that the PhoPQ system controls the expression of several additional virulence factors and that they may be involved in interactions with other regulatory system (Venkatesh et al. 2006).

The regulatory locus, *pehSR* in *Ralstonia solanacearum* governs the PG production and other virulence functions. The endoPG activity of *pehSR* mutants was negligible, whereas the exoPG activity was reduced by 50%. Despite the reduction in virulence of the mutant, its growth in culture remained normal. The expression of pehSR in planta increased, as the pathogen proliferated. Involvement of the global virulence gene regulator PhcA in the negative regulation of the *pehSR* locus was revealed by the reporter gene studies (Allen et al. 1997). The extracellular plant cell wall-degrading enzymes (CWDEs) and other proteins are secreted via the highly conserved type II protein secretion system (T2SS). *R. solanacearum* with a defective T2SS is weakly virulent. Mutants of wild-type strain GM100 lacking either T2SS or up to six CWDEs were generated. In soil-drench inoculation tests, virulence of mutants lacking only pectic enzymes (PehA, PehB, PehC and Pme) was not significantly different from the wild-type. However, all the mutants lacking one or both cellulolytic enzymes (Egl or CbhA) wilted plants significantly slowly compared with the wild-type. The GMI-6 mutant lacking all six CWDE was more virulent than the mutant lacking only its two cellulolytic enzymes and both were significantly more virulent than the T2SS mutant GMI-D, since the T2SS mutant was much less virulent than the six fold CWDE mutant. It can be concluded that other secreted proteins contribute subsequently to the ability of *R. solanacearum* (wild-type) to systemically colonize tomato plants (Liu et al. 2005a).

Erwinia chrysanthemi requires additional factors such as flagellar proteins and exopolysaccharides (EPS) for efficient colonization of host tissues, in addition to CWDEs. Involvement of the protein H-NS associated with the nucleoid of

E. chrysanthemi in pathogenesis was assessed. The *hns* mutant exhibited reduced growth rate and motility. However, there was increase in EPS synthesis and sensitivity towards high osmomolarity. Furthermore, drastic reduction in Pel production and consequent reduced virulence of the mutant were also discernible, indicating that H-NS protein has a vital role in the expression of the virulence genes and pathogenicity of *E. chrysanthemi* (Nasser et al. 2001). The aggressiveness (virulence) of several *E. chrysanthemi* strains on potato was shown to be negatively correlated with their osmotic tolerance. Disruption of the *ousA* encoding the major osmoprotectant uptake system dramatically increased the virulence of mutants on potato tubers. In addition, *ousA* disruption also increased the efficiency of maceration on potato tubers under anaerobic conditions. In the absence of oxygen, Pel production showed significant increase in the tissue macerated with *ous* (^-) strain than the wild-type. As the oxygen contents in infected and healthy potato tissues, *ousA* might be a contributory factor in disease development within host tissues. The results reveal a relationship between *E. chrysanthemi* pathogenicity factors and the activity of *ousA* under low oxygen status. Further, *ousA* and osmoprotectant effects on Pel are not linked by osmoregulation, since complex regulations may be involved in Pel production, *ousA* and osmoprotection via compounds that are liberated during plant infection (Gloux et al. 2005).

A mutant of *R. solanacearum* defective in pectin methyl esterase (*pme*) gene did not show detectable Pme activity in vitro and it could not utilize methylated pectin (93%) as a carbon source. Its virulence, on tomato, egg plant and tobacco was, however, not reduced, indicating Pme was required for growth, but not virulence (Trans-Kersten et al. 1998). In *Xanthomonas oryzae* pv. *oryzae* (*Xoo*), several proteins including a xylanase required for virulence are secreted through type II secretion system (T2SS). The *xynB* gene encoding for secreted xylanase and a paralog *xynA* were cloned. These genes *xynA* and *xynB* are adjacent to each other in *Xoo*, as well as in *X. axonopodis* pv. *citri* (*Xac*). The secreted xylanase activity was affected by mutations in *xynB*, but not in *xynA*. Western blot analysis using anti-XynB antibodies on exudates from infected rice leaves indicated that xylanase was expressed during growth in planta. Mutations in either *xynB* or *lipA* (encoding a lipase / esterase (LipA) secreted through T2SS), partially affected virulence. In contrast a *lipA-xynB* double mutant showed significant reduction in virulence. Complementation of virulence-deficient phenotype of the *lipA-xynB* mutant with the pRR7 clone containing an intact *xynB* gene restored the virulence of the double mutant to the levels of wild-type. It seems that there is functional redundancy among the T2SS secreted proteins *Xoo* in promoting virulence on rice (Rajeswari et al. 2005).

Xylella fastidiosa (*Xb*) a xylem-limited bacterium causes the Pierce's disease of grapevine accounting for significant losses. *Xf* systemically colonizes the xylem elements of infected grapevine plants breaching the pit pore membranes and consequently separating the xylem vessels. Based on the detection of genes involved in the plant cell wall degradation in *Xf* genome, it was considered that the cell wall grading enzymes produced by *Xf* may be involved in the cell wall degradation. Several beta-1,4-endoglucanases xylanases and xylosidases and one PG encoding genes were identified in *Xf* genome. A functional PG was shown to be encoded by the *pglA*. A mutant defective in *pglA* became nonpathogenic and was compromised in its ability to colonize the tissues of grapevine systemically. This study indicated the requirement of PG for successful infection of grapevine by *Xf* and its function as a critical virulence factor for *Xf* pathogenesis in grapevine (Roper et al. 2007).

Cyclic di-GMP is considered a second messenger with a role in the regulation of a range of cellular functions in diverse bacteria including virulence of pathogens. The role of the putative elements of cyclic di-GMP signaling in the virulence and environemtnal adaptation of *X. campestris* pv. *campestris* (*Xcc*) causing black rot disease of cruciferous crops was investigated. A two-component signal transduction system comprising the HY-GYP domain regulator RpfG and cognate sensor kinase RpfC is required for a number of factors including synthesis of extracellular enzymes and extracellualr polysaccharide and also for formation of biofilm for bacterial dispersal. The HD-GYP domain of RpfG was shown to be a cyclic di-GMP phosphodiesterase implicating cyclic di-GMP in the control of virulence, virulence factor synthesis and biofilm formation in *Xcc*. The genome of *Xcc* encodes 37 proteins with GGDEF, EAL or HD-GYP domains. The contribution of these domains to virulence and virulence factor production in *Xcc* was assessed. The RpfG controlled expression of a subset of virulence genes, one of which was under the control of HrpG, the master regulator of the *Xanthomonas* hrp regulon. The results indicated that regulatory network allowed *Xcc* to integrate information from diverse environmental inputs to modulate virulence factor synthesis as well as cyclic di-GMP signaling systems controlling other specific functions (Ryan et al. 2007).

Bacterial pathogenesis is affected by contents of available iron in the apoplastic fluids of plants. Ferric citrate is the major iron carrier present in the conducting elements of plants. *E. chrysanthemi* produces a catechol type siderophore, chrysobactin which aids in the sequestration of free iron. *E. chrysanthemi* also produces two receptors capable of interacting with bacterial iron carrier molecules. Pel activity registers significant increases under low iron conditions. A regulatory *cbrAB* positively regulates the expression of *pelB*, *pelC* and *pelE* (Sauvage and Expert 1994). The role of iron in the pathogenicity of *E. chrysanthemi* was investigated by generating mutants by insertional mutagenesis. The iron transport pathway is mediated by desferrioxamine (DFO) which is required for iron utilization by *E. chrysanthemi* during pathogenesis. DFO may also have a role in the oxidative burst elicited by the bacteria (Dellagi et al. 1998). Following transposon mutagenesis technique, a novel mutant 74M913 of *Xanthomonas oryzae* pv. *oryzae* (*Xoo*) with reduced virulence and ability to cause hypersensitive response in leaf-blight resistant rice and tomato, was produced. The transposon in 74M913 was inserted in a gene homologous to the phosphoglucose isomerase (*pgi*) gene of *X. axonopodis* pv. *citri*. The transposon was demonstrated to disrupt *pgi* function by the reduction in the growth of *Xoo* in a synthetic medium containing fructose or xylose as a sole carbohydrate. The results indicated that *pgi* and the *hrp* gene (*hrcU*) were expressed independently in synthetic medium and suggested that PGI was involved in pathogenicity of *Xoo* (Tsuge et al. 2004).

Phytobacterial pathogens have to overcome plant defenses and gain access to nutrients trapped inside the plant cell in order to colonize the host tissues and multiply resulting in the expression of different symptoms characteristic of various diseases of plants. Many bacterial pathogens require type III secretion system (TTSS) to mount a successful invasion of the susceptible host plant species. The TTSS was first discovered in the mammalian pathogen *Yersinia* and it has been reported in taxonomically diverse, Gram negative bacterial pathogens of plants later (Lindgren 1997; He 1998). The TTSS acts as a syringe and virulence proteins are injected by the bacteria into the host cells. The type III effector proteins are considered to facilitate disease development by altering the normal physiology of the plant in favor of the pathogen. The TTSS of the bacterial pathogens belonging to the genera *Erwinia, Pseudomonas* and *Xanthomonas* possess *hrp* (for hypersensitive response (HR) and pathogenicity) and *hrc* (HR and conserved) genes. The *hrp* genes are involved both in the functions of pathogenicity and elicitation of resistance by bacterial pathogens such as *E. amylovora, P. syringae* pv. *tomato* and *X. campestris* pv. *vesicatoria* inducing necrosis, and these genes have been found to be homologous (van Gijsegem et al. 1995a, b). Depending on the bacterial species or pathovar, the number of hrp genes may vary. Three biochemical functions such as gene regulation, protein secretion and induction of HR are ascribed to the *hrp* genes (Gopalan and He 1996). The TTSS delivers into plant cell a wide array of proteins known as effectors which are translocated into the plant cell cytoplasm. The best-studied TTSS effectors are Avr proteins and others are Hop (*H*rp *o*uter *p*rotein) in *Ralstonia* (earlier included in the genus *Pseudomonas*) (Table 2.3).

Virulence of bacterial pathogens is frequently abolished / reduced, if mutations occur affecting type III protein secretion. When the pathogens come into contact with host cells, the TTSS is activated in vivo and the virulence proteins are considered to accelerate the leakage of plant nutrients in the extracellular space (apoplast) in plant tissues infected by bacteria (Lindgren 1997; He 1998). Host specific symptoms may be induced by the expression of single gene as in the case of *X. axonopodis* pv. *citri* (*Xac*) (citrus canker). The *pthA* gene controls division, enlargement and death of host cells. As elicitation of specific symptoms depend on a functional type III protein secretion system in *Xac*, it was suggested that PthA protein could be a specific plant signal whose site of action may be inside the plant cell (Duan et al. 1999).

The protein–protein interactions involving subunits, regulators and substrates of the type III secretion system encoded by the *hrp* cluster of *X. axonopodis* pv. *citri* (*Xac*) were identified using a yeast two-hybrid system (Alegria et al. 2004). In *Xac*, the *hrp* cluster is composed of 25 genes extending from *hpa2* and *hpaF* (da Silva et al. 2002). A region of the *hrp* cluster of *Xac* was isolated and the interaction of mutants for the type III secretion system with host and nonhost plants was characterized. Mutants for the operons *hrpB, hrpD* and *hrpF* were constructed. These mutants failed to induce canker in citrus plants or HR in cotton plants (a nonhost) but the wild-type could do both. In addition, the mutants did not induce HR in bean, tomato, pepper and *Nicotiana benthamiana*, whereas the wild-type strain induced HR in all these plant species. This may be due to the requirement of the gene *pthA* for elicitation of HR. It has been demonstrated the PthA is secreted by TTSS and

Table 2.3 Genes of type III secretion system (TTSS) in bacterial plant pathogens and the predicted functions of effectors

Name of genes	Predicted functions of effectors	References
Erwinia amylovora		
avrRpt2 (EA)	AvrRpt 2(EA)-virulence factor, inducing systemic infection	Zhao et al. (2006)
avrE1	constitute Hrp pathogenicity island	Badel et al. (2006)
hopM1	conserved effector locus (CEL)	
hop AA1-1		
dspA	DspA, virulence factor	Barny et al. (1999)
dspA/E	DspA/E virulence factor, induction of necrotic symptoms	Boureau et al. (2006) and Meng et al. (2006)
hrpN	HrpN-virulence factor	Barny et al. (1999)
hrpW	HrpW-a negative factor	Barny et al. (1999)
E. chrysanthemi		
hrp/hrc	Encoding a putative mixture of virulence factors	Rojas et al. (2004)
hrpN	HrpN (harpin) facilitating colonization of host tissues	Rojas et al. (2004)
hrpC	Viulence factors	Yang et al. (2002)
hrpG		
P. syringae pv. phaseolicola (Psp)		
hrc C	encoding core structural components	Thwaites et al. (2004)
hrc J		
hrc N		
hrc V		
hrp F		
hrp J avrPto	AvrPto suppressing a cell wall based defense	Hauck et al. (2003)
hrp P		
hrpR avrPphF	encoding secreted effectors	Thwaites et al. (2004)
hrpL virPphA		
hrpY virPphC	allows evasion of HR-based immunity	Tsiamis et al. (2000)
hrpZ virPphF		
***P. syringae* pv. *syringae* (*Pss*)**		
syrB	Synthesis of syringomycin	Cao et al. (2005)
sal A	LuxR regulatory protein family, production of syringopectin	Lu et al. (2005)
syrB1	syringomycin synthetase	Wang et al. (2006a)
syr F	syrF-signal molecules	Wang et al. (2006b)
***Pseudomonas syringae* pv. *tomato* (Pst)**		
avrPto	Necrosis associated with disease; enhancement of virulence	Lin and Martin (2005) and Anderson et al. (2006)
avrPtoB	Functions as cell death inhibitor / programmed cell death (PCD)	Lin and Martin (2005)
avrPto and avrPtoB	Regulation of *LeACO1* and *LeACO2* host genes involved in biosynthesis of ethylene forming enzyme ACC oxidase	Cohn and Martin (2005)
avrRpt2	Virulence factor Avirulence factor	Lim and Kunkel (2004a, b)

Table 2.3 (continued)

Name of genes	Predicted functions of effectors	References
hrpA	HrpA, production of pilus	Brown et al. (2001)
hrpZ	HrpZ, production of pilus	Brown et al. (2001)
hrpW	HrpW, production of pilus	Hu et al. (2001) and Badel et al. (2002)
hopPtoA1 *hopPtoA2*	Putative effector genes	
hop Pto M	Induction of symptoms	Badel et al. (2003)
hopPtoN	Reduction in symptom intensity	López-Solanilla et al. (2004)
Ralstonia solanacearum		
hrpB	HrpB-transcription activator acting as a master regulatory gene	Zolobowska and van Gijsegem (2006)
hrpG	homologous to HrpG from *Xcv* activating *hrp* genes	Brito et al. (1999)
prhJ	Plant inducible, regulatory protein	Brito et al. (1999)
popA	Putative effector expressed in early stages	Kanda et al. (2003)
***X. axonopodis* pv. *citri* (*Xac*)**		
pthA	pthA-division, enlargement and death of plant cells	Duan et al. (1999)
***Xanthomonas campestris* pv. *vesicatoria* (*Xcv*)**		
avrBs1	unknown function	Ronald and Staskawicz (1988) and Escaler et al. (2001)
avrBs2	Putative glycerophosphoryl diester phosphodiesterase	Swords et al. (1996)
avrRxv	YopJ/AvrRxv-family, putative cysteine protease	Whalen et al. (1993); Ciesiolka et al. (1999) and Bonshtien et al. (2005)
XopB	Homology to HopD1 (*Pseudomonas syringae* pv. *tomato*)	Noël et al. (2001)
XopC	Unknown function	Noël et al. (2003)
XopD	SUMO (small ubiquitin-like modifier) cysteine protease	Noël et al. (2002) and Hoston et al. (2003)
XopF1	Unknown function	Roden et al. (2004b)
XopF2	Unknown function	Roden et al. (2004b)
XopJ	AvrRxy/YopJ. family, putative cysteine protease	Roden et al. (2004b)
XopN	Unknown function	Roden et al. (2004b)
XopO	Homology to HopK1 and AvrRps4 (*P. syringae*)	Roden et al. (2004b)
XopP	unknown function	Roden et al. (2004b)
XopQ	HopQ1-1 family protein, putative inosine-uiridine nucleoside N-ribohydrolase	Roden et al. (2004b)
XopX	Unknown function	Metz et al. (2005)
Ecf	Early chlorosis factor, homology to HopAE1 (*P. syringae* pv. *syringae*)	Morales et al. (2005)

Table 2.3 (continued)

Name of genes	Predicted functions of effectors	References
avrBs3	AvrBs3 – nuclear localization singal (NLS), recognition	Bonas et al. (2000) and Szurek et al. (2002)
AvrXv4	SUMO (small ubiquitin-like modifier) protease localized to plant cytoplasm	Roden et al. (2004a)
avrBs4	AvrBs4-triggers hypersensitive response (HR)	Schornack et al. (2005)
***X. oryzae* pv. *oryzae* (*Xop*)**		
avrXa7	AvrXa7-virulence factor and avirulence specificity	Yang et al. (2000) and Yang and White (2004)
avrBs3	virulence factor, interaction with host DNA and transcriptional machinery	Yang et al. (2000)

Source: Gürlebeck et al. (2006)

the loss of HR in the nonhost cotton may be due to the requirement for a functional secretion system that translocates PthA in to the plant cell. The results indicated that genes present in operons *hrpB* and *hrpD* and the *hrpF* gene are required for pathogenicity in susceptible plant species and also for induction of HR in nonhost plant species (Dunger et al. 2005).

Elucidation of an essentially complete inventory of TTSS effectors that are present in several phytopathogenic bacterial pathogens has been a significant achievement in understanding the virulence strategies of bacterial pathogens. The TTSS effectors may be required to promote bacterial growth in apoplast by defeating host defenses and releasing nutrients from plant cells. To accomplish this goal, two levels of host defenses have to be disrupted. Basal defenses are triggered from the outside of plant cells by nonpathogens, TTSS mutants, heat-killed bacteria and factors bearing pathogen-associated molecular patterns (PAMPs) (Medzhitov and Janeway 2002; Nurnberger and Brunner 2002). These factors are also known as general elicitors which include lipopolysaccharides (Dow et al. 2000), flagellin (Gomez-Gomez and Boller 2002) and cold shock proteins (Felix and Boller 2003). Basal resistance to nonpathogenic bacteria is recognized by the production of a callose-rich papilla beneath the plant cell wall at the site of bacterial contact, localized induction of reactive oxygen species (ROS) and enhancement of expression of defense-related genes (Jones and Takemoto 2004). Basal defences are triggered early (<10 min after contact) in the plant–pathogen interaction in response to the perception by plant pathogen recognition receptors (PRRs) of extracellular pathogen-associated molecular patterns (PAMPs). Plants have evolved PRRs that can recognize certain PAMPs which are important molecules for the microbial life style. PAMPs contain a conserved structural feature that is recognized by a PRR. Recognition of a PAMP accelerates several early frontline defenses against bacterial pathogens (Gomez-Gomez and Boller 2002). The second level of host defenses, the HR, in contrast, is elicited inside of plant cells by many bacterial pathogens effectors during gene-for-gene interactions involving matching *avr* and *R* genes.

The programmed cell death (PCD) is a resultant of recognition and signal transduction events and the surrounding plant tissues exhibit resistance following a complex physiological activities of the host tissues. Very little is clearly known about the cell death associated with lesions that develop in later stages of compatible interactions involving all kinds of pathogens with a TTSS (Alfano and Collmer 2004).

Many of the TTSS effectors seem to have acquired by horizontal gene transfer and they are commonly associated with mobile genetic elements (Kim and Alfano 2002; Arnold et al. 2003). In various bacterial strains, of over 150 effector genes have been collectively identified and in the model *Pseudomonas syringae* pv. *tomato* (Pst) strain DC 3000, more than 40 TTSS substrates have been confirmed experimentally (Schechter et al. 2004; Wehling et al. 2004). The effectors injected into host cells have a pivotal role in promoting the virulence of pathogenic bacteria infecting plants and animals. The effectors delivered by *Pst* (over 30) have diverse enzymatic activities such as cysteine protease (Shao et al. 2002; López-Solanilla et al. 2004; Coaker et al. 2005), ubiquitin-like protease (Hoston et al. 2003; Roden et al. 2004), E3 ubiquitin ligase (Abramovitch et al. 2006b; Janjusevic et al. 2006) and protein phosphatase activity (Espinosa et al. 2003). The TTSS secretory accessors proteins designated translocators which form pores in the eukaryotic plasma membrane are essential for the translocation of effector proteins. The HrpF was suggested to act as a translocator in *X. campestris* (Rossier et al. 2000; Büttner et al. 2002). HrPF shares similarity with *P. syringae* HrpK and *R. solanacearum* PopF1/F2 (Alfano and Collmer 2004). Most of the effectors do not show sequence similarity to known proteins and their functions are not known entirely (Abramovitch et al. 2006a).

The host proteins and signaling pathways that are targeted by TTSS effectors to suppress PAMP signaling, have been identified in certain pathosystems. Two effectors AvrRpt2 and AvrRmp1 have been demonstrated to inhibit flag 22 (a 22-amino acid epitope of flagellin)-induced defenses in *Arabidopsis* and promote the growth of TTSS-deficient bacteria (Kim et al. 2005c). The effector AvrB, as well as Avr Rpt2 and Avr Rmp1 were each shown to interact with RIN4, a negative regulator of resistance (R) protein-mediated defenses in *Arabidopsis.* The overexpression or absence of RIN 4 in *Arabidopsis* resulted respectively in the inhibition or enhancement of flg 22-stimulated callose deposition and growth of TTSS-deficient bacteria. RIN4 was cleaved by AvrRpt2 having cysteine protease activity (Kim et al. 2005a; Coaker et al. 2005). Furthermore, induction of phosphorylation of RIN4 by AvrRmp1 was also demonstrated (Mackey et al. 2002). Ultrastructural analyses and suppression of expression of genes associated with basal defenses may probably depend on the strength of recognition by PRRs (Li et al. 2005).

Bacterial effectors AvrPto not only suppress basal defenses, but also contribute to the virulence of *Pst* (Shan et al. 2000). Some effectors have been shown to suppress HR-based immunity (Espinosa and Alfano 2004; Chisholm et al. 2006). The gene *VirPphA* enabled *P. syringae* pv. *phaseolicola* (Pph) to evade HR-based resistance (Jackson et al. 1999). Likewise, the studies with *Pst* DC 3000 effector protein Avr PtoB indicated that the effectors might suppress HR-based immunity by suppressing PCD (Kim et al. 2002b; Abramovitch et al. 2003). The enzymatic activities of several HR-suppressing effectors indicate that these proteins are modifying or degrading targets in signaling pathways that are associated with the HR. The activity of E3 ubiquitin, detected in AvrPtoB, was required to suppress HR-based PCD and plant immunity (Abramovitch et al. 2006a, b; Janjusevic et al. 2006).

A functional Hrp TTSS is essential for the pathogenicity of *Erwinia amylovora*. Harpin, a protein playing a major role in virulence, is exported in vitro via the TTSS machinery. This protein could not be detected inside the host cells by using a specific antiserum, but it was found to be associated with the pathogen cells and secreted. The extracellular localization of harpin is in agreement with the physiological effects induced by purified harpin applied as an exogenous elicitor (Banar et al. 1994; Perino et al. 1999). The *dsp* (disease specific) region located next to the *hrp* gene cluster is essential for *E. amylovora*, but not for elicitation of HR. The operon *dspEF* contains genes *dspE* and *dspF* and it is positively regulated by *hrpL* gene. Similarity in the *dspE* to a partial sequence of the *avrE* locus of *Pseudomonas syringae* pv. *tomato* was observed. When the *dspEF* was introduced into a plasmid, *P. syringae* pv. *glycinea* race four became avirulent on soybean (Bogdanove et al. 1998). Three proteins secreted by *E. amylovora* through functional Hrp-secretion pathway were identified. Harpin (HrpN), a glycine-rich heat stable protein capable of eliciting HR on tobacco, DspA/E homologous and functionally equivalent to AvrE and HrpW with structural similarity to harpin and homologous to class III pectate lyases were purified and their functions were determined by examining the mutants defective in *hrpN, dspA* and *hrpW*. HrpN was shown to be a virulence factor, rather than a pathogenicity factor. DspA was an essential pathogenicity factor and HrpW was not required for pathogenicity of *E. amylovora*. It was further shown that although harpin was the main HR elicitor, DspA also participated in HR elicitation on tomato. HrpW acted as a negative factor, since hrpW mutants could induce stronger HR than the wild-type strain (Barny et al. 1999).

The TTSS has been demonstrated to be essential for *E. amylovora* to induce disease on hosts and HR on nonhost plant species. The DspA/E protein, essential for pathogenicity, was secreted via the TTSS in vitro. *DspA/E* belongs to a type III effector family that is conserved in several bacterial pathogens of plants. An in situ immunogold labeling protocol was developed for visualization of specific localization of TTSS effector proteins of *E. amylovora*, using monoclonal antibodies generated against recombinant Ea 110 HrpA and DspA/E proteins. The pili, but not the flagella were densely labeled by the antibodies specific to Ea110 HrpA protein. The HrpA antibody attached directly to Hrp pili (Fig. 2.9(a)). The locations of secreted HrpN and DspA/E were found specifically along Hrp pili indicating that HrpN and DspA/E were apparently secreted at Hrp pilus assembly site, but not from other channels / structures on the cells surface of *E. amylovora* (Fig. 2.9(b)). This study also showed that protein secretion system was not associated with pilus. These results provide direct evidence for the secretion of TTSS at the site of Hrp pilus assembly and for Hrp pilus being the conduit / guiding filament for the transfer of effector proteins outside bacterial cell (Jin et al. 2001).

In *E. amylovora* DspA/E has been implicated in the generation of an oxidative stress during disease and the suppression of callose deposition. The fate of DspA/E in planta was investigated by delivering it artificially to apple or tobacco

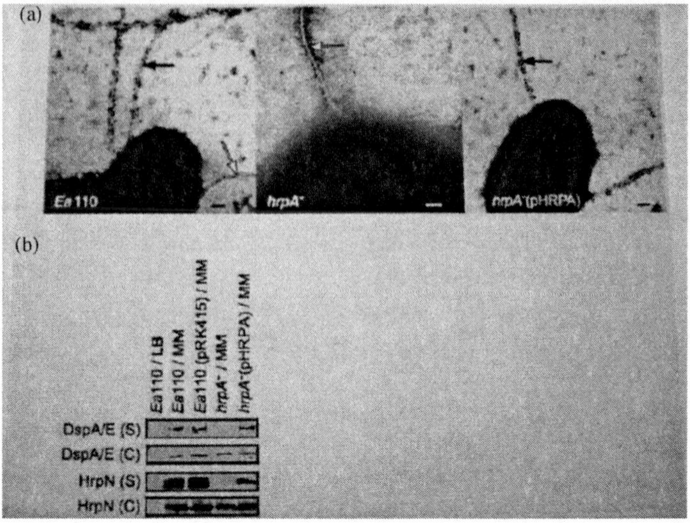

Fig. 2.9 Immunogold labeling of Hrp pili of *Erwinia amylovora* (*Ea*) with Hrp antibody
(a) Surface appendages of Ea 110 immunogold labeled with Hrp antibody as seen in transmission electron microscope (TEM) images Open arrow indicates flagellum (∼15 nm in diameter); filled arrow indicates pilus (∼8 nm diameter); **(b)** Immunoblot of HrpN and DspA/E in the supernatant (S) and cell-associated (**C**) fractions of different bacterial strains grown in either LB medium or hrp-inducing MM. (Courtesy of Jin et al. 2001; Blackwell Science Ltd, UK)

cells by agro- infection. Dsp/E induced necrotic symptoms indicating its possible delivery via TTSS. DspA/E was inferred to act as a major cell-death inducer during disease and HR, as the dspA/E mutant had severely impaired ability to induce electrolyte leakage in apple and tobacco leaves. When dspA/E was transiently expressed in tobacco, expression of the defense marker gene *PR1* was delayed, suggesting that DspA/E–mediated necrosis may be associated with an alteration of defense responses (Boureau et al. 2006). In another study, the functions of the gene dspA/E encoding an essential pathogenicity effector of 198-kDa have been studied. DspA/E interacted physically and specifically with four similar putative leucine-rich (LRR), receptor-like serine/threonine kinases (RLK), from apple, an important host of *E. amylovora*. The genes encoding those four Dsp A/E interacting proteins of *Malus* x *domestica* (DIPM 1–4) are conserved in a wide range of cultivars representing highly susceptible to highly resistant categories. All the four DIPMs were expressed constitutively in host plants. These proteins may act as susceptibility factors required for disease progression (Meng et al. 2006). The expression of *dspE* (a putative effector) and pelD (a major pectin-degrading enzyme) in *E. chrysanthemi* (Ech 3937) was studied using green fluorescent protein gene as a reporter under different cultural conditions and on inoculation on potato tubers and Chinese cabbage. Wide variations in the expression of *dspE* and *PelD* in culture and in plants were observed (Peng et al. 2006).

Erwinia amylovora (*Ea*) requires type III secretion pathway to infect various host plant species. The TTSS allows extracellular bacteria to inject virulence proteins into the cytosol of plant host cells. The *hrp* genes, involved in disease induction or HR, are located on a 61-kb chromosomal pathogenicity island (Oh et al. 2005). The proteins secreted via the TTSS, the T3 secretome, include effector and helper proteins. The effectors are virulence proteins interfering with host metabolism and defense, while helper proteins are predicted to function extracellularly in effector delivery (Collmer et al. 2002). To have a more clear understanding of the role of T3 secretion in interactions between *Ea* and its hosts, an inventory of T3-secreted proteins was made by comparing the extracellular proteins of a wild-type strain with those of mutants defective in functions of structural or regulatory gene required for T3 secretion. The secretome of a mutant deficient in production of two harpins, HrpN and HrpW, and a mutant deficient in the production of Hrp J was also investigated. Six effector-like or helper-like proteins [Eop1, Eop2 Eop3, Eop4 (*Erwinia* outer protein) HrpJ and HrpK] were identified. These proteins were detected earlier as protein secreted by TTSS of *Ea.* In addition, HrpJ was required for extracellular accumulation of harpins for pathogenicity and for wild-type elicitation of the HR (Nissinen et al. 2007).

A total of 12 proteins secreted through TTSS of *Ea* in a defined inducing minimal medium was identified. Only four of these proteins, Eop1, Eop3, Eop4 and DspA/E exhibited similarity to known effectors. Analysis of the secretome of a nonpolar *hrpJ* mutant revealed that HrpJ was required for accumulation of wild-type levels of secreted harpins. HrpJ was found to be essential for pathogenesis and also to have an important role in elicitation of the HR in tobacco. The proteins were resolved by gel electrophoresis and identified using mass spectrometry and a draft sequence of *Ea* genome. The wild-type strain Ea 273 included ten protein bands that were absent or extremely faint in the *hrp* secretion and *hrp* regulation mutants Ea273-hrcN:Tn10 and Ea273-hrpL: Tn10 respectively (Fig. 2.10) (Nissinen et al. 2007).

Erwinia amylovora (*Ea*) infects various aerial plant parts including immature fruits of pear and apple. As it is difficult to maintain required number of trees in the greenhouse for various experiments, immature fruits are used to study infection by *Ea.* A genome-wide examination of gene expression patterns during host infection to uncover pathogenesis strategies was undertaken to gain a better understanding of the molecular mechanisms governing plant–bacterial pathogen interactions. A modified in vivo technology was employed to identify the genes of *Ea* that are activated during infection of immature pear tissue. The unique pear fruit-induced (*pfi*) genes (394) were identified based on the sequence similarity to known genes and they were classified into nine putative function groups including host–pathogen interaction. The virulence genes including *hrp/hrc* components of TTSS, the major effector gene *dspE*, type II secretion, levansucrase (lsc) and regulators of levan sucrase and amylovoran biosynthesis, were upregulated during pear tissue infection. The virulence factors, the presence of which was known already in *E. carotovora* and *P. syringae* were detected for the first time in *Ea*. The other virulence factors identified were HecA hemagglutinin family adhesion, Peh polygalacturonase, new effector HopPtoC$_{EA}$, and membrane-bound lytic murein transglycosylase MltE$_{EA}$.

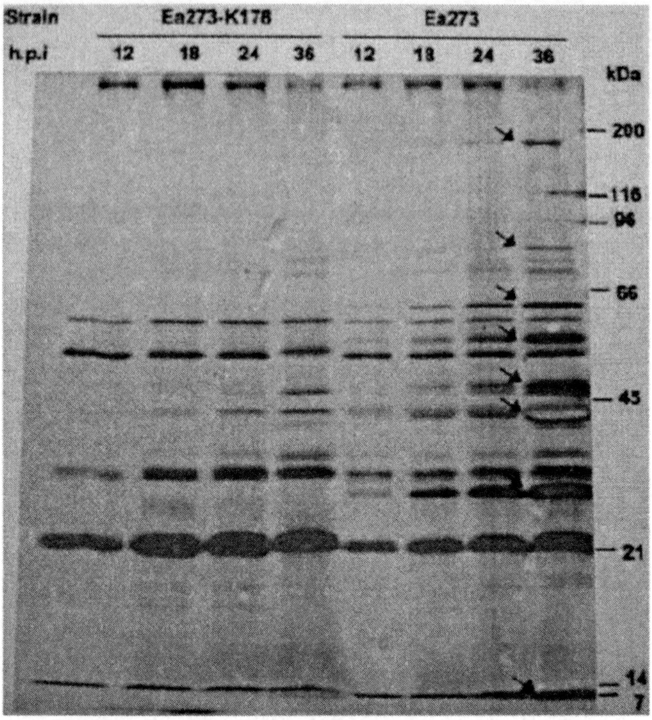

Fig. 2.10 Secreted proteins of *Erwinia amylovora* (Ea) hrp secretion mutant Ea 273-hrcN:Tn19 (Ea 273-K178) and wild-type strain 273 present in culture supernatant (SN) resolved in 12.5% SDS-PAGE and visualized by silver staining
Arrows indicate Hrp-specific proteins. Molecular weights (k-Da) are marked at right extremity. (Courtesy of Nissinen et al. 2007; Blackwell Publishing Ltd, Oxford, UK)

The results suggested that *Ea* could utilize a range of strategies during plant infection and overcome nutritional and environmental stresses (Zhao et al. 2005).

In another investigation aimed to identify genes induced during infection of pear fruit by *Ea*, a putative effector gene, avrRpt2$_{EA}$ was identified and cloned. The deduced amino acid sequence of the translated AvrRpt2$_{EA}$ protein was found to be homologus to the effector protein AvrRpt2 reported earlier in *P. syringae* pv. *tomato* (*Pst*). However, the secretion and translocation domain was different. The presence of *avrRpt2*$_{EA}$ was detected in all strains of *Ea* tested, but this gene was not present in other closely related *Erwinia* spp. The ability to cause systemic infection on immature pear fruits was reduced in an *avrRpt2*$_{EA}$ deletion mutant, indicating that *avrRpt2*$_{EA}$ acts as a virulence factor on its native host. The gene *avrRpt2*$_{EA}$ promoted virulence of *Pst* DC 3000 on *Arabidiopsis* similar to *Pst* expressing avrRpt2. The signal sequence and promoter of avrRpt2$_{EA}$ may affect its expression, secretion and translocation or in combination, in *Pst* DC 3000. The results indicated that the *RPS2* disease resistance gene in *Arabidopsis* when expressed in *Pst* DC 3000 recognized genetically avr Rpt2$_{EA}$. The expression and secretion of similar effector genes present in different bacterial pathogens, may be under specific regulation by the native pathogen (Zhao et al. 2006).

Erwinia chrysanthemi 3937 requires *pelA*, *pelD*, *pelE* and the *pemA* and *pemB* (pectin methylesterase) genes for full virulence on African violet (*Saintpaulia ionantha*), whereas the *pem* genes and *pelA*, *pelB*, *pelC*, *pelD*, *pelI*, *pelL and pelZ* are essential for full virulence on chicory leaves (Beaulieu et al. 1993; Robert-Baudouy et al. 2000). On the other hand, mutations in the type III *hrp* gene secretion do not seem to drastically reduce the virulence of *E. chrysanthemi* (Barras et al. 1994; López-Solanilla et al. 2001) or *E. carotovora* (Rantakari et al. 2001). The perceived minimal role for *hrp* genes in *E. chrysanthemi* may possibly be due to lack of stringent host plant species available for virulence assays. Hence, improved virulence assays for *E. chrysanthemi* 3937 using African violet varieties and a new method for the construction of precise bacterial gene knockouts are needed. The virulence of the *hrpG* and *hrcC* mutants was substantially reduced on leaves of semi-tolerant African violet varieties, whereas an *hrpN* mutant strain produced delayed symptoms on African violet leaves. Furthermore, an *hrpN* Δpel (deletion of five major pectate lyase genes) double mutant was entirely nonpathogenic (Table 2.4), whereas the wild-type strain could produce HR in tobacco rapidly. The *hrcC* and *hrpG* mutants did not induce HR and the *hrpN* mutant gave only a reduced HR. This investigation reveals the importance of using appropriate host plant species for virulence assays and *hrp* genes on the virulence of *E. chrysanthemi* and their ability to elicit HR on nonhosts (Yang et al. 2002).

The DNA sequencing of the entire *E. chrysanthemi hrp/hrc* gene cluster and approximately 12 kb of the flanking regions showed that *E. chrysanthemi* TTSS genes were systemic and had similarity (>50 amino acid sequence identity) with their *E. amylovora* orthologs. But the *hrp/hrc* cluster was interrupted by a cluster of four genes, only one of which, a homolog of lytic trans-glycosylases, is implicated in TTSS functions. The results of experiments for generation of mutants defective in all

Table 2.4 Levels of virulence of wild-type and mutant strains of *E. chrysanthemi* 3937 as determined by virulence assays on African violet varieties

African violet varieties[a]	Wild-type	hrpN mutant		hrpG mutant	
	DI[b]	DI	DI ratio[c]	DI	DI ratio
Irene	4.5	3.8	0.84	2.9	0.64
Maki	3.9	3.9	1.00	1.7	0.44
Katja	4.6	3.5	0.76	0.7	0.15
Rosalie	3.0	1.9	0.63	0.3	0.10
Van Gogh	2.6	1.5	0.58	0.1	0.04
Patsy	1.9	0.7	0.37	0.0	0.00

[a] Ten leaves in 5 plants inoculated with each bacterial isolate

[b] Diseases index (DI) based on the scale representing 0 = no maceration symptom; 1 = less than 20% maceration of the total leaf; 2 = 20%–39%; 3 = 40%–59%; 4 = 60%–79% and 5 = 80%–100%

[c] DI ratio = DI of mutant / DI of wild-type strain

Source: Yang et al. (2002)

of the genes in the regions flanking or interrupting the *hrp/hrc* cluster, indicated that the 44-kb region of *E. chrysanthemi* EC16 genome that is centered on the *hrp/hrc* cluster encodes a mixture of virulence factors, but none of them seem to be a TTSS effector (Rojas et al. 2004).

The necrotrophic bacterial plant pathogens such as *E. chrysanthemi* (strain 3927) produces pectinolytic enzymes that play a key role in inducing soft rot symptoms. Additional factors are however, required for efficient colonization of plants by *E. chrysanthemi*. These factors include HrpN (harpin), a heat-stable, glycine-rich hydrophilic protein, secreted by TTSS. The expression of hrpN in *E. chrysanthemi* 3937 in various environmental conditions and regulatory backgrounds was studied. The gene *hrpN* was repressed when pectinolysis started and negatively regulated by the repressors of pectate lyase synthesis, PecS and PecT. The results suggested a coordinated regulation of HrpN and pectate lyases by PecS and PecT (Nasser et al. 2005).

By applying an in situ immunogold labeling procedure using specific monoclonal antibodies, the majority of AvrPto and HrpW proteins of *Pst* were found to be localized specifically along the Hrp pili, but not along the flagella or randomly distributed in this intercellular space (Fig. 2.11) (Jin et al. 2001). The *hrp* locus of *P. syringae* strains examined, consists of 27 open reading frames (ORFs) and is located in the chromosome. These ORFs are organized into six operons and their

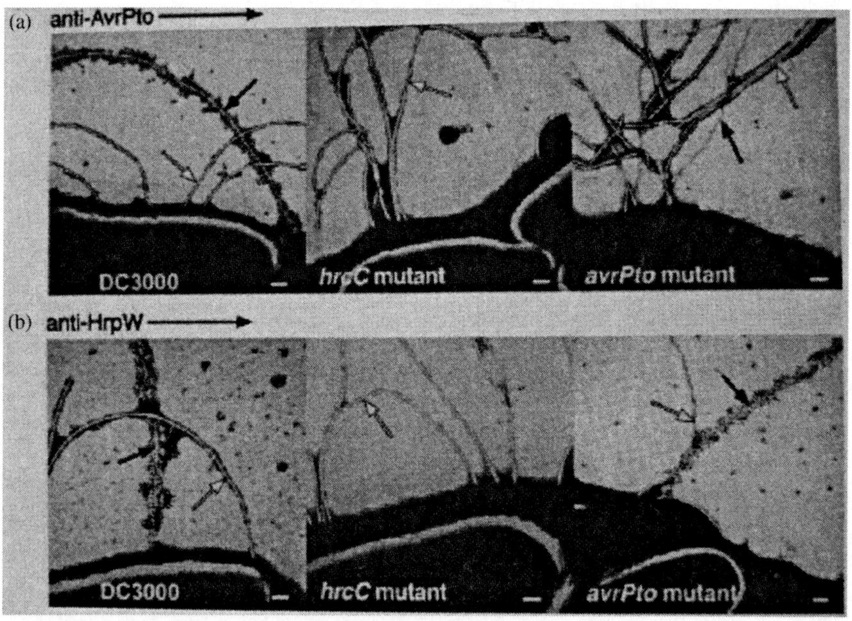

Fig. 2.11 Immunogold labeling of *Pseudomonas syringae* pv. *tomato* (*Pst*) strains with antibodies raised against AvrPto (**a**) or HrpW (**b**) proteins visualized by electron microscopic observations Open arrows indicate flagella. Filled arrows indicate pili. (Courtesy of Jin et al. 2001; Blackwell Publishing Ltd, Oxford, UK)

expression is coordinately regulated in planta and in an *hrp* – inducing minimal medium (Xiao et al. 1992). Three classes of *hrp* genes within *hrp* locus have been recognized. One of them encodes core components of the TTSS, including eight *hrc* (*hrp* genes conserved) genes that exhibit significant sequence similarities with the flagellum assembly genes and they are conserved among all bacteria that contain a TTSS. The regulatory proteins such as HrpL, HrpR, HrpS and HrpV controlling the expression of all TTSS-associated genes in planta or in *hrp*-inducing minimal medium included in the second class of *hrp* genes. The induction of TTSS genes is mediated by HrpL and the transcription of *hrpL* is activated by HrpR and HrpS, two homologous DNA binding proteins encoded by the *hrpRS* operon . In order to evaluate *Pst*DC 3000 genes regulated by *hrpL* and *hrpRS* in minimal medium (MM), microarray analysis was performed. A number of novel *hrpL*-activated genes with a putative TTSS-independent function, was identified. The genes regulated by *hrpL* were mostly regulated by *hrpRS* in the same manner. However, many genes regulated by *hrpRS* were found to be *hrpL* independent, indicating that *hrpL* represents one branch of the regulatory pathways downstream of *hrpRS*. The activation of the TTSS apparently has a cost on the basic cellular activities (Lan et al. 2006).

Another class of hrp genes encodes secreted proteins, some of which are extracellular components of the TTSS, as in the case of Hrp. The conserved effector locus and the exchangeable effector locus flank the *hrp* locus in *P. syringae*. The effector genes in the loci are respectively conserved and variable. All the three loci are linked to form a large pathogenicity island (Alfano et al. 2000). In *P. syringae, hrp/hrc* genes encode the Hrp secretion system, whereas avirulence (*avr*) and Hrp-dependent outer protein (*hop*) genes encode effector proteins. A tripartite mosaic Hrp pathogenicity island (Pai) is formed by the *hrp/hrc* genes of *P. syringae* pv. *syringae* 61 and *P. syringae* pv. *tomato* DC 3000 flanked by exchangeable effector locus and conserved factor locus. Cosmid PH1R11 carrying a protein of the strain 61 Hrp pathogencitiy island could direct *E. coli* and *P. fluorescens* to inject HopPsyA into tobacco cells, eliciting an HR normally triggered only by pathogenic bacteria. The conserved locus has been found to be essential for pathogenicity, while the exchangeable effector locus has only a minor role in the growth of bacteria in tomato (Collmer et al. 2000).

The HrpA subunits constitute the Hrp pilus which is an essential component of the type III secretion system in *P. syringae*. The production of pilus in vitro from *P. syringae* pv. *tomato* (*Pst*) strain DC 3000, was examined by applying electronmicroscopy and immuno-cytochemistry techniques. Following labeling with antibodies to HrpA pili developed rapidly in a nonpolar manner soon after the detection of the *hrpA* transcript and extended upto 5 μm into the surrounding media. The basal bodies of flagella could be clearly differential from the structures at the base of the pilus. Another protein HrpZ also secreted via TTSS was found by immunogold labeling to be associated with the pilus in vitro. After inoculation of the wild-type DC 3000 and *hrpA* and *hrpZ* mutants into leaves of *Arabidopsis,* the accumulation and secretion of HrpA and HrpZ were monitored and estimated quantitatively. The functional pilus crossed the plant cell wall to generate tracks of immunogold labeling for

HrpA and HrpZ. The mutants did not assemble pili, though HrpA was produced and they were nonpathogenic. The mutants were compromised for the accumulation of HrpZ (Brown et al. 2001). The newly formed effector proteins were observed to exit from the tip of the Hrp pilus, suggesting a conduit function for the Hrp pilus (Jin and He 2001; Li et al. 2002). The Hrp pilus elongated by adding the new HrpA subunit at the distal end which is similar to flagellum assembly mechanistically (Li et al. 2002). In *Pst* DC 3000, the Hrp system secretes HrpA, HrpZ, HrpW and AvrPto and assembles a surface appendage Hrp pilus in hrp-gene-inducing minimal medium. The antibody against HrpA-efficeintly labeled Hrp pili. In constrast, antibodies against HrpW and HrpZ did not label Hrp pili. By applying immunogold labeling protocol, a characteristic lineup of gold particles was disernible around bacteria and / or at the bacterium-plant contact sites in *A. thaliana* leaves inoculated with *Pst.* The results showed clearly that HrpA is the major structural proteins of the Hrp pilus and provide evidence for assembly of Hrp pili both in vitro and in planta (Hu et al. 2001; Büttner and Bonas 2003).

P. syringae TTSS secretes, in addition to translocating effector proteins, several extracellular accessory proteins or helper proteins in the extracelular spaces of plant leaves (the apoplast), which help in the translocation of effectors and helper proteins are designated HOP proteins(*H*rp *o*uter *p*roteins). The confirmed Hop inventory of *Pst* DC 3000 is more than 40 (Alfano and Collmer 2004). Many TTSS substrates utilize intracellular accessory proteins known as type III chaperones (TTCs) that facilitate their secretion and translocation via the TTSS (Feldman and Cornelis 2003; Alfano and Collmer 2004). Specific chaperones are essential for the secretion of some effector proteins in mammalian pathogenic bacteria (Cornelis and van Gijsegem 2000). The chaperones are generally small acidic proteins with no sequence similarities to each other. Sequence-based examination has indicated that at least seven putative chaperones of type III effectors in *P. syringae*. The Shc A protein was shown to be a chaperone specific to HopPsyA of *P. syringae* pv. *syringae* (*Pss*). This chaperone is essential for the secretion of HopPsy A in vitro and for its translocation into plant cells. ShcA possesses certain general features of chaperons present in animal pathogens (Van Dijk et al. 2002). The TTCs seem to have a multidimensional roles, such as preventing interactions between TTSS substrates in the bacterial cell and stabilizing effectors in addition to regulatory functions (Feldman and Cornelis 2003; Yip et al. 2005).

The presence of TTCs has been confirmed in *Erwinia amylovora* (DspB) (Gaudriault et al. 2002), *Xanthomonas campestris* pv. *vesicatoria* (HpaB) (Büttner et al. 2004) and several TTCs in *P. syringae* (Shc A, ShcF, Shc M, ShcN and ShcV (Badel et al. 2003; López-Solanilla et al. 2004; Shan et al. 2004; Van Dijk et al. 2002; Wehling et al. 2004)). The operons *hop01-1, hopS1* and *hopS2* in *Pst* DC 3000 were characterized. These operons encoded three homologs TTCs, Shc01, ShcS1 and ShcS2 which facilitated the type III secretion and / or translocation of their cognate effectors Hop01-1, HopS1 and HopS2 respectively. Two novel Hops, Hop01-2 and HopT1-2 were also shown to be secreted via the TTSS. ShcS1 and ShcS2 could substitute for Shc01 and enhance the type III secretion and translocation of the Hop01-1 effector. A DC 3000 *hop01-1* operon deletion mutant was able to induce symptoms of disease as the wild-type strain. But its ability to proliferate in *A. thaliana* was reduced (Guo et al. 2005).

It is important to define the complete effector inventory in order to understand the pathogenicity and host specificity of *Pst* DC 3000. The pathogenicity island (Pai) that encodes the TTSS apparatus was examined. The *hrp-hrc* genes encoding the TTSS apparatus is present in the central region of the pathogenicity island and this central region is flanked by the conserved effector locus (CEL) which contains many effector and helper genes. *Pst* DC 3000 CEL mutants exhibited reduced ability to grow in plants and cause disease symptoms. On the other hand, CEL mutants showed only subtle reduction in the production of disease symptoms and growth in planta (Alfano et al. 2000). The *hrpK* gene is located in the conserved *hrp-hrc* cluster at the border of the CEL in *P. syringae*. The *hrpK* mutants are able to secrete the HrpZ harpin in culture, suggesting that HrpK is not an essential component of *P. syringae* TTSS apparatus (Collmer et al. 2000). The transcriptional expression of the ORFs within the CEL of *Pst* DC 3000 was characterized. One of the ORFs PSPTO 1406 (*hopB1*) was expressed in the same transcriptional unit as hrpK. Both HopB1 and HrpK were secreted in culture and translocated into plant cells via the TTSS. HrpK acted as a translocator for the *P. syringae* TTSS. The *Pst* DC 3000 hrpK mutants were significantly reduced in virulence. HrpK defective strains were less efficient at effector translocation compared with wild-type strains, based on their reduced HR phenotypes. However, *hopB1* mutants produced phenotypes similar to the wild-type strain. In addition, *hrpK* mutants exhibited reduced ability to elicit HR, a PCD-associated plant defense. The results indicate that HopB1 is a type III effector, whereas HrpK has an important role in the TTSS and is a putative type III translocator (Petnicki-Ocwieja et al. 2005).

The transcription of selected *hrp* and effector genes in *P. syringae* pv. *phaseolicola* (*Psp*) strains 1448A (race 6) and 1449B (race7) was examined by quantitative real-time PCR with specific TaqMan probes. Transcripts from the genes encoding regulators *hrpR* and *hrpL, hrcN, hrcU* and *hrpA*; the first ORF of each *hrp* operon including *hrpF, hrpJ, hrpP* and *virPphA* were detected. The rates and levels of transcript accumulation within the first 2 h after inoculation were considerably higher in planta than in vitro and indicated that plant cell wall contact may enhance transcription of TTSS and effector genes in *Psp* (Thwaites et al. 2004). A novel TTSS-dependent effector that functions as a host-species-specific virulence factor in Psy61 was identified in *P. syringae*. Most genes encoding effectors are dispersed in the genome of *P. syringae* and all are regulated coordinately by the alternative sigma factor HrpL. Hop PsyL was structurally similar to other secreted effectors and carried a putative chloroplast-targeting signal and two predicted transmembrane domains (Losada et al. 2004). The number of predicted effector proteins in *P. syringae* seems to be unexpectedly high. It is apparent that the plants use TTSS effectors as a main source of recognition to activate innate defense and turn virulence-related effector gene into *avr* genes. For its survival, *P. syringae* has to mutate these *avr* genes to evade recognition or evolve additional effector genes to mask the function of the *avr* genes, resulting in rapid production of effector genes in the pathogen genome (Jin et al. 2003).

Two regulatory *hrpG* and *prhJ* genes of *Ralstonia solanacearum* located at the right hand end of the *hrp* gene cluster were identified. These genes are required for full pathogenicity of *R. solanacearum.* The hrpG was found to be homologous to HrpG which activates *hrp* genes in *Xanthomonas campestris* pv. *vesicatoria* (*Xcv*). PrhJ is a novel *hrp* regulatory protein and shares homology with LuxR / UhpA family of transcriptional activators; it is specifically involved in *hrp* gene expression in the presence of plant cells. Transcription of *hrpG* and *prhJ* genes is plant inducible through the *PrhA*-dependent pathway (Brito et al. 1999). The effector protein popA of *R. solanacearum* OE1-1 (OE1-1) (pathogenic to tobacco) showed 97.6% identity to popA of GM1 1000 strain (not pathogenic to tobacco). RT-PCR analysis showed that popA in OE1-1 was expressed at 3 h after inoculation (hai) in infiltrated tobacco leaves. A popABC operon-deleted mutant of OE1-1 (Delta ABC) indicated that pop ABC is not involved in pathogenicity of OE1-1 directly. The results suggest that tobacco plants can recognize PopA, when it is expressed early in disease development and respond with an effective defense in the intercellular spaces (Kanda et al. 2003).

The repertoire of genes under the control of the transcription activator HrpB that governs the pathogenesis of *Ralstonia solanacearum* were examined by a 70-mer oligonucleotide-based DNA microarray system. Among the 5074 genes generated, 143 hrpB up-regulated genes and 50 hrpB down-regulated genes were identified. It was demonstrated that the hrpB regulon extends beyond type III secretion system-related functions to include a number of genes governing chemotaxy, biosynthesis and catabolism of various low-molecular-weight compounds and siderophore production and uptake. The results indicate that *hrpB* may act as a master regulatory gene governing a physiological swing associated with the shift from saprophytic to parasitic life (Occhialini et al. 2005). The effects of infection by *R. solanacearum* GMI 1000 and *hrp* mutants on the root system of petunia plants were assessed. Inhibition of lateral root elongation and induction of swellings of the roots were the adverse effects caused by both the wild-type and the mutants. Further, the production of new root lateral structures (RLS) was seen due to infection by the wild-type, but not by the mutants. The RLS have been found to be efficient colonization sites allowing extensive bacterial mutliplication. It is possible that the RLS may play a role in rhizosphere-related stages in the life cycle of *R. solanacearum* leading to successful infection by the wild-type strain (Zolobowska and van Gijsegem 2006).

Two novel Hrp-secreted proteins, PopF1 and PopF2 from *R. solanacearum* exhibited similarity to one another and to putative translocators, HrpF and NopX from *Xanthomonas* spp. and rhizobia respectively. Both *popF1* and *popF2* belong to the HrpB regulon and are essential for interaction with plants. They are not required for secretion of effector proteins, but they are needed for translocation of AvrA proteins into tobacco cells. It is concluded that PopF1 and PopF2 are type III translocators belonging to the HrpF/NopX family. The *hrpF* gene of *Xcc* partially restored HR-inducing ability to *popF1* and *popF2* mutants of *R. solanacearum*, suggesting that translocators of *R. solanacearum* and *Xanthomonas* are functionally conserved (Meyer et al. 2006). The hrpB-regulated gene *lrpE* (*hpx5/brg24*) in *R. solanacearum* encodes a PopC-like leucine-rich repeat (LRR) protein that carries 11 tandem LRR in the central region. Mutant defective in *lrpE* had slightly reduced virulence on host plants and changed morphology resulting in the aggregation of bacterial cells in a minimal medium. In the *ΔlrpE* mutant, bundled Hrp pili were observed on the cell surface even in late growth phase, in contrast to the disappearance of Hrp Pili much earlier in the wild type strain. Such bundles were entangled and anchored the mutant cells in the aggregate preventing dispersal of cells. LrpE accumulated in bacterial cells and did not translocate into plant cell as an effector protein. In Delta *lrpE* mutant, the expression levels of *hrp* genes attained a 3–5 fold increase in comparison to those in the wild-type strain. It was proposed that LrpE may negatively regulate production of Hrp pili on the cell surface of *R. solanacearum* to disperse the bacterial cells from aggregates leading to rapid pathogen movement in the vascular elements of host plants (Murata et al. 2006).

Ralstonia solanacearum encodes a large number of putative TTSS effectors (upto 80) (Cunnac et al. 2004; Occhialini et al. 2005). Among these effector proteins, a group of seven genes that have homologies with plant-specific leucine-rich repeats (LRR) has been identified and they are designated “GALA” proteins after a conserved GAXALA sequence in their LRR. GALA6 was translocated into plant cells (Cunnac et al. 2004). *R. solanacearum* has effectors that contain both a LRR and an F-box domain. The F-box is a short domain (48aa) characteristic of the eukaryotic F-box proteins, allowing them to interact directly with the SKP1-like proteins. The Cullin 1 and SKP1-like proteins interact to form the SCF-type E3 ubiquitin ligase complex and control specific ubiquitinylation. The ubiquitin-tagged proteins have either modified properties or are doomed for degradation by the 26S proteasome. A *R. solanacearum* strain in which all of the seven GALA effector genes have been detected or mutated lost the pathogenicity entirely on *Arabidospsis* or became less virulent on tomato. In addition, GALA7, a host specificity factor was essential for pathogenicity on *Medicago truncatula* plants. Since the F-box domain is required for virulence function of GALA7, it was hypothesized that these effectors may act by hijacking their host SCF-type E3 ubiquitin ligases to interfere with their host ubiquitin / proteasome pathway to promote disease (Angot et al. 2006).

Pseudomonas syringae pv. *tomato* (*Pst*) DC 3000 elaborates a phytotoxin designated coronatine (COR) which induces necrotic lesions on susceptible tomato plants. COR was also produced in vitro by mutants of DC 3000 defective in *hrcC* which encodes an outer-membrane protein required for type III mediated secretion. The *hrcC* mutants did produce COR in planta, which induced chlorotic lesions that did not turn necrotic as in the case of lesions induced by wild-type strain. The results suggested that the *hrp/hrc* secretion system was not needed for COR production, but mutations in this system may have regulatory effects on the production of virulence factors such as COR (Bender 2000). Two distinct components viz., coronafacic acid (CFA) and coronamic acid (CMA) are present in the COR molecule and they are intermediates in the COR biosynthetic pathway. In *P. syringae* pv. *glycinea* PG 4180, *corR* encoding a response regulator is reported to positively regulate CFA and CMA synthesis. But in *Pst* DC 3000 the role of *corR* is not clear. In a later study, a *corR* mutant of *Pst* DC 3000 defective in the production of COR, CFA and CMA induced less severe symptoms on tomato compared with wild-type strain. In addition, mutation in *hrpL* which encodes an alternate RNA polymerase sigma factor needed for the expression of genes encoding components of TTSS inhibited production of COR in *Pst* DC 3000. The RT-PCR experiments revealed that the upstream effector gene *holPtoAA* which is associated with the *hrp* box, was cotranscribed with *cor*. In addition, the *corR* mutant of *Pst* DC 3000 exhibited both reduced and delayed expression of *hrpL* and was impaired in its ability to elicit a HR on *Nicotiana benthamiana*. The results showed that *corR* impacted directly the expression of the *hrp* regulon in *P. syringae* (Sreedharan et al. 2006).

A neutral leucine aminopeptidase (LAP-N) and an acidic LAP (LAP-A) are expressed constitutively in tomato plants during floral development and in leaves in response to infection by *Pst*, in addition to insect infestation and wounding. The mechanism underlying the strong induction of LAP upon *Pst* infection was investigated. Following infection of tomato by COR-producing and-deficient *Pst* strains, *LapA* expression was monitored. LapA RNA and activity were detected only with COR-producing *Pst* strain. COR treatment of excised shoots induced increases in RNAs for jasmonic acid (JA)-regulated PR protein gene *PR-1*. The results indicated that COR mimicked the wound response. However, it was insufficient to activate JA-regulated PR genes (Pautot et al. 2001).

A family of structurally closely-related peptide derivatives designated syringolines is secreted by strains of *P. syringae* pv. *syringae* (*Pss*). Among these compounds, syringolin A was shown to be a major variant. Syringolin A induced hypersensitive death of cells colonized by powdery mildew in wheat, reprogramming a compatible interaction into an incompatible one. The genes involved in the biosynthesis of syringolin A encode proteins consisting of modules typical of nonribosomal peptide synthetases and type I polyketide synthetases (Amrein et al. 2004). The expression of the *syrB* gene that controls synthesis of syringomycin, a nonhost-specific phytotoxin produced by *Pss* was determined in vitro by using aqueous extracts of bark tissues of peach trees. The *syrB* gene expression was quantified by a β-galactosidase activity expressed by *Pss* B3AR-132 containing a syrB:: lacZ fusion. The expression of *syrB* was significantly less in plants receiving N fertilization compared with nonfertilized plants. Nitrogen fertilization appeared to decrease host susceptibility to *Pss* (bacterial canker) by producing or reducing compounds that may induce or antagonize syrB expression (Cao et al. 2005).

The *salA* gene in *Pss* is a key regulatory element for production of syringomycin and encodes a member of the LuxR regulatory protein family. In addition, *salA* a member of the GacS/GacA signal transduction system has been shown to be essential for bacterial virulence, syringomycin production and expression of the *syrB1* synthetase gene. By using oligonucleotide microarray analysis of the *SalA* regulon controlling phytotoxin production, it was observed that expression of 16 genes was significantly higher (>2 folds) in the wild-type for some toxin biosynthesis genes. Except for the *sylD* synthetase gene for syringolin production, all ORFs controlled by *SalA* were located in the syr-syp genomic island and were associated with biosynthesis, secretion and regulation of syringomycin and syringopeptin. The subgenomic oligonucleotide microarray has the potential for use as power tool for defining the

SalA regulon and its relationship to other genes important to plant pathogenesis (Lu et al. 2005).

Production of syringomycin and expression of *syrB1*, a syringomycin synthetase gene in *Pss* are induced by specific plant signal molecules. Likewise, syringopeptin production is activated by plant signal molecules and GacS, SalA and SyrF regulatory pathway mediates transmission of plant molecules to the *syr-syp* biosynthesis apparatus. Addition of arbutin (100 μM) and D-fructose (0.1%) to syringomycin minimal medium (SRM) increased syringopeptin production by 2-folds in BR 132 strain of *Pss*. The *salA* mutant did not produce syringopeptin, indicating the involvement of SalA/SyrF regulon in its synthesis. The subgenomic analysis of transcriptional expression with a 70-mer oligonucleotide microarray confirmed that the *syr-syp* genes are induced 2.5–10.5 fold by addition of arubtin and D-fructose to SRM. Thus plant signal molecules are found to be transmitted through the GacS, SalA/SyrF pathway to activate the coordinated transcriptional expression of the *syr-syp* genes (Wang et al. 2006a). Transcriptional analysis with 70-mer oligonucleotide microarrays, along with β-glucuronidase (GUS) assays and real-time RT-PCR analysis confirmed that all of the *syr-syp* genes belong to the SalA regulon. The synthetase genes for syringomycin (*syrB1* and *syrE*) and syringopeptin (*sypA, sypB* and *sypC*) and four regulatory genes (salA, syrF, syrG and syrP) and nine putative secretion genes dedicated to the production of two toxins were activated by the phenolic plant signal molecule arbutin (Lu et al. 2005). In a later study, the 132-kb *syr-syp* genomic island was shown to be organized into five polycistronic operons along with eight individual genes based on RT-PCR and bioinformatics analysis. In addition, the conserved promoter sequences of the *syr-syp* genes were indicated to contribute to the coregulation of syringomycin and syringopeptin production (Wang et al. 2006b).

Pseudomonas syringae pv. *syringae* (*Pss*) causing mango apical necrosis disease, elaborates an antimetabolite mangotoxin, production of which is controlled by the chromosomal region of 11.1 kb. Six complete ORFs including a large gene (ORF5) with a modular architecture characterisitic of nonribosomal peptide synthetases (NRPS) designated *mgoA* were identified in this chromosomal region. A Tn5 mutant disrupted in *mgoA* was defective in the biosynthesis of mangotoxin, indicating the involvement of the putative NRPS gene in the biosynthesis of mangotoxin. In addition, the virulence of the defective strain was also reduced as revealed by severity of necrotic symptoms on tomato leaflets. Complementation of this NRPS mutant with plasmid pCG2-6 that contained an 11,103-bp chromosomal region cloned from the wild-type strain of *Pss* that included the putative NPRS gene (*mgoA*) restored both virulence and the ability to synthesize mangotoxin. The results revealed the involvement of *mgoA* in the virulence of *Pss* strain infecting mango (Arrebola et al. 2007).

Xanthomonas albilineans causing leafscald disease in sugarcane, is a systemic, xylem-invading bacterial pathogen. Albicidin, a major toxic compound produced by the pathogen has a critical role in pathogenicity. Albicidin-deficient mutants cannot induce disease symptoms and transgenic sugarcane plants expressing an albicidin-detoxifying gene exhibit resistance to the disease (Zhang et al. 1999). The 49-kb albicidin biosynthesis gene cluster and two additional 3 kb-genomic regions are essential for albicidin production. Variability in biosynthesis in relation to pathogenicity was examined by RFLP analysis. Albicidin genetic diversity showed high similarity to the genetic diversity of *X. albilineans* based on whole genome. There was no relationship between variability in albicidin biosynthesis genes and the quantity of albicidin produced in vitro by *X. albilineans.* The strains of *X. albilineans* (137) tested, were placed in three different pathogenicity groups based on symptom severity and pathogen population density in the stalk (Champoiseau et al. 2006).

Streptomyces turgidiscabies causing potato scab disease, produces the phytotoxin, thaxtomin A. The biosynthetic genes for thaxtomin A are present on the large mobile pathogenicity island (PAI), the acquisition of which makes the saprophytic genus *Streptomyces* pathogenic to plants. The Nec1 protein induces necrosis on excised potato tuber tissue. The G+C content of *nec1* indicated a lateral transfer from unrelated taxon. An active 16-kDa form of Nec1 was shown to be secreted by *S. turgidiscabies.* The 151-amino acid C-terminal region of the Nec-1 was shown to be secreted by *S. turgidiscabies.* The 151-amino acid C-terminal region of the Nec-1 proteins conferred the necrogenic activity, as revealed by deletion analysis of *nec1* gene. The *nec1* deletion mutant exhibited significantly compromised virulence on *A. thaliana, N. benthamiana* and radish (*Raphanus sativus*). Aggressive colinization and infection of root meristem of radish by the wild-type Car8 strain was observed, whereas the *Δnec1* mutant Car811 did not infect the radish root tissues (Joshi et al. 2007).

The ability to sense the surrounding environment and regulate gene expression accordingly is an important feature of microorganisms including *P. syringae*. The tetracycline repressor (TetR) family of bacterial regulatory proteins that allow bacteria to sense and respond quickly to the environment TetR-like regulators bind DNA via highly conserved amino-terminal helix-turn-helix (HTH) motif (Hillen and Berens 1994). The TetR-like AefR (*A*HL and *e*piphytic *f*itness *r*egulator) protein was demonstrated to regulate the synthesis of *N*-acyl homoserine lactone (AHL) and contribute to the epiphytic fitness of *P. syringae* pv. *syringae* strain B728a on soybean leaf surfaces (Quinones et al. 2004).

It is likely that TetR-like regulators may have a role in pathogen adaptation to plant host environment, in addition to their ability to modulate the expression of genes in response to environmental signals. A mutant strain DB4H2 of *Pst* DC 3000 contains a single Ta5 insertion in PSPT 03576, an ORF that is predicted to encode protein belonging to TetR family of transcriptional regulators. The ORF PSPT03576 was shown to be essential for the virulence in *Pst* DC 3000 and the encoded protein was designated TvrR (*T*etR-like *v*irulence *r*egulator). The gene *tvrR* was required for virulence in *Pst* DC 3000 on *A. thaliana* plants. Plants inoculated with tvrR mutant strains did not produce the characterisitic disease symptoms of water-soaked lesions and chlorosis which are the physical hallmarks of DC 3000 infection. Further, *tvrR* mutants had significantly impaired growth within plant tissues. Complementation of *tvrR* mutant strain restored wild-type growth and symptom development. The *tvrR* gene was constitutively expressed in the artificial media and in planta. The *tvrR* mutant strains were able to synthesize coronatine (COR) required for virulence of *Pst* DC 3000 on *A. thaliana.* Since *tvrR* mutants are not defective for type III secretion or COR production, *tvrR* seems to be a novel virulence factor required for an unknown process that may be essential for pathogenesis (Preiter et al. 2005).

The pathogenicity of *Xanthomonas* spp. as in the case of *Pseudomonas* spp. depends on a specialized type III secretion system (TTSS) which secretes some helper and accessory proteins into the extracellular milieu that support the injection of effector proteins, the major substrate class of the TTSS system, into the host cell. The specialized protein system is encoded by a 23-kb hrp chromosomal *hrp* gene cluster which contains six operons *hrpA* to *hrpF*. The *hrp* mutants are unable to infect susceptible plants and induce symptom and also fail to eilicit HR response in resistant plants. A few nonconserved proteins essential for pathogenicity of *X. campestris* pv. *vesicatoria* (*Xcv*) have been identified. HrpB2 and HrpE were secreted and essential for secretion of other proteins by the TTSS (Rossier et al. 2000). HrpE has been found to be a major subunit of the Hrp pilus, an extracellular appendix considered to be associated with the TTSS apparatus (Koebnik 2001; Weber et al. 2005), whereas HrpF is essential for pathogenicity, but not for secretion in *Xcv* (Rossier et al. 2000). On the other hand, *hrpF* mutants of *X. oryzae* pv. *oryzae* (*Xoo*) were only reduced in virulence (Sugio et al. 2005). HrpF may possibly be the major component of the *Xcv.* TTSS translocon, a predicted protein channel in the plant membrane mediates the transport of the effector proteins into the plant cell cytosol (Büttner et al. 2002).

The formation of filamentous structures, the Hrp pili, at the surface of cells of *Xcv* was visualized by immunoelectron microscopy by using antisera generated against HrpE, HrpF and AvrBs3. The Hrp pili were heavily labeled by HrpE-specific antiserum, revealing that HrpE was the major pilus subunit. Flagella were occasionally seen, but they were not labeled, indicating the specificity of the Hrp pilus labeling. The other type III secreted proteins HrpE and AvrBs3 were also in close contact with Hrp pilus during and / or after their secretion. Analysis of proteins in the supernatant of the pilus preparation using Tricine SDS-PAGE technique showed that the HrpE protein was the predominant component of the pilus preparations from *Xcv* (Weber et al. 2005). The gene *hrpE* is unique to the genus *Xanthomonas* and exhibits no sequence similarity to other pili genes. Bacteria with mutations in *hrpE* are unable to either cause disease or elicit HR in plants. In addition, HrpE is essentially required for all the TTSS for secretion of HrpF, AvrBs1 and AvrBs3. The Hrp pilus, is therefore, an indispensable component of a functional TTSS. The 9-kDa HrpE protein forms a slender pilus 8–10 nm in diameter and up to 4 μm in length (Weber et al. 2005). The Hrp pilus acts as a cell surface appendage of the TTSS and serves as a conduit for the transport of bacterial effector proteins into the plant cell cytosol. The functional domains of HrpE were mapped by linker-scanning mutagenesis and by a reporter protein fusions to an N-terminally truncated avirulence protein (AvrBs3Δ2). The N-terminal 17 amino acid residues act as an efficient TTS signal (Weber and Koebnik 2005).

The *hpa* (*hrp*-associated) genes that contribute to, but are not essential for the pathogenic interaction of *Xcv* with host plants were identified by the analysis of nonpolar mutants in the *hrp* gene cluster (Noël et al. 2002; Büttner et al. 2004). The gene *hpaA* encodes a protein needed for the HR and full pathogenicity. HpaA secreted by the TTSS, is localized to the plant cell nucleus when expressed in planta (Huguet et al. 1998). In contrast, HpaB remains in the pathogen to accomplish its role in the control of type III protein export (Büttner et al. 2004). XopA (*Xanthomonas* outer protein) secreted by TTSS has some features of an Hpa. The *XopA* gene located adjacent to the left border of the *hrp* region and its expression is HrpG-and HrpX-dependent. XopA exhibits homology to Hpa1 from *Xoo* and the homolog HpaG from *X. axonopodis* pv. *glycines* (Kim et al. 2003, 2004a). The protein HpaB, showing typical features of type III chaperones was identified. It physically interacts with the first 50 amino acids of AvrBs3 and probably is needed for the translocation of AvrBs3 (Büttner et al. 2004). The *XopA* mutants were reduced in virulence, but not in type III-mediated secretion (Noël et al. 2001, 2002). HrpG and HrpX are the key regulatory protein in *Xanthomonas* which are encoded outside of the *hrp* gene cluster (Wengelnik and Bonas 1996). HrpG belonging to the ompR-family controls the expression of HrpX, whereas HrpX, an AraC-type transcriptional activator, regulates the expression of most genes of the HrpG regulon including the *hrp* operons hrpB to hrpF (Wengelnik and Bonas 1996; Noël et al. 2001, 2002, 2003).

Xanthomonas oryzae pv. *oryzae* (*Xoo*) causing rice bacterial leaf blight (BLB) disease is considered as a model organism for analysis of plant–pathogen interaction, since more than 30 races differing in virulence and 25 resistance genes in rice have been identified. The complete genome sequence of *Xoo* strain MAFF 311018 has been defined (Zhu et al. 2000; Ochiai et al. 2005). The genome sequences of *X. axonopodis* pv. *citri* (*Xac*) and *X. campestris* pv. *campestris* (*Xcc*) have also been reported earlier (da Silva et al. 2002). The deduced genome of *Xoo* strain MAFF311018 is a circular chromosome of 4,940,217 bp and the average G+C content is 63.7%. The *hrp* genes encoding a TTSS are the most important of the genes related to pathogenicity of *Xoo*. The *hrp* gene cluster contains 27 genes extending from hpaZ to hpaF and the structure has similarity to those of other characterized *Xanthomonas hrp* gene clusters (Kim et al. 2003) the exception being the *hpaB* (*hrpE2*) – *hrpF* region. Three novel genes were detected in the region between *hpaB* and *hrpF*. In addition, four tandem transposase homologs were present between hpaB and hrpF (Ochiai et al. 2005).

Successful infection by bacterial pathogens commences with attachment of pathogen cells to the host tissue surface. Many Gram-negative bacteria have type IV fimbriae (also known as pili) for adhesion. The presence of type IV fimbriae genes distributed in several loci on the genome of *Xoo* has been observed. Further, a non-fimbrial adhesion designated XadA detected in *Xoo*, may also have a role in virulence. Two *xadA* homologs at different loci have been identified in *Xoo* MAFF 311018 strain (Ray et al. 2002; Ochiai et al. 2005). *Xoo* has homologs of two sets of characteristic two-component regulators involved in microbial interactions with plants. One set is the *virA/virG* two-component system of *Agrobacterium tumefaciens* and other one is the nodV/ nodW two-component system of *Bradyrhizobium japonicum* (Stachel and Zambryski 1986; Göttfert et al. 1990). The similarity to the *vir* and *nod* gene regulators suggests that they may play a role in regulation of pathogenicity in response to plant or environmental signals. Comparison of the *hrp* cluster among *Xoo, Xac* and *Xcc* revealed that the gene orders were similar but the clusters were located in different regions on the genomes (Ochiai et al. 2005).

Pantoea stewartii subsp. *stewartii* causes wilting and leaf blight, preceded by production of water-soaked lesions on young leaves of sweet corn and maize. The Group I *hrp* clusters are present in this bacterial pathogen as in *Pseudomonas* and *Erwinia*. The *hrp* genes are activated by complex regulator pathways encoding HrpS and HrpL (Frederick et al. 2001). The hrp type III secretion regulon of *P. stewartii* subsp. *stewartii* is regulated by a cascade involving the HrpX / HrpY two-component system, the HrpS enhancer binding protein and the HrpL alternate sigma factor. The gene *hrpXY* is both constitutive and autoregulated. HrpY controls *hrpS* and HrpS activates *hrpL*. These regulatory genes are sequentially arranged in the order *hrpL*, *hrpXY* and *hrpS* and they constitute three operons. A novel autoregulatory loop involving HrpS has been detected in this bacterial pathogen. HrpS initiates this regulatory loop caused by read-through transcription from *hrpL* into *hrpXY* (Merighi et al. 2005). Each bacterial pathogen appears to respond differently to external signals and growth conditions so that regulation of *hrpS* by specific and global regulators may be fine-tuned for each pathogen's niche. The genetic analysis showed that the HrpX sensor kinase and the HrpY response regulator are at the head of a complex cascade of regulators controlling *hrp/hrc* secretion and *wts* effector genes. Further study showed that only some of the multiple signals controlling the *hrpS* promoter are mediated by HrpY. The *hrpS* promoter was characterized by locating its transcription start site and identifying the region required for HrpY-dependent regulation and determining the sequences to which HrpY binds. HrpY has a single phsophorylation site and phosphorylation increases its binding affinity to P_{hrpS} (Merighi et al. 2006).

Plants possess mechanisms, evolved over time, to detect the presence of microbial pathogens and activate defenses that delay or arrest induction of disease symptoms, as culmination of successful infection. In the case of bacterial pathogens, plants may activate basal defenses upon detection of bacterial flagellin or lipopolysaccharides (Zipfel et al. 2004). Cell wall reinforcements and expression of defense-related proteins are induced later, contributing to the development of effective defense system. Presence of intracellular resistance (R) proteins that can detect the presence of specific type III effectors inside the plant cells has been detected in certain pathosystem (Martin et al. 2003). Hypersensitive response (HR) is the key phenomenon associated with R protein-mediated resistance (immunity). HR is a process involving rapidly induced programmed cell death (PCD) in cells near the site of infection (Greenberg and Yao 2004).

Pathogenicity of *Xanthomonas campestris* pv. *vesicatoria* (*Xcv*) on susceptible plants and induction of HR on resistant plants are controlled by *hrp* genes, most of which are clustered in a 23 kb chromosomal region. Nine *hrp* genes encode different components of the type III secretion system (TTSS). *Xcv* secretes proteins in a *hrp*-dependent manner. Further, pathogenicity factors and avirulence proteins such as AvrBs3 are secreted via TTSS. The AvrBs3 protein is considered to govern recognition resulting in HR induction in pepper plants carrying the corresponding *Bs3* resistance gene. The AvrBs3 protein contains a nuclear localization signal (NLS) and acts inside the plant cell. It is hypothesized that AvrBs3 may be transferred into the plant cell viz the Hrp type-III pathway and recognition of AvrBs3 may take place in the plant cell nucleus (Bonas et al. 2000). The AvrBs3 protein belonging to a large family of conserved proteins in *Xanthomonas* spp. is considered to play a role in both virulence and avirulence (Bai et al. 2000; Yang et al. 2000a; Marois et al. 2002). The 17.5 repeats of the 122-kDa AvrBs3 protein are required for recognition by the Bs3 resistance gene and induction of cell hypertrophy in susceptible plants (Herbers et al. 1992; Marois et al. 2002).

By applying immuno-cytochemistry technique, direct evidence for TTSS-dependent translocation of AvrBs3 from *X. campestris* pv. *vesicatoria* (*Xcv*) was provided. Immunochemistry protocols for pepper cells expressing avrBs3 via *Agrobacterium*-mediated gene transfer under the control of the constitutive CaMV 35S promoter were developed. The subcellular localization of AvrBs3 was determined in serial transverse sections of susceptible pepper (ECW) leaf tissue expressing avrBs3, using an AvrBs3-specific PAB and alkaline phosphatase-coupled secondary antibody. The DNA-specific staining with 4′,6-diamidino-2-phenylindole (DAPI) showed that the AvrBs3 protein was detectable in the nucleus of some, but not all mesophyll cells, probably because all cells might not have been transformed by *Agrobacterium.* AvrBs3 translocation could be seen in situ in native tissues of both susceptible and resistant plants. AvrBs3 was detected in the nucleus as early as 4 h post-infection which was dependent on a functional TTSS and the putative translocator HrpF. Deletion of the nuclear localization signals (NLS) in the AvrBs3 C-terminus, which are required for the AvrBs3-mediated induction of the HR in resistant pepper plants abolished AvrBs3 localization to the nucleus (Szurek et al. 2002). This report seems to be the first providing a direct evidence for translocation of a native type III effector protein from bacterial pathogen to the host plant cell.

Both animal and plant bacterial pathogens have homologs of the *Yersinia* virulence factor YopJ, indicating several bacterial pathogens may use YopJ-like proteins to regulate interactions with host plants during infection. These proteins are hypothesized to functionally mimic small ubiquitin-like modifier (SUMO) proteases in eucaryotic cells. *Xcv* strains possess four Yop-J like proteins, *AvrXv4, AvrBsT, AvrRxv* and *XopJ.* Characterization of AvrXv4 showed that *Xcv* secreted and translocated the AvrXv4 protein into plant cells during infection in a type III-dependent manner. Once inside the plant cell, AvrXvr is localized to the plant cytoplasm. Deletion and mutational analysis helped identification of amino acids required for type III delivery and for host recognition. Expression of AvrXv4 in planta led to a reduction in SUMO-modified proteins, demonstrating that AvrXv4 has SUMO-isopeptidase activity. It was concluded that the YopJ-like effector AvrXv4 encodes a type III SUMO protease effect that could be active in the cytoplasmic compartment of plant cells (Roden et al. 2004a).

The avirulence gene *avrRxv* in *Xcv* localized predominantly to the cytoplasm and possibly in association with plasma and nuclear membranes in both resistant and susceptible tomato plants. For host recognition, the AvrRxv cysteine protease catalytic core was shown to be required, since introduction of mutations in this domain effected the ability of AvrRxv to elicit HR and the inhibition of bacterial growth in resistant plants. Analysis of expression profiles revealed that 420 of approximately

8600 tomato genes, were identified as differentially modulated by the expression of a functional AvrRxv, including over 15 functional classes of proteins and a large number of transcription factors and signaling components (Bonshtien et al. 2005). In another investigation, characterization of *Bs4* (disease resistance gene) promoter and its application for low level expression of bacterial type III effector proteins in planta were taken up. Real-time PCR assay revealed that Bs4 was constitutively expressed at low levels and transcript abundance did not alter significantly following infection with avrBs4-containing *Xcv* strains. The gene avrBs3 did not trigger Bs4-dependent HR when expressed under the control of the weak Bs4 promoter. On the other hand, avrBs4 was able to trigger HR. However, the pepper Bs3 gene which mediates recognition of AvrBs3, but not AvrBs4-delivering xanthomonads, retained its recognition specificity, even when *avrBs4* was expressed in planta from the strong 35S promoter (Schornack et al. 2005).

Xanthomonas oryzae pv. *oryzae* (*Xoo*) incitant of rice bacterial blight disease, produces AvrXa7 protein, the product of the pathogen gene *avrXa7* and this protein is a virulence factor. The gene *avrXa7* is included in the *avrBs* avirulence gene family which encodes proteins targeted to plant cells by the type III secretion system. The specificities of virulence and avirulence are associated with the central repeat domain, which in *avrXa7* consists of 25.5 direct repeat units. Mutations in three C-terminal nuclear localization motifs abolish both avirulence and virulence activities in rice. The products of the *avrBs3*-related genes are the virulence factors targeted to the host cell nuclei and have the potential to interact with host DNA and transcriptional machinery (Yang et al. 2000a). Avr Xa7, a member of the avBs3/pthA gene family, is the only known type III secretion system effector from *Xoo* with a major contribution to bacterial growth and lesion formation in bacterial blight disease. The effectors from diverse strains of *Xoo* were identified in order to assess the general requirement for effectors for the AvrBs3/pthA family. The growth of *Xoo* in plants was severely inhibited following inactivation of a single effector gene in representative strains from Japan, Korea and Philippines. The results appear to lend support to the hypothesis that the bacterial disease of rice is highly dependent on a single class of type III effectors and also indicated that avrXa7 aviruelnce specifically is separable from virulence activity (Yang and White 2004). The endogenous avr/R interactions have not been identified for rice leaf streak disease caused by *X. oryzae* pv. *oryzicola*. The genes *avrXa7* and *avrXa10* from *Xoo* failed to elicit disease resistance-associated HR and failed to prevent development of leaf streak in rice cultivars with the corresponding *R*genes after introduction into *X. oryzae* pv. *oryzicola*, despite the ability of this pathovar to deliver an Avrxa10: Cya fusion protein into rice cells. In addition, coinoculation of *X. oryzae* pv. *oryzicola* inhibited the HR of rice cv. IRBBIO to *Xoo* carrying *avrXa10*. Inhibition was quantitative and dependent on the TTSS of *X. oryzae* pv. *oryzicola*. The results suggest that one or more *X. oryzae* pv. *oryzicola* type III effectors interfere with avr/R-gene-mediated recognition or signaling and subsequent defense response in host (Makino et al. 2006).

The core *hrp* cluster of xanthomonads contains six operons from *hrpA* to *hrpF*. *X. oryzae* pv. *oryzicola* also possesses clusters of *hrp* genes that controls elicitation of HR in nonhost tobacco and pathogenicity in host rice. Ten *hrp* nine *hrc* (*hrp* conserved) and eight *hpa (hrp*-associated) genes and seven regulatory plant-inducible promoter boxes were identified in a 27-kb region of *X. oryzae* pv. *oryzicola* genome. The *hpa2* to *hpa B* and the hrpF operons in *X. oryzae* pv. *oryzicola* showed resemblance to those in other xanthomonads, the *hpaB-hrpF* region incorporated into an *hrpE3* gene that was absent in *Xoo*. An *hrpF* mutant could not induce HR in tobacco and was unable to infect adult rice as well, although it could cause water-soaked symptoms in rice seedlings. The Hpa1 was found to be an HR elicitor in nonhost tobacco, whose expression was controlled by an *hrp* regulator, HrpX. A small hrp cluster containing the *hrpG* and *hrpX* regulatory genes, separated from the core *hrp* cluster, was also identified in this pathogen. Furthermore, a gene, *prhA* (plant-regulated *hrp*) with a key role in Hrp phenotype was also detected. Thus *X. oryzae* pv. *oryzicola* appears to be the first bacterial pathogen reported to posses three separate DNA regions for HR induction in nonhost tobacco and pathogenicity in host rice, providing the fundamental base to understand pathogenicity determinants of *X. oryzae pv. oryzicola* in comparison with *Xoo* (Zou et al. 2006).

Pseudomonas syringae pv. *tomato* (*Pst*) DC 3000 uses the TTSS for delivering directly the bacterial effector proteins into the host cell (Galan and Collmer 1999). Mutations in the TTSS caused loss of the ability to induce disease, indicating that the effectors are essential agents of *P. syringae* pathogenesis (Collmer et al. 2000). Many effectors have been isolated based on their ability to trigger host immunity. The concept of plant immunity based on the gene-for-gene model is based on the recognition of a pathogen avirulence (Avr) effector protein by a plant resistance (R) protein (Cohn and Martin 2003). The tomato R-protein Pto, a serine / threonine protein kinase recognizes and directly interacts with DC 3000 effector proteins AvrPto and AvrPtoB and uncharacterized signaling mechanisms (Tang et al. 1999). The availability of the genome sequence of *Pst* facilitated the genome wide surveys of TTSS based on the presence of *hrp* box motif in the promoter and other properties. These studies indicated that more than 30 putative effector genes may be present in *Pst* (Guttman et al. 2002; Petnicki-Ocwieja et al. 2002). The role of effectors as virulence factors has been intensively, investigated. However, the molecular basis of how effectors function as virulence factors during compatible plant–pathogen interactions is being examined in the recent years. Many TTSS effectors have been demonstrated to enhance the ability of bacterial pathogens to develop rapidly and induce disease symptoms on respective host plant species (Abramovitch and Martin 2004).

The functions of different effectors of *P. syringae* pv. *tomato* DC 3000 have been investigated intensively. Decrease in intensity of symptoms, without marked reduction in pathogen growth in tomato leaves was observed, following deletion of the TTSS effector gene *hopPtoM* in *Pst* DC 3000 (Badel et al. 2003). In contrast, dramatic increase in disease symptoms occurred due to mutation of another effector gene, *hopPtoN* in *Pst* DC 3000 which did not affect the pathogen development (López-Solanilla et al. 2004). Necrosis associated with disease was increased when both AvrPto and AvrPtoB were delivered by TTSS of *Pst* DC 3000 (Lin and Martin 2005). The AvrPtoB type III effector protein is conserved among diverse genera of plant bacterial pathogens, suggesting that it plays an important role in pathogenesis. AvrPtoB acts inside the plant cell to inhibit PCD initiated by the Pto and Cf disease resistance proteins and remarkably, the pro-apoptotic mouse protein Box. In addition, AvrPtoB could suppress PCD in yeast, confirming that AvrPtoB functioned as a cell death inhibitor across kingdoms. Distinct N-and C-terminal domains of AvrPtoB that are sufficient for host recognition and PCD inhibition respectively could be differentiated.

A novel resistance phenotype, *Rsb*, that is triggered by an AvrPtoB truncation disrupted in the anti-PCD domain, was identified. A *Pst* DC 3000 strain with a chromosomal mutation in the AvrPtoB C-terminus elicited *Rsb*-mediated immunity in previously susceptible tomato plants and disease was restored, when full-length AvrPtoB was expressed in trans. The result show that TTSS effector can induce plant susceptibility to bacterial infection by inhibiting host PCD (Abramovitch et al. 2003). The AvrPto protein from *Pst* DC 3000, following delivery into plant cells by TTSS, either promotes host susceptibility or in tomato plants expressing the Pto kinase elicits disease resistance. Two dimensional electrophoresis revealed evidence that AvrPto was phophorylated when expressed in plant leaves. Three Ser residues clustered in the C-terminal 18 amino acids of AvrPto were identified in vitro as putative phosphorylation and one site at S149 was confirmed by mass spectrometry, as an in vivo phosphorylation site. When Ala was substituted for S149, the ability of AvrPto to enhance disease symptoms was reduced significantly, in addition to reduction in growth of *Pst* in susceptible leaves. Furthermore, the avirulence activity of AvrPto in resistant tomato plants was decreased by S149A. Based on the results, a model in which AvrPto has evolved to mimic a substrate of a highly conserved plant kinase to enhance its virulence activity has been proposed. Residues of AvrPto that promote virulence are also monitored by plant defenses (Anderson et al. 2006).

The type III effector AvrPtoB from *Pseudomonas syringae* elicit differential host responses depending on the resistance / susceptibility of the inoculated plants. In resistant plants, AvrPtoB is recognized by the Pto resistance (R) protein and elicits PCD to limit pathogen invasion, whereas in susceptible plants, in contrast, AvrPtoB suppresses PCD associated immunity, permitting pathogen proliferation. AvrPtoB targets a conserved PCD pathway, since AvrPtoB inhibits PCD induced by diverse agonists in plants (like mouse BAX) and also suppresses PCD in yeast (Abramovitch et al. 2003). The AvrPtoB coterminal domain (CTD) exhibits remarkable homology to the RING-finger and U-box families of proteins involved in ubiquitin ligase complexes in eukaryotic E3 ubiquitin ligases. The AvrPtoB, ubiquinated in vitro, exhibited E3Ub ligase activity in the presence of recombinant E1 activating enzyme and specific E2 U6-conjugating enzymes. Mutation of Avr PtoB lysine residues in the C-terminus disrupted AvrPtoB-Ub interactions, decreased AvrPtoB-mediated anti-PCD activity and arrested disease development in susceptible tomato plants. Furthermore, quantitative decreases in AvrPtoB anti-PCD activity were correlated with reduction in AvrPtoB ubiquitination and E3 Ub ligase activity. The data establish that *Pst* DC 3000 exhibits a unique pathogenesis strategy where AvrPtoB manipulates the host Ub system and requires intrinsic E3 Ub ligase activity to suppress plant immunity. The pathogen employs a mimic of host E3 ubiquitin ligases to inactivate plant defenses (Abramovitch et al. 2006b; Janjusevic et al. 2006).

In addition to the ability of AvrPtoB to suppress HR-based PCD, it may possibly have other virulence functions. The DC 3000 :: mut5 mutant lacking cell death suppressor (CDS) activity showed a 10-fold decrease in growth on susceptible RG-*prf* tomato plants (Abramovitch et al. 2003). Since the CDS activity might also act as a quantitative virulence factor in the absence of HR-based immunity, plants with prf3 mutations cannot mount *Pto*-or *Rsb*-mediated immunity. AvrPtoB may promote disease symptom production, possibly by inducing speck-associated cell death, as indicated by the observation that DC 3000 :: *Δ*avr PtoB could cause reduced speck number on susceptible tomato leaves without reducing pathogen growth (Lin and Martin 2005).

AvrRpt2, an effector protein from *Pseudomonas syringae* pv. *tomato* (*pst*) is able to function as an virulence factor activating resistance in *A. thaliana* lines expressing the resistance gene *RPS2*. In addition, AvrRpt2 can also enhance pathogen fitness by promoting the ability of the bacteria to grow and cause disease on susceptible *A. thaliana* lines that lack functional *RPS2* (Leister and Katagiri 2000). AvrRpt2 has been shown to induce the disappearance of the *A. thaliana* protein RIN4 (Axtell and Staskawicz 2003; Mackey et al. 2003). Later it was shown that the ability of six novel AvrRpt2 mutants to induce RIN4 disappearance correlated well with their avirulence activities, but not with their virulence activities. Furthermore, the virulence activity of wild-type AvrRpt2 was detected in an *A. thaliana* line lacking *RIN4*. The virulence activity of *Avrhpt2* in *A. thaliana* is likely to relay on the modification of host susceptibility factors other than or in addition to RIN4 (Lim and Kunkel 2004a). In a further study, the functional relationship between the avirulence and virulence activities of *avrRpt2* was examined by analyzing a series of six avrRpt mutants deficient in eliciting the RPS2-dependent HR. The aviruelnce and virulence activities of avrRpt2 could be genetically uncoupled. The reduced AvrRpt2 virulence activity was correlated with reduced efficiency in the induction of RIN4 disappearance (Lim and Kunkel 2004b).

The effector protein VirPphA has been shown to allow *P. syringae* pv. *phaseolicola* (*Psp*) to evade HR based immunity in bean (Jackson et al. 1999). In addition, the other effectors of *Psp* also help the pathogen to avoid triggering host immunity, including Avr PphC and Avr Pph F (Tsiamis et al. 2000). The AvrPto was reported to suppress a cell wall-based defense in transgenic *Arabidopsis* lines (Hauck et al. 2003). Likewise, several effector proteins such as AvrPtoB, are considered to act as general suppressors of host programmed cell death (PCD) which is the hallmark of HR-based immunity in plants depending on the genetically controlled and regulated process (Abramovitch et al. 2003; Jamir et al. 2004). In some pathosystems, expression of genes associated with defense is reportedly suppressed by TTSS effectors (Chen et al. 2004; Jamir et al. 2004).

The role of host plant in the development of disease has been indicated to be significant. The phytohormones like ethylene, jasmonates and salicylic acid (SA) which have been demonstrated to be involved in plant defense responses, are also implicated in facilitation of disease progression. The gaseous hormone ethylene is a vital component of plant responses to pathogen invasion. The molecular roles of TTSS effects of *Pst* DC 3000 in induction of disease symptom was investigated, using a DNA microarray protocol. The expression of about 8600 random tomato genes in response to inoculation with wild-type strain and a mutant lacking a functional TTSS was studied. Many of the differentially expressed genes identified, encode proteins associated with hormone response or hormone biosynthesis pathways. The effector proteins AvrPto and AvrPtoB (important virulence factors) induced a set of host genes involved in ethylene biosynthesis and signaling. They regulated particularly two genes, *LeAC01* and *LeACO2* encoding the ethylene-forming enzyme ACC oxidase. Ethylene production by the transgenic tomato lines with diminished ACC oxidase activity was found to be essential for full virulence activity of both AvrPto and AvrPtoB. By upregulating genes involved in ethylene production, AvrPto and AvrPtoB seemed to enhance the disease symptoms in tomato leaves (Cohn and Martin 2005).

The host plant innate immune system responds to two general classes of pathogen-derived molecules-pathogen-associated molecular patterns (PAMPs)-the type III effector proteins in the case of bacterial pathogens. Lipopolysaccharides and flagellin from bacteria stimulate plant encoded PAMP receptors (Nurnberger et al. 2004). Bacterial pathogens deploy type III effector proteins that promote their virulence. Collectively, the type III effectors make vital contributions to the virulence of bacteria. Some of the type III effectors can inhibit basal defenses of the plants (Keshavarzi et al. 2004). A second system employs disease-resistance (R) proteins to recognize type III effector proteins that are delivered into the plant cell by the bacterial TTSS. RIN4 from *Arabidopsis* is a host target of type III effectors that is "guarded" by R proteins. The effector AvrRpt2, a protease targets RIN4 and induces its post- transcriptional disappearance. The disappearance of RIN4 activates RPS2, the cognate R protein of AvrRpt2 (Axtell and Staskawicz 2003; Mackey et al. 2003). The type III effectors AvrRpt2 and AvrRpm1 inhibited PAMP-induced signaling and weakened basal defense system of *Arabidopsis.* RIN4 is a protein targeted by AvrRpt2 and AvrRmp1 for degradation and phosphorylation respectively. RIN4 itself was found to be a regulator of PAMP signaling. The effector-induced perturbations of RIN4 were sensed by the R proteins RPS2 and RPM1. RIN4 negatively regulates PAMP signaling and PAMP-induced defense responses are inhibited or enhanced in plants overexpressing or lacking RIN4 respectively (Kim et al. 2005a).

The virulence factors introduced by the bacterial pathogens, render the host more hospitable for pathogen proliferation. The gene *avrRpt2* from *Pst* JL1065 has been shown to promote virulence of heterologous *P. syringae* strains on *A. thaliana.* A mutant derivative of JL1065 carrying disruption in *avrRpt2* was found to be impaired in its ability to induce disease on tomato, indicating that avrRpt2 acts as a virulence gene in its native strain on a natural host also. In addition, the virulence activity of *avrRpt2* could be detected on tomato lines that were defective in either ethylene perception or accumulation of SA, but could not be detected on tomato mutant insensitive to JA. It was further shown that the enhanced virulence conferred by the experession of *avrRpt2* in JL1065 was not associated with the suppression of several defense-related genes induced during infection of tomato (Lim and Kunkel 2005). *P. syringae* has developed strategies to counteract host resistance by injecting virulence proteins into the host cell. A conserved *P. syringae* virulence protein HopM1 targets an immunity-associated protein, AtMIN7 in *A. thaliana*. Hop M1 mediates the destruction of AtMIN7 via the host proteasome. The results focus on a strategy by which a bacterial pathogen exploits the host proteasome to subvert host immunity and to cause infection in plants (Nomura et al. 2006).

Introduction of AvrPto into a *Pst* strain lacking AvrPto was found to enhance bacterial growth and development of visible symptoms of infection on susceptible tomato plants (Chang et al. 2000; Shan et al. 2000). Suppression of the expression of cell wall-associated defense genes, inhibition of callose deposition in response to *Pst* infection and enhancement of bacterial proliferation were observed in *Arabidopsis thaliana* due to over expression of AvrPto (Hauck et al. 2003). Furthermore, the ability of AvrPto to suppress cell death in response to nonhost pathogens was also demonstrated (Kang et al. 2004). Some of the bacterial effectors appear to mimic the substrates of eukaryotic enzymes and as a result, they are posttranslationally modified within the host cell. By applying the two dimensional gel electrophoresis, evidence for the phosphorylation of AvrPto, when expressed in leaves of host plant species, was obtained (Fig. 2.12). In vitro phosphorylation of AvrPto by plant extracts occurred independently of Pto and was due to a kinase activity that is conserved in tomato, tobacco and *A. thaliana*. One of the three putative phosphorylates sites, S149 was identified as an in vivo phosphorylation site by using mass spectrophotometry. The ability of AvrPto to enhance disease symptoms and promote growth of *Pst* in susceptible tomato leaves was reduced significantly following substitution of Ala for S149. A model for evolution of AvrPto for mimicking a substrate of a highly conserved plant kinase to enhance its virulence activity was proposed by Anderson et al. (2006).

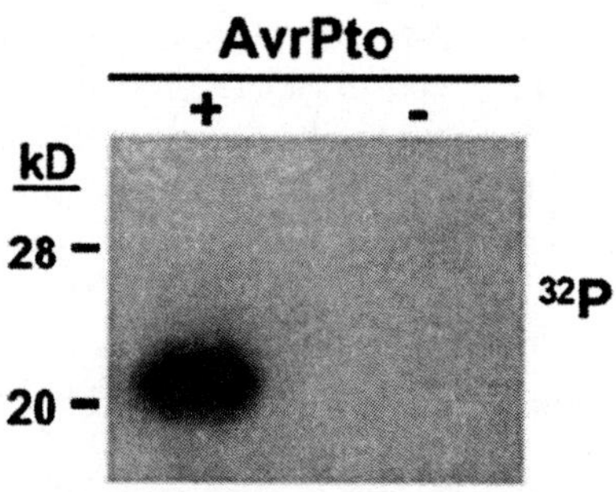

Fig. 2.12 Detection of a protein (~21 kD) corresponding to the expected size of AvrPto-dHA in protoplasts expressing AvrPto(+) compared with vector only control(−) by direct labeling of AvrPto in the presence of ^{32}P-orthophosphate; leaf protoplasts were transfected with either 35S:avrPto-HA or vector only
Note the presence of a single ^{32}P-labeled band corresponding to AvrPto in protoplasts expressing AvrPto(+) only, but not in the vector only (−) lane. (Courtesy of Anderson et al. 2006; American Society of Plant Biologists, Rockville, MD, USA)

The *Arabidopsis-Pseudomonas syringae* pv. *phaseolicola* (*Psp*) pathosystem has been demonstrated to be a model genetic system for nonhost resistance investigations. The *nho1* mutant generated by a recessive mutation, nonspecifically supports the development of both nonhost and avirulent *Pseudomonas* bacteria. *NHO1* has been found to be ineffective against virulent strain and the *nho1* mutant does not exhibit higher level of susceptibility (Lu et al. 2001). In a later investigation, *Arabidopsis* NHO1 has been shown to encode a glycerol kinase that is required for resistance to and induced by *P. syringae* isolates from bean and tobacco. The resistance due to *NHO1* is also needed for resistance to the fungal pathogen *Botrytis cinerea*. *Pst* DC 3000 was fully virulent on *Arabidopsis* in which *NHO1* expression was actively suppressed. However, the presence of *avrB* in *Pst DC 3000* restored the induction of *NHO1* expression and activation of a cultivar-specific gene-for-gene resistance in *Arabidopsis*. The gene *NHO1* deployed for both general and specific resistance in *Arabidopsis* is targeted by *Pst* DC 3000 for parasitism (Kang et al. 2003).

The elicitors of pathogen origin are termed as pathogen-associated molecular patterns (PAMPs). The flagellin from *Pseudomonas* which activates innate immunity has been characterized. A conserved N-terminal peptide of flagellin, flg22, is a potent elicitor of defense responses in tomato and *Arabidopsis* (Gomez-Gomez and Boller 2002). The *NHO1* gene in *Arabidopsis* is required for resistance to multiple strains of nonhost *P. syringae*. In contrast, it could not induce resistance against *Pst* DC 3000, as the *NHO1* was induced in a flagellin-dependent manner. A *Ptab* strain lacking the flagellin gene *fliC* elicited disease symptoms and multiplied revealing the contribution of flagellin signaling to nonhost resistance. The wild type *Pst* DC 3000 strain rapidly suppressed the *NHO1* induction. However, the DC 3000 mutants defective in TTSS had reduced ability to suppress *NHO1*. Direct expression of DC 3000 effectors in protoplasts indicated that at least nine effectors, HopS1, HopAI1, HopAF1, HopT1-1, HopT1-2 and HopAA1-1, HopF2 and HopC1 and AvrPto were able to suppress the flagellin-induced *NHO1* expression. The effector HopAI1 (conserved in both animal and plant pathogenic bacteria), when expressed in transgenic *Arabidopsis* plants, promoted growth of the nonpathogenic hrpL$^-$ mutant bacteria. The purified phytotoxin coronatine, a known virulence factor of *P. syringae* suppressed the flagellin-induced *NHO1* transcription. The results showed that a large number of *Pst* DC 3000 virulence factors act in the plant cell by suppressing the species level defenses and that contributed to the specialization of *Pst* DC 3000 on *Arabidopsis* (Li et al. 2005).

Plants respond to an array of microbe associated molecular patterns (MAMPs) from both pathogenic and nonpathogenic microbes (Ausubel 2005). In the absence of specialized immune system as in animals, the plant cells seem to have developed the ability to respond to MAMPs such as flagellin, harpin (HrpZ), lipopolysaccharides (LPS), chitin and necrosis-inducing proteins (NPP) and activate defense gene transcription and MAPK signaling (Asai et al. 2002; Fellbrich et al. 2002; Navarro et al. 2004; Ramonell et al. 2005). The LRR receptor kinase that perceives a conserved 22-amino acid peptide (flg22) from bacterial flagellin and activates MAPK cascades and WRKY transcription factor in *Arabidopsis* (Asai et al. 2002; Gomez-Gomez and Boller 2002). Activation of the flg22-mediated MAPK cascade confers resistance to both bacterial and fungal pathogens (Asai et al. 2002). Many type III effectors such as AvrPto, AvrRpt2, AvrRpm1, and HopAI1. suppress defense responses elicited by either TTSS-defective mutants or flg22 (Hauck et al. 2003; Kim et al. 2005a; Li et al. 2005). The molecular mechanisms underlying nonhost immunity in plant has been studied in *Arabidopsis–Pss* strains interactions. Pathogenic strains suppress early MAMP marker-gene activation whereas the nonhost/nonpathogenic strains sustain this phenomenon. A cell-based genetic screen of virulence factors was performed and AvrPto and AvrPtoB were identified as potent and unique suppression of early-defense gene transcription and MAP kinase (MAPK) signaling. These two proteins intercepted multiple MAMP-mediated signaling upstream of MAPKKK at the plasma membrane linked to the receptor. AvrPto blocked early MAMP signaling and also enabled nonhost *P. syringae* to grow in transgenic *Arabidopsis.* Reduction in virulence in *P. syringae* following deletion of AvrPto and AvrPtoB, was observed. A fundamental role of MAMP signaling in nonhost immunity and a novel action of type III effectors from pathogenic bacteria were elucidated by the studies of He et al. (2006).

The pathogenicity and elicitation of hypersensitive response have been demonstrated to depend largely on the TTSS and effector proteins that this system translocates into plants. In addition to these proteins, additional factors may contribute to the virulence of *P. syringae* pv. *tomato* (*Pst*) and other bacterial pathogens dependent on TTSS. The invovlement of the twin-arginine translocation (Tat) system in physiological functions of *Pst* DC 3000 was investigated. The Tat pathway operates in the inner membrane of many Gram-negative bacteria (Berks et al. 2003). Tat substrates are synthesized with cleavable N-terminal signal peptides that are characterized by a highly conserved twin-arginine motif in the positively charged N-terminal region, a weakly hydrophobic core region and a positively charged Sec pathway-avoidance signal in the C-terminal region (Cristobal et al. 1999). Tat system has remarkable ability to transport proteins that have already folded in the cytoplasm (Berks et al. 2000). Inactivation of the Tat system in *Pst* DC 3000 led to multiple complex phenotypes, including loss of motility on soft agar plates, deficiency in siderophore synthesis and iron acquisition, sensitivity to copper, loss of extracellular phospholipase activity and attenuation in host plants leaves. Furthermore, localization of a Tat-specific protein (a disulfide bond protein), the powerful periplasmic disulfide catalyst DsbA was found to be essential for the full virulence of both *Pst* DC 3000 and *P. aeruginosa* on *A. thaliana* and for efficient type III secretion in *Pst* DC 3000 (Kloek et al. 2000). The *tatC* mutants exhibited impaired localization of *DsbA* protein to the periplasm compared with wild-type *Pst* DC3000. Two putative virulence factors PlcA1 and PlcA2 that carry Tat-specific targeting signals were also identified. For the first time, inactivation of either Tat or type II secretion systems has been shown to result in a significant attenuation of virulence. The results of this investigation suggest that the two-step secretion might function as a complementary virulence mechanism to the predominant TTSS of *P. syringae* (Bronstein et al. 2005).

The role of the twin arginine translocation (Tat) pathway in the virulence of two pathovars *P. syringae* pv. *tomato* (*Pst*) DC 3000 and *maulicola* (*Psm*) ES4326 was studied. The deletion mutants in Tat pathway exhibited a range of pleiotrophic phenotypic changes such as defects in fluorescent siderophore production, a decrease in sodium dodecyl sulfate (SDS) and copper resistance, in addition to a significant loss in fitness using *A. thaliana* or factors in *Pst* were predicted to be translocated via the Tat pathway. These virulence factors include several proteins involved in iron scavenging (two siderophore receptors) and aminotransferase involved in siderophore biosynthesis. The candidates for Tat-dependent pathogencity determinants may include the homologs of a cell wall amidase, an enzyme involved in periplasmic glucan biosynthesis and two putative phospholipases. The strains deleted for the genes encoding the probable aminotransferase and amidase enzymes had reduced infectivity in comparison to wild-type strain (Caldelari et al. 2006).

The genes induced during infection of plants by *Erwinia amylovora* were identified. A putative effector gene, avrRpt2 (EA) was cloned and the deduced amino acid sequence of the translated AvrRpt(EA) protein was found to be homologous to the effector protein AvrRpt2 previously reported in *Pst*. Though the protein exhibited 70% similarity in the functional domain, the secretion and translocation domain varied. The *avrRpt2* (EA) was detected in all *E. amylovora* strains tested, but not in other closely related *Erwinia* spp. A deletion mutant showed reduced ability to cause systemic infection on immature pear fruits in comparison to the wild-type strain, indicating that *avrRpt2*(EA) acts as a virulence factor on its native host. The expression of avrRp2 (EA) by *Pst* DC3000 in *Arabidopsis rps2* mutant resulted in acceleration of pathogen growth (10 folds). But a weak HR was induced in *Arabidopsis,* when avrRpt2 (EA) was expressed in *Pst* DC 3000 in its native form. The results suggest that although similar effector genes are present in distinct pathogens, *E. amylovora* and Pst DC 3000 the expression and secretion of these effectors may be under specific regulation by the native pathogen (Zhao et al. 2006). In another investigation, the role of the Hrp pathogenicity island conserved effector locus (CEL) was studied. The Delta CEL mutation affected three putative or known effector genes viz., *avrE1, hopM1* and *hopAA1-1.* Comparison of genomic sequences showed that only these effector genes are present in the CEL of *Pst, Psp* 1448A, and *PssB* 728a. A DC 3000 Delta *avrE1* mutant was reduced in its ability to produce lesions but not in its ability to grow in host tomato leaves. Deletion of both *avrE1* and *hopM1* resulted in significant reduction in growth and lesion phenotype of Delta CEL mutant. Further, both the *avrE1 / hopM1* and Delta CEL mutants were partially impaired in their ability to elicit the hypersensitive response in nonhost *N. benthamiana* leaves. The results suggest that AvrE1 acts within plant cells and promotes lesion formation and the combined activity of AvrE1 and HopM1 is particularly important in promoting bacterial growth in planta (Badel et al. 2006).

The pathogenicity of *Xanthomonas oryzae* pv. *oryzae* (*Xoo*) depends on a functional TTSS, while members of the *avrBs3 / pthA* gene family of TTSS substrate effectors are required for its ability to induce bacterial leaf blight (BLB) in rice (Yang and White 2004). The products of *avrBs3/pthA* have characteristics of nuclear transcription activation factors and are designated transcription activator-like (TAL) effectors (White et al. 2000). The TAL effectors are included in a large family of closely related proteins. They can be individually differentiated by the nature and number of direct repeats within a central repetitive region and they contribute to virulence of bacterial pathogens (Schornack et al. 2006). The TAL effectors, secreted through TTSS, are targeted to host cell nuclei and each effector has a C-terminal transcription activation-like domain that is essential for both virulence and avirulence activities (Szurek et al. 2002). The TAL effector AvrBs3, when introduced ectopically, was shown to be associated with the induction of pepper genes (Marois et al. 2002). Likewise, AvrXa27, a TAL effector of XooPXO99^{A} was found to be associated with induction of R gene *Xa27* (Gu et al. 2005). In a later investigation, whole genome-based microarray of rice genes was employed to identify genes that were expressed following challenge by *Xoo*. The expression of the rice gene *Os8N3*, a member of the *MtN3* gene family from plants and animals was enhanced upon infection by *Xoo* strain PX099^{A} and it depended on the type III effector gene *PthXo1*. *Os8N3* resides near *R* gene *Xa13*. The strain PXO99^{A} did not induce *Os8N3* in rice lines with *Xa13*. Rice plants with *Os8N3* silenced by inhibiting RNA were resistant to infection by strain PX099^{A} only, but not to other strains. Further the effector gene *avrXa7* from strain PXO86 made PXO99^{A} to be compatible with *Xa13* or *Os8N3*-silenced plants. The results indicated that *Os8N3* is a host susceptibility gene for BLB targeted by the type III TAL-effector *PthXO1* (Yang et al. 2006). *Xa27* is a dominant R gene whose expression is specifically elevated in the presence of the TAL effector AvrXa27 (Gu et al. 2005). Likewise, Pth X01 is the second TAL effector demonstrated to be associated with the enhanced expression of a host gene that has known consequences for the outcome of the host–pathogen interaction (Yang et al. 2006).

The soil bacterium *Agrobacterium tumefaciens* infects and transfers a piece of its tumor-inducing (Ti) plasmid, the transferred DNA (T-DNA), to several dicotyledonous plants, thereby modifying their genome and inducing hyperplastic response resulting in crown gall symptom. The *A. tumefaciens*–plant interaction appears to be the only verified example of natural inter-kingdom DNA transfer. Hence, this bacterium is widely employed to genetically engineer plants and to generate insertional disruptions in genes, facilitating functional genomics of plants (Azpiroz-Leehan and Feldmann 1997). The interaction between *A. tumefaciens* and plants involves a complex series of chemical signals communicated between the pathogen and the host. The molecular basis of this interaction is discussed hereunder.

As the first step in the process of tumor induction, the bacterium attaches to the plant surface. The cDNA transfer process is initiated, when *Agrobacterium* perceives the presence of certain phenolic and sugar compounds released from the wounded plant cells. These phenolic compounds can serve as inducers (or coinducers) of bacterial *vir* genes, although these compounds may be normally involved in inducing production of defense-related substances, thus subverting part of the host's defense mechanism. In addition, the pathogen uses these compounds to signal the presence of a potentially susceptible plant (Bolton et al. 1986). Acetosyringone and other related compounds are perceived by *Agrobacterium* via the VirA sensory protein (Stachel et al. 1986; Dye et al. 1997). Autophosphorylation of VirA protein and the subsequent transphosphorylation of VirG protein leads to the activation of *vir* gene transcription (Jin et al. 1990a, b). The *vir* genes present in one half of Ti plasmid (Vir region) control the virulence and tumor formation, whereas the genes for replication, opine catabolism and conjugation are located in the other half of *Ti* plasmid. The octopine *Ti* plasmid contains eight operons (VirA to VirH) in the Vir-region. Mutations in the *virA, virB, virO* and *virG* operons result in loss of tumor formation function. On the other hand, host range of *Agrobacterium* is affected following mutations in *virC, virE* and *virH*. The functions of *Ti*-plasmid genes encoding virulence proteins in the pathogen and the host plant have been determined to some extent.

A highly motile strain of *A. tumefaciens* showed significantly greater level of chemotaxis toward phenolic compounds that strongly induce *vir* genes compared to poorly motile strain. This pTi-dependent chemotoxis depends on *virA* and virG genes (Shaw et al. 1988; Hawes et al. 1988). For the bacterial cell attachment to plant surface, the genes *chvA, chvB, pscA* (or *exoC*) *att* located in the chromosome are required. Mutation in these loci may result in loss of virulence to many plant species. The *att* (*attA1* to *attH*) genes involved in the bacterial cell attachment to the plant cell surface, appear to be essential for signaling between the bacteria and host plant (Matthysee 1994). Att A1-H deletion mutants may be defective in sensing the plant signal or in responding to it, whereas att$^-$ mutants may lack bacterial adhesin or blocked steps prior to adhesion (Matthysee et al. 1996; Reuhs et al. 1997). The attachment of bacteria to the plant cell surface is strengthened by cellulose fibrils, whose synthesis is controlled by *cel* genes located on the bacterial chromosomes near but not contiguous with *att* genes (Matthysee 1983). The *cel* mutants have reduced virulence, indicating the requirement of *cel* genes for full virulence of this pathogen (Matthysee and McMahan 1998). Mutations in *chvA* and *chrB* genes results in reduced motility and failure to bind to plant cells (Zorreguieta et al. 1988; Cangelosi et al. 1989). The *exoC*$^-$ mutants are defective in synthesis of EPS, whereas defective cellulose synthesis and aggregate formation are noted in *cel* mutants which bind loosely to plant wound sites (Matthysee 1983; Uttaro et al. 1990).

The Vir-regions of octopine and nopaline *Ti* palsmids direct the transfer of oncogenic T-DNA of the nuclei of host plant cells. The *vir* genes form a regulon consisting of a set of operons coregulated by the same regulatory proteins (Duban et al. 1993). A two-component regulatory system consisting of VirA and VirG proteins mediates the expression of the *vir* genes. The *virA* gene, which is constitutively expressed, produces a protein located in the inner membrane that responds to plant wound metabolites. VirA is a membrane-spanning protein with an N-terminal periplasmic "sensor" domain (capable of sensing acetosyringone and related phenolics) a "linker" domain (responding to pH changes and interacting with ChVE, a sugar-binding protein) a "kinase" domain and a "receiver" domain. Autophosphorylation of VirA protein results in the activation of the intracellular VirG which in turn is phosphorylated to become the transcriptional activator for all *vir* genes (Hooykaas and Beijersbergen 1994; Sheng and Citovsky 1996). After induction of the *vir* genes, the bacterial cells generate a linear single-stranded DNA designated T-DNA or T-strand. The T-strand is the coding (bottom) strand of the T-DNA region of the *Ti*-plasmid. The *virD2* gene encodes and enzyme VirD2 which liberates the linear T-DNA (Stachel et al. 1986). The enzyme VirD2 is a cleaving-joining enzyme that cuts the lower strand of the border sequences in a site-specific manner. The excised T-DNA is removed and the resulting single-stranded gap is repaired probably by replacement DNA-strand synthesis (Yanofsky et al. 1986). The Vir D2 effector may be covalently bound to the T-DNA and pilots the transfer of TDNA into plant cell. The Vir D4 protein functions as the coupling protein located at the inner membrane which recognizes and directs the substrates to the translocation machinery (Atmakuri et al. 2004; Cascales and Christie 2004b).

The transfer of T-DNA and the effectors appears to occur due to the action of the type IV secretion system. But it is not known clearly how they cross the host plasma membrane. Following the induction of the transfer machinery *A. tumefaciens* forms a pilus which is composed of the cyclized VirB2 protein (Fullner et al. 1996; Eisenbrandt et al. 1999; Lai et al. 2002). The gene products of the complex *virB* locus of the Ti plasmid regulate the transfer of the T-complex. An operon of the *VirB* locus encodes eleven proteins (VirB1-VirB11). The VirB membrane proteins are considered to form a pore for the transport of the T-DNA across the bacterial membranes. The nature of the transport pore complex was studied using immunofluorecence (IF) and immunoelectron microscopy (ISEM) techniques. The VirB8, VirB9 and VirB10 proteins localized primarily to the inner membrane, outer membrane and periplasm respectively, may form the components of the transport pore. In a *virB8* deletion mutant, VirB9 and VirB10 proteins were randomly distributed on the cell membrane, indicating the requirement of VirB8 protein for assembly of the transport pore complex (Kumar et al. 2000). Since the VirB2 protein is required for virulence, the pilus was proposed to either pierce the plasma membrane to deliver the T-DNA and the effectors into the host cell cytoplasm or to mediate intimate contact with the plant cell (Gelvin 2003; Christie 2004). However, studies with VirB uncoupling mutants suggested that wild-type T-pilus is not required for substrate transfer (Jakubowski et al. 2005). Further, the ability of the effector VirE2 protein to integrate in black lipid membranes (BLM) and to form large anion-selective channels that transfer ssDNA across membranes in vitro also suggested a role for VirE2 in assisting T-DNA passage across the eukaryotic plasma membrane (Dumas et al. 2001; Duckely and Hohn 2003).

The multiple functions of VirE2 have been revealed by different reports. Its ability to bind to ssDNA (Gietl et al. 1987), to protect the T-DNA from nuclease digestion (Citovsky et al. 1988) and to assist in nuclear import (Citovsky et al. 2004; Lacroix et al. 2005) have been demonstrated. The T-DNA after insertion into the host plant cell, is transcribed, resulting in the synthesis of opines required for bacterial growth and the growth regulators that accelerate host cell enlargement and division leading to formation of characteristic tumors (Burr et al. 1998). The proteins VirE2 and VirF are secreted directly into the plant cells via the same VirB / VirD4 transport system. Both proteins have been shown to assist in the transformation of normal cells into tumor cells. Deletion of *virF* led to reduced virulence of *A. tumefaciens*. It could be complemented by expression of *virF* in the host plant, indicating the intracellular functioning of the VirF protein which may be involved in the targeted proteolysis of specific host proteins during early stages of transformation of plant cells (Schrammeijer et al. 2001).

The VirE2 protein has an important role in the transformation process. The *virE2* mutants exhibited drastically reduced virulence although they had low level of virulence on some host plants species (Dombek and Ream 1997). Among the Vir proteins, VirE2 was found to be the most abundant in acetosyringone-induced *Agrobacterium* cells (Engstrom et al. 1987). The VirE2 protein and T-DNA appear to be transferred to plant cells separately or VirE2 protein can form a complex with incoming T-DNA in the plant cytoplasm. The critical role of VirE2 protein in the transfer of ssDNA from *A. tumefaciens* to the nucleus of the plant host cell because of its ssDNA binding activity and assistance in nuclear import and putative ssDNA channel activity has been studied recently. VirE2 is associated with its specific chaperone, the VirE1 protein which is a small, acidic, 7-kDa protein and capable of preventing the VirE2 protein to form aggregates and increasing its half life by three folds (Zhao et al. 2001b). The native form of VirE2 in the bacterial cytoplasm is complex with its chaperone, VirE1. Electronmicroscopic observations revealed that upon binding of VirE1 VirE2 to ssDNA, helical structures similar to those reported for the VirE2-ssDNA complex, were formed. The VirE1 VirE2 complex associated with different kinds of lipids. The black lipid membrane tests showed the ability of the VirE1 VirE2 complex to form channels (Duckely et al. 2005).

Agrobacferium tumefaciens (*At*) infection of plants is viewed traditionally as a process of T-DNA transport. A multitude of bacterial proteins such as VirD2, VirF and VirE3 are also exported into the host cell most of them separately from the T-DNA itself (Cascales and Christie 2004b). VirE3 is encoded by other soil bateria such as *A. rhizogenes* and *Rhizobium etli* (Winans et al. 1987). As the biological functions of VirE3 were not known, investigations were taken up. It was demonstrated that VirE3 was transferred from the bacterium *At* to plant cell and then imported into its nucleus via the karyopherin α-dependent pathway. In addition, VirE3 interacted with VirE2, a major bacterial protein that was directly associated with T-DNA and facilitated its nuclear import. The nuclear import of VirE2, in turn, was mediated by a plant protein VIP1 and acted as an "adapter" molecule between VirE2 and karyopherin α and piggy-backing' VirE2 into host cell nucleus. It was proposed that *Agrobacterium* has evolved to produce and export to the host cells its own virulence protein VirE3 that, atleast partially complements the VIP1 function (Lacroix et al. 2005). In a later study, double mutation of virF and virE3 strongly reduced the ability of *At* to form tumors on tobacco, tomato and sunflower. The VirE3 protein was translated from a polycistronic mRNA containing the *virE1, virE2,* and *virE3* genes in *At*. The VirE3 protein had nuclear localization sequences, suggesting that it may be transported into the plant cell nucleus upon translocation. VirE3 was shown to interact in vitro with importin-α and a VirE3-GFP fusion protein was found localized in the nucleus. VirE3 was also able to interact with two other proteins viz., pcsn5, a component of the CoP9 signalosome and pBrp, a plant cell specific general transcription factor belonging to the TFIIB family. VirE3 could induce transcription in yeast, when bound to DNA. The translocated effector protein VirE3 was transported into the nucleus, where it may interact with the transcription factor pBrP to induce the expression of genes required for tumor formation (García-Rodríguez et al. 2006).

The nature of opines such as octopine, nopaline, succinamopine and leucinopine present in the tumors, forms the basis of classifying the strains of *A. tumefaciens*. The chromosomal gene *acvB* isolated from strain A208 with nopaline Ti plasmid was also shown to be essential for virulence (Wirawan et al. 1993). Another gene, *virJ*, homologous to *acvB* was also detected in an octopine Ti plasmid (Pan et al. 1995). A transposon (Tn5) insertion in the *acvB* gene abolished virulence, indicating the requirement of this gene for virulence. The *acvB* gene is involved in the transfer of T-DNA to the tobacco cell nuclei (Fujiwara et al. 1998). The *virJ* mutants and *acvB* mutants are avirulent indicating their requirement for virulence of *A. tumefaciens*. VirF may also be exported to the plant cells. Nopaline type strains lacking a functional *virF* gene are either aviruelnt or could induce very small tumors on *Nicotiana glauca* (Melchers et al. 1990).

The nuclear targeting of the T-DNA in plant cells is likely to be mediated by the agrobacterial proteins. Active import of proteins and nucleoprotein complexes in eukaryotic cells needs specific nuclear localization signals (NLS) that are recognized by nuclear import cytosolic factors such as importins. The VirD2 and VirE2 proteins have plant-active nuclear localization signals. When fused to β-galactosidase, VirD2 tightly associated with 5′ end of the T-DNA, a peptide containing NLS could target the chimeric protein to plant nuclei (Herrera-Estrella et al. 1990). On the other hand, VirE2 protein containing two separate bipartite NLS region may target linked reporter proteins to plant cell nucleic (Citovsky et al. 1997).

The bacterial T-DNA gets integrated into the plant genome in cell nucleus by illegitate recombination, a mechanism that joins two DNA molecules which do not share extensive homology. Very little is known about the process of T-DNA integration with host genome (Gheysen et al. 1991). The possible role for VirD2 protein in T-DNA integration was suggested by Tinland et al. (1995). However, the in vitro analysis of the potential function of VirD2 in ligation of 5′ end of T-DNA to the 3′ end of plant DNA indicated its inability to perform T-DNA ligation. The T-DNA ligase activity could be detected in extracts from tobacco BY-2 cells and pea axes (Ziemienowicz et al. 2000). The extent of involvement of host components in recognition, transfer and integration of T-DNA remains largely unknown. A plant cyclophilin (Deng et al. 1998) and a plant karyopherin α-protein (Ballas and Citovsky 1997), in addition to a plant histone H2A (Mysore et al. 2000) and a plant DNA-ligase (Ziemienowicz et al. 2000) appear to have a role in the process of T-DNA integration.

Numerous other genes-in addition to *vir* genes on the Ti plasmid known as *chv* genes located on other replicons, are essential for the successful interaction of *Agrobacterium* with its hosts. The *chv* genes serve dual functions, whereas *vir* genes appear to be dedicated solely to the interaction of *Agrobacterium* with its host plants. The *chv* genes are required for the physiological functions of *Agrobacterium* developing in the absence of host plants as well as in the interaction of this pathogen with its host plants. The gene *chvE* is involved in the transport of specific sugars required for bacterial cells as a carbon source, in addition to its role in activating *vir* genes (Cangelosi et al. 1990). Likewise, the gene *katA* assists the pathogen in overcoming the host plant defenses and probably plays a role in the stress response of the cell (Zupan et al. 2000). In order to identify genes important in the interaction of *Agrobacterium* with plants, insertion mutagenesis using Tn5 or derivative such as TnPhoA was applied to randomly mutate the bacterial genome. This approach has been successful in identifying new *chv* genes. Citrate synthase (CS) governs the entry of carbon into the TCA cycle and it is important in *Agrobacterium* because of its requirement as the main pathway for generation of energy. A CS deletion mutant of *A. tumefaciens* C58 was found to be highly attenuated in virulence. The mutation in CS led to a 10-fold decrease in *vir* gene expression which could be a reason for reduced virulence. When a plasmid containing a constitutive *virG* [virG (con)] locus was introduced into this mutant, induction of *vir* gene occurred to the level comparable to that of wild-type. In addition, the size and number of tumors induced by the *virG* (con)-complemented CS mutant strain were similar to the wild-type strain (Suksomtip et al. 2005).

The cell-to-cell communication in diverse species of Gram-negative bacteria is achieved by quorum-sensing gene regulation. It is characterized by the presence of specific proteins that function as transcriptional regulators (LuxR homologs) and N-acetylhomoserine lactone (AHL) synthases (LuxI homologs) (Whitehead et al. 2001). Quorum sensing process has been shown to be involved in the regulation of many important physiological functions of the bacteria such as symbiosis, conjugation and virulence. *Agrobacterium vitis* causes tissue-specific necrosis, in addition to formation of tumors on grapes and HR-like symptom on nonhosts like tobacco (Herlache et al. 2001). The *luxR* homolog *aviR* in *A. vitis* strain F2/5 was associated with induction of a HR on tobacco and necrosis on grapevine plants indicating that the responses were regulated by quorum sensing. The mutant M1320 with disruption in a second luxR homolog, *avhR* exhibited HR-negative and reduced grape necrosis characteristics. The expression of *aviR*, avhR and the *clpA* (encoding one of two ATPase subunits of the ClpAP protease of *E. coli*) in F2/5 and mutants M1154 and M1320 and their complemented derivatives was determined by RT-PCR. The gene *avhR* was expressed in F2/5 and in complemented M1320 and also in the *aviR* mutant M1154. Further, *aviR* expression in F2/5 and M1320 could be detected indicating that *aviR* and *avhR* were likely to be expressed independently and not in a hierarchical manner. The expression of clpA was observed in F2/5 and in all mutants (Fig. 2.13) (Hao et al. 2005).

As the first step in the perception of pathogens by plants, perception of pathogen-associated molecular patterns (PMAPs) or general elicitors by the host plant results in rapid activation of defense mechanisms such as cell-wall reinforcement by callose deposition, production of reactive oxygen species (ROS) and induction of several defense-related genes. On the other hand, virulence factors produced by pathogens may inhibit these PAMP-elicited basal defenses (Nomura et al. 2005; Kim et al. 2005c). The structures characteristic of bacterial pathogens such as lipopolysaccharides (LPS), bacterial cold-shock protein (CSP), flagellin and ET-Tu may be perceived by plants (Nurnberger et al. 2004; Zipfel and Felix 2005). The pathogens are recognized by the plants through the activity of an array of

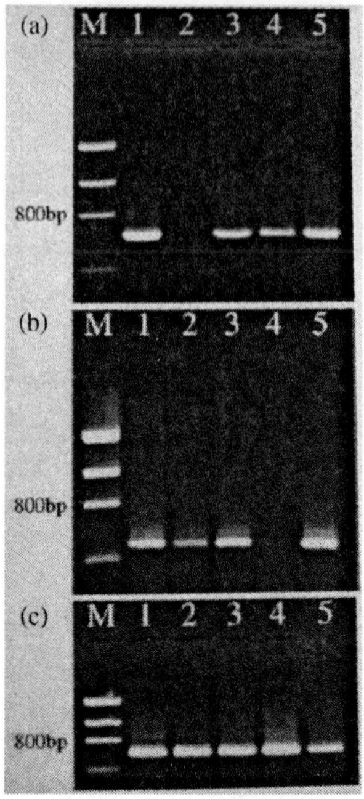

Fig. 2.13 RT-PCR assay to assess the expression of *aviR* (**a**), *avhR* (**b**) and *clpA* homolog (**c**) in F2/5, M1154 (avi R mutant), M1320 (avhR mutant) and complemented M1154 and M1320 Lane M: Low molecular-size DNA markers (standard); Lane 1: F2/5; Lane 2: M1154; Lane 3: Complemented M1154; Lane 4: M1320; Lane 5: Complemented M1320. (Courtesy of Hao et al. 2005; The American Society for Microbiology, Washington, USA)

pattern recognition receptors (PRRs) which are capable of recognizing characteristic molecular structures, the PAMPs. *Arabidopsis* plants can detect a variety of PAMPs including conserved domains of bacterial flagellin and EF-Tu. *Agrobacterium tumefaciens* has an EF-Tu that is fully active as an elicitor in *Arabidopsis* (Kunze et al. 2004). Flagellin and EF-Tu activate a common set of signaling events and defense responses. A targeted reverse-genetic approach was adopted to identify a receptor kinase essential for EF-Tu perception designated EFR. Upon transient expression of EFR *Nicotiana benthamiana* (a plant unable to perceive EF-Tu) acquired EF-Tw binding sites and responsiveness. Suceptibility to *A. tumefaciens* was increased in *efr* mutants of *Arabidopsis* into which T-DNA transformation occurred with higher efficiency. The results indicated that EFR was the EF-Tu receptor and that plant defense responses induced by PAMPs like EF-Tu reduced transformation by *A. tumefaciens* (Zipfel et al. 2006).

Many virulence factors, in addition to *hrp* controlled pathogenic factors (secreted via TTSS) are essential for the induction of different types of symptom in susceptible host plant species. Water-soaking symptom associated with several bacterial diseases in the early stage of infection, is due to the action of extracellular polysaccharides (EPS) produced by the bacterial pathogen. Large amounts of EPS are produced by many bacterial pathogens both in vitro and in vivo. The amounts of EPS produced appears to be related to the level of virulence in *Xanthomonas oryzae* pv. *oryzae* (*Xoo*). The most virulent strain BX08 of *Xoo* produced higher amount of EPS (43.62 μg/ml) than the strain BX01 which could produce only 32.5 μg/ml. In addition, the compositional variation of EPS may have some influence on the virulence of *Xoo* (Singh et al. 2006). The *eps* genes are required for EPS synthesis and the ability to produce EPS has been found to be positively correlated with virulence of the bacterial species. In *Burkholderia (Pseudomonas) solanacearum*, the biosynthesis and transport of the acid EPS (high MW) are encoded by at least nine structural genes located in the *eps* operon. Inactivation of the *eps* operon by transposon tagging resulted in reduction in the ability to produce EPS in planta and consequently the EPS-mutants had reduced virulence. Acidic EPS is an essential wilt-inducing factor produced by *B. solanacearum* (Denny and Schell 1994). In another study on *Clavibacter michiganensis* which produces two distinct types of EPS (A and B), no difference in the virulence of EPS mutants was discernible. Hence, wilting was not considered to be due to EPS (Bermpohl et al. 1996). The pathogenicity locus *pat-1* located in the plasmid pCM2 was shown to be essential for virulence of *C. michiganensis* subsp. *michiganensis* (*Cmm*). The endophytic plasmid-free isolates of *Cmm* became virulent following introduction of the *pat-1* region (Dreier et al. 1997).

The capsular EPS of *Erwinia amylovora* is composed of complex EPS amylovoran. The gene clusters *cms* and *cps* controlling the biosynthesis of levan and amylovoran have been sequenced. Amylovoran negative mutants could not induce any symptoms in inoculated pear seedlings (Bernhard et al. 1993). Proteins with transport functions were tagged by fusions with β-lactamase. Expression of levan sucrose is influenced by *rlsA*, located in the *dsp/hrp* region. Although no signal peptide could be detected in the amino acid sequence of levan sucrose, its fusion with β-lactamase produced antibiotic resistant cells, confirming its transport in *E. amylovora.* A protease deficient strain was retarded in colonization of host plants (Geider et al. 1999). In the case of *E. chrysanthemi* EC 3937, the *eps* genes are clustered on the chromosome and they are repressed by a regulator of pectate lyase synthesis designated PecT. The reduction in the efficiency of tissue maceration in an *eps* mutant indicated that full expression of virulence in *E. chrysanthemi* required production of EPS (Condemine et al. 1999). The effects of multiple mutations involving three virulence systems operating in strains of *E. chrysanthemi* EC 16 viz., *pel* (coding for the major pectate lyases pel A, B, C, E), *hrp* and *sap* (sensitivity to antimicrobial peptides) were assessed. The *sap* mutations, tested on potato tubers, exhibited greater reduction of virulence than *pel* mutations. Analysis of the strains affected in Pel-Sap, Hrp-Sap and Pel-Hrp-Sap suggested that effects of these mutations were additive and that the mutations in the *hrp* locus can be complemented in vivo by coinfection, while mutation in *pel* and *sap* cannot (López-Solanilla et al. 2001).

Erwinia amylovora requires pathogenicity factors for the ability to produce the capsular exopolysaccharide (EPS) amylovoran and to cause a hypersensitive response (HR) in nonhost plants. The synthesis of the virulence factors such as the effector protein harpin and the synthesis of amylovoran are tightly regulated by environmental conditions. The global regulation of bacterial transcription is a general response to the environmental changes depending on many factors including histone-like proteins like H-NS (Dorman and Deighan 2003). H-NS is mainly a multifunctional gene regulator, predominantly negative, but H-NS may also activate genes by repressing a repressor (Dorman 2004). Attenuation of virulence of animal pathogens by the small basic histone-like protein H-NS is known. An *hns* homolog from *E. amylovora* was identified by complementing an *Escherichia coli hns*-mutant strain with a cosmid library from *E. amylvora*. A difference in the functions of the two *hns*-like genes in *E. amylvora* was noted for the production of EPS amylovoran. Levan production was significantly increased by *hns* mutations. But synthesis of the capsular EPS amylovoran and of levan was reduced, if *hns* from *E. amylovora* plasmid was over- expressed. Increase in amylovoran synthesis was seen due to a mutation in chromosomal *hns* of *E. amylovora*. Both mutations affected development of symptoms adversely on immature pear fruits (Hildebrand et al. 2006).

Xanthomonas axonopodis pv. manihotis (*Xam*) causing cassava bacterial blight (CBB) disease induces water-soaked leaf lesions that develop into extensive necrotic areas as the bacteria spread systemically. *Xam* produces copious amounts of fibrillar extracellular material during colonization of intercellular spaces of cassava leaves and xylem vessels as revealed by electron microscopic observations (Cooper et al. 2001). The antibodies specific to EPS from *Xanthomonas campestris* pv. *campestric* (*Xcc*) labeled the material in advance of invading *Xam* cells and it was associated closely with degraded cell walls of cassava indicating the nature of the extracellular material as EPS. The EPS produced by xanthomonads is designated xanthan with multifunctions such as protection to bacterial cells, enhancement of attachement of bacterial cells to surfaces, assisting water-soaking of intercellular spaces (Denny 1995). In order to determine the ability of EPS as a virulence determinant of *Xam*, EPS-mutants were created by targeted disruption of the biosynthetic gene *gumD*. Pathogenicity of mutants was severely reduced. Although two EPS^- mutants could induce water-soaking and chlorosis, symptoms were restricted to inoculated spots, while the wild-type I 56 rapidly spread and developed into flaccidity. The wild-type population produced EPS as high as 200 folds of the EPS^- mutants. Petioles injected with EPS did not exhibit any visible symptom, whereas severe wilting was seen in petioles inoculated with I 56. Furthermore, EPS conferred resistance to H_2O_2 and UV irradiation to which the mutants were vulnerable. The reduced epiphytic survival of the mutants correlated with their sensitivity to UV irradiation (Kemp et al. 2004a).

Mutants of *Xanthomonas campestris* pv. *campestris* (*Xcc*) produced less or altered xanthan in culture and were less virulent compared with wild-type strain. Another xanthan mutant produced by transposon tagging was comparatively less virulent and the ability to multiply in mesophyll tissue was markedly reduced (Newman et al. 1994). The *pigB* gene has been shown to govern the production of EPS xanthomonadin pigements and the diffusible signal molecule diffusible factor (DF). Following application of the DF extracellularly, the *pigB* mutants were able to synthesize EPS and xanthomonadin. They could infect cauliflower through wounds but not through natural openings like hydathodes. A functional *pigB* gene may be essential for epiphytic survival and natural host infection (Poplawsky et al. 1998; Poplawsky and Chun 1998). Production of extracellular enzymes (including proteases, pectinases and cellulases and EPS) by *Xcc* is regulated by the products of the *rpfABFCHG* cluster of genes. The *rpFB* gene encoding a putative enoyl

CoA hydratase and *rpfB* gene encoding a long chain fatty acyl CoA ligase have been implicated in the synthesis of a small diffusible signal factor (DSF) (Barber et al. 1997; Slater et al. 2000). The DSF, as the cell–cell communication signal of *Xcc* plays an important role in regulation of virulence factor production and biofilm dispersal. The core members of DSF regulon include 165 genes in *Xcc* (He et al. 2006). Exogenous addition of DSF restored the ability to produce extracellular enzymes and polysaccharides by the rpfF transposan DSF defifcient mutant to the levels of wild-type *Xcc* strain (Barber et al. 1997). Furthermore, the cell aggregates formed by *rpfF* mutants could be dispersed by external addition of DSF or endo-β-1,4-mannanase encoded by *man4* (Dow et al. 2003).

The homologs of RpfF, the enzyme implicated in DSF biosynthesis have been detected in several bacterial genomes. DSF-like activity was recorded in more than 20 bacterial strains belonging to 10 bacterial species (Wang et al. 2004b). By applying microarray technique, several new functions of *Xcc* coordinated by DSF cell–cell communication signals have been demonstrated. *Xcc* has been shown to recruit a DSF which has been structurally characterized as *Cis*-11-methyl-d-dodecenoic acid, as a cell–cell communication signal to synchronize virulence gene expression and biofilm dispersal. The DSF-mediated functions were in general conserved in different *Xcc* isolated, although phenotype variations were evident. Of the 165 genes included in DSF regulon, the majority (133) was activated by the signal. This feature as "gene inducer" of DSF is similar to AHL-type quorum-sensing systems. The DSF-activated genes belong to many functional groups and most of the genes repressed by the signal are located in three functional groups such as chemotaxis, HR and pathogenicity (*hrp*) in addition to certain antiporter and efflux protein homologs (pha DEF). The genomic profiles of DSF regulon were investigated by oligomicroarray analysis by comparison of gene expression patterns of wild-type strain XC and its DSF-deficient mutant XC1dF. New functions mediated by DSF such as flagellum synthesis, resistance to toxins and oxidative stress were revealed by microarray analysis. The results indicated that DSF cell–cell signaling is essentially required for coordination of expression of virulence as well as for keeping up the general competence of the pathogen in ecosystem (He et al. 2006).

The operation of cell–cell signaling in *Xcc* and *X. axonopodis* pv. *citri* (*Xac*), mediated by diffusible signal factor (DSF) was investigated. Production of pathogenesis factors such as synthesis of extracellular plant cell wall-degrading enzymes and extracellular polysaccharide (EPS) xanthan is regulated by a cluster of genes (A-I) known as *rpf* (for *r*egulation of *p*athogeneic *f*actors) in *Xcc* (Slater et al. 2000). Two of these genes *rpfH* and *rpfI* were absent in *Xac,* although the cluster of *rpf* genes in *Xac* has synteny with the corresponding gene cluster in *Xcc*. Two strains of *Xac*, one with a mutation in the *rpfF* gene resulting in inability to produce DSF and another mutant defective in *rpfC* with DSF overproducing ability were constructed. These mutants also produced lower levels of extracellular cyclic β-1,2-glucans, endoglucanase and protease extracellular enzymes. These mutants were severely compromised in the ability to induce canker symptoms in lemon leaves compared to the wild-type strain. The results indicate that *rpf* genes in *Xac* are involved in controlling virulence factors mediated by DSF (Siciliano et al. 2006).

2.2.3 Symptom Expression

Numerous proteins known as effectors are delivered into the host plant cell via TTSS of bacterial pathogens. These proteins are considered to suppress plant defenses or alter the plant metabolic machinery to promote the synthesis of nutrients required for the proliferation of bacteria. The hrp-dependent factors and certain toxins-host-specific/host nonspecific may be involved in the suppression of host defense gene expression and induction of symptoms characteristic of the bacterial pathogen concerned. Expression of some genes has been shown to be essential for the production of symptoms of the disease(s). Many bacterial pathogens induce water-soaked symptom and chlorosis in different host plant. The water-soaking symptom is due to the action of extracellular polysaccharides (EPSs). Large amounts of EPS are produced by many bacterial pathogens both in vivo and in vitro. The *eps* genes govern the synthesis of EPS and the ability to produce EPS is positively correlated to the virulence of the strains of *Erwinia amylovora*, *E. chrysanthemi*, *Burkholderia solanacearum*, *Clavibacter michiganensis* and *Xanthomonas campestris* (Narayanasamy 2002).

Chlorosis is the visible result of the loss of chlorophyll which may occur by a variety of mechanisms, including a response to the activity of toxins resulting in cell / tissue collapse and ultimate death of plants. In tomato, *Xanthomonas campestris* pv. *vesicatoria* (*Xcv*) causes bacterial spot disease whch induces chlorosis as the early visible symptom and infected tissues turn necrotic later. A novel locus from *Xcv* that induces early chlorosis in tomato and several nonhost has been identified and characterized. The gene, chlorosis factor (*ecf*) encodes a hydrophobic protein with similarity to four other proteins in plants including HolPsyAE. The 2.4 kb fragment containing *ecf* was capable of conferring the chlorosis-inducing activity on *Xcv* 89-1 strain. The *ecf* gene product seemed to be translocated to hosts and the associate phenotype was *hrp* dependent. There was no influence of *ecf* on electrolyte leakage or on the bacterial growth in planta in response to infection (Morales et al. 2005).

Damage to plant tissues due to biotic or abiotic stress may result in the release of chlorophyll from thylakoid membranes. The chlorophylls have to be degraded rapidly to prevent cellular damage by the photodynamic action (Takamiya et al. 2000). In the absence of chlorophyll degradation, the amount of reactive oxygen species (ROS) produced may exceed the antioxidant capacity of plant system and the toxic molecules may cause cell death. In *A. thaliana*, of the two genes encoding chlorophyllases *AtCLH1* and *AtCLH2*, only the former is rapidly induced in response to wounding, methyl jasmonate (MeJA) and the bacterial jasmonate-mimicking toxin coronatine. Further, the expression of *AtCLH1* was reduced in the JA-insensitive *Coi1* mutant plants (Benedetti and Arruda 2002). By using RNA interference (RNAi), the specific silencing of *AtCLH1* encoding the first enzyme in the chlorophyll degradation pathway, chlorophyllase 1, led to accumulation ROS. In the *A. thaliana* plant overexpressing *AtCLH1*, disease symptoms such as tissue maceration were clearly seen even from 24 h after inoculation with *Erwinia carotovora* subsp. *carotovora* (*Ecc*). In contrast, the inoculated leaves RNAi silenced plants exhibited no visible symptom of infection. At 72 h after inoculation, infection had spread in the wild-type and *AtCLH1* overexpression lines and the treated leaves were entirely macerated. The *AtCLH1*- silenced lines became highly susceptible to infection and susceptibility was further enhanced following SA treatment resulting in infection of almost all the local leaves. RNAi silencing of AtCLH1 enhanced, the resistance to *Ecc* and in contract, increased susceptibility to the fungal pathogen *Alternaria brassicola* (Kariola et al. 2005).

Hypersensitive response is a defense response of the nonhost or host plants, involving rapid, localized cell death leading to inhibition of further development of the pathogen. The cell death associated with HR is a genetically controlled and regulated process. Programmed cell death (PCD) is a commonly of observed feature in hosts exhibiting HR and such a symptom is considered to reflect high levels of resistance of immunity. The effector AvrPtoB from *Pseudomonas* acts inside the plant cell to inhibit PCD initiated by the Pto and Cf9 disease resistance proteins (Abramovitch et al. 2003). Inactivation of salicylic acid (SA)-mediated resistance is a critical step in pathogenesis. The Δ CEL mutant of *Pst* DC3000 reached higher population levels and caused more pronounced chlorosis and necrosis in NahG plants than in wild-type Col-o *Arabidopsis* plants. The NahG plants cannot accumulate SA, because of the degradation of SA by *nahG* encoded SA hydroxylase. The effector locus CEL of *Pst* DC 3000 contains atleast four characterized effector genes (avrE, hopPtoM, hrpW and hopPtoA1) that are conserved among various *P. syringae* pathovars (DebRoy et al. 2004).

Erwinia amylovora causing fire blight of apple induces plasmolysis and deterioration of parenchyma cells due to electrolyte leakage (Sjulin and Beever 1978). *E. chrysanthemi* produces pectic enzymes and siderophore-dependent iron uptake systems which are recognized to be essential major pathogenicity factors. At least eight endo-pectate lyases (PelA, PelB, PelC, PelD, PelE, PelI, PelL and PelZ), two exo-pectate lyases (PelX and PelW), two pectin methyl esterases (PemA and PemB) and a pectin acetyl-esterase (Pae Y) have been examined. The genes encoding these enzymes were essential for full expression of virulence of *E. chrysanthemi* (Franza et al. 1999; Robert-Baudouy et al. 2000). Rapid maceration of tissues occurs due to the activity of pectic and other enzymes. The susceptibility of various African violet varieties was related to the expression of the virulence factors (Yang et al. 2002). Falgellin from *Pseudomonas syringae* in the pathogen-associated molecular pattern (PAMP) that has been characterized well. A conserved N-terminal peptide of flagellin, flg22 is a potent elicitor of defense responses in tomato and *Arabidopsis*. The flg22 strongly induced transcription of NHO1 (nonhost resistance gene) in *Arabidospsis* in a flagella-dependent manner. But a *Ptab* strain lacking the flagellin gene *fliC* induced disease symptoms and multiplied in *Arabidopsis.* In contrast, the wild-type strain did not cause any visible symptom. The absence of flagellin results in symptom induction, while its presence elicits induction of the NHO1 gene in *Arabidopsis* (Li et al. 2005).

The bacterial effector proteins contribute to induction of disease symptoms without substantially affecting pathogen growth, suggesting that the molecular mechanisms underlying these two phenotypes are different. The ability of *P. syringae* pv. *syringae* to induce disease is abrogated by a mutation in *gacS*, without affecting its proliferation in bean leaves (Willis et al. 1990). Likewise, reduction in symptom intensity results, due to deletion of the TTSS effector gene *hopPtoM* in *Pst* DC3000 and the growth of the pathogen in leaves, however, is not reduced significantly (Badel et al. 2003). In contrast, enhancement of disease symptom is seen with no adverse effect on the development of pathogen, due to mutation in another effector gene *hopPtoN* from *Pst* DC3000 (López-Solanilla et al. 2004). The effectors AvroPto and AvrPto B, when delivered, enhance necrosis associated with the disease induced by *Pst* DC 3000 (Lin and Martin 2005).

The phytohormones appear to have dual function in plant pathogenesis. They are involved in both plant defense response and disease progression. The gaseous hormone ethylene is a critical component of plant responses to pathogen attack, in addition to being essential for developmental processes such as fruit ripening and senescence. In tomato, ethylene perception is needed for later stages of symptom development following inoculation with *Pst* and *Xanthomonas* spp. as characterized by extensive necrosis of infected tissue, after initial lesion formation (Lund et al. 1988). In addition, ethylene has been reported to promote salicylic acid (SA) production in tomato in response to *X. campestris* pv. *vesicatoria* (*Xcv*) which is critical for development of necrosis induced by this pathogen (O'Donnell et al. 2003).

The induction of disease symptoms in tomato by *Xcv* requires the cooperative and sequential actions of jasmonate, ethylene and salicylic acid (SA). The tomato lines insensitive to ethylene (*Never ripe*) or inhibited in ethylene synthesis [ACC deaminase (ACD)] and tomato NahG line that fails to accumulate SA were inoculated with *Xcv*. The disease development was attenuated, with no measurable effect on bacterial growth (O'Donnell et al. 2001). When tomato plants were inoculated with *Pst* DC 3000 which produces COR, no difference in bacterial growth or disease development was noted in *defl,* AS-DOC mutants and their wild-type controls (O'Donnell et al. 2003). The effectors AvrPto and AvrPtoB appear to enhance symptom intensity in tomato leaves inoculated with *Pst* DC 3000, in part, by upregulating genes involved in ethylene production (Scofield et al. 1996; Cohn and Martin 2005). A group of T-DNA genes in *Agrobacterium tumefaciens*, directs the synthesis of plant growth hormones that are required for the enlargement and proliferation of transformed plant cells resulting in the production of characteristic galls or tumors in the infected plants. The *iaaM* and *iaaH* gene products are involved in the conversion of tryptophan via indoleacetamide to indole acetic acid (auxin) (Bins and Costantino 1998). Two other T-DNA genes, gene 5 and gene 6b (tml) are also thought to play ancillary roles in tumorigenesis (Hooykaas et al. 1988).

Some bacterial pathogens produce phytotoxins which are considered to induce some symptoms of the disease. *P. syringae* pv. *tomato* (*Pst*) DC 3000 produces a phytotoxin designated coronatine (COR) in susceptible tomato plants causing necrotic lesions. Coronatine is composed of coronafacic acid (CFA) and coronamic acid (CMA). The *hrp/hrc* secretion system does not appear to be essential for COR production, since the *hrp* mutants can produce COR (Bender 2000). The results of experiments conducted on tomato inoculated with *Pst* DC3000 indicated that intact COR molecule was required for both suppression of salicylic acid (SA)-mediated defense responses and full disease symptom expression in tomato (Uppalapati et al. 2007). *P. syringae* pv. *syringae* (*Pss*) produces syringomycin; *P. syringae* pv. *tabaci* elaborates tabtoxin, while phaseolotoxin is known to be formed by *P. syringae* pv. *phaseolicola* (*Psp*). The genes involved in the production of these toxins have been cloned and characterized (Xu and Gross 1988; Willis et al. 1991; Kamdar et al. 1991). These genes may be located in one or more clusters. Tissue necrosis is the symptom induced by the phytotoxin. The *lemA* gene in *Pss* strain B728a is required for the production of extracellular protease and syringomycin and consequently for lesion formation, whereas *lemA* is not essential for the pathogenicity of *Psp* (Barta et al. 1992). *Burkholderia glumae* (causing rice seedlings and grain rot disease) produces the toxin called as toxoflavin composed of two proteins viz., TRP-1 and TRP-2. The TOX^- mutants are nontoxigenic and also nonpathogenic to rice (Suzuki et al. 1998; Yoneyama et al. 1998).

Burkholderia glumae causes rice seedlings rot in the nursery boxes and grain rot in the field after heading in Japan (Uematsu et al. 1976). *B. glumae* elaborates a phytotoxin, toxoflavin that retards growth of leaves and roots, in addition to induction chlorotic symptoms on panicles in the grain rot phase. Two toxoflavin biosynthesis-related proteins viz TRP-1 and TRP-2 from wild-strains were identified. TRP-1 was encoded by *toxA* gene. An 8.2 kb region in the pathogen DNA containing the *tox* operon (*toxABCDE*) involved in toxoflavin biosynthesis was cloned and sequenced. The sequence downstream of *toxA* had four ORFs – *toxB, toxC, toxD* and *toxE*. Mutants disrupted in the *tox* operon did not produce the toxoflavin and did not induce typical chlorotic lesions on infected rice panicles. The *tox* operon was indicated to be polycistronic by RT-PCR assay. The results suggested that toxoflavin might be synthesized in part through a biosynthetic pathway common to the synthesis of riboflavin with which the toxin exhibited similarity in certain physical and chemical properties (Suzuki et al. 2004).

Ralstonia solanacearum infecting many crops produces extracellular polysaccharides (EPS) in culture and correlation between EPS production and virulence of the pathogen strains has been observed (Buddenhagen and Kelman 1964). The crude cell-free extracts induced wilting in tomato cuttings. Using specific EPS mutants of *R. solanacearum*, the ability of the acidic EPS of inducing wilting was noted. Hence, this form of EPS was considered as the wilting factor of *P. solanacearum* (Denny and Schell 1994). The inactivation of *eps* operon by transposon tagging resulted in reduction in production of acidic EPS in planta and consequently the EPS mutants had reduced virulence (Denny and Schell 1994). *Xanthomonas campestris* pv. *campestris* (*Xcc*) produces a range of extracellular enzymes and one extracellular polysaccharide, xanthan which may be required at early stages of infection. However, xanthan is produced abundantly at later stages of pathogenesis in tissues undergoing necrosis (Vojnov et al. 2001). Xanthan polymers have been implicated in several symptoms, including wilting induced by vascular pathogens and the water soaking symptoms associated with foliar pathogens (Denny 1995).

Suppression of host defenses is emerging as a key pathogenesis-related mechanism. The role of xanthan in the pathogenicity and symptom production of *Xcc* in *Arabidopsis* and *Nicotiana benthamiana* has been investigated in detail. Two

Xcc mutants, strains 8397 and 8396, defective in xanthan production were used. Leaves of *N. benthamiana* and *Arabidopsis* preinfiltrated with xanthan exhibited symptoms in response to inoculation with *Xcc* strain 8397 and 8396. The bacterial population in the infected leaves was approximately 30 or 25-fold higher than control (water treated) and infection reached the levels similar to that of wild-type strain. The xanthan effect was dose-dependent and the minimum xanthan concentration required for restoring disease symptoms in 8397 strain was 50 μg/ml. The role of callose deposition and its suppression in host defense and susceptibility was determined by monitoring callose deposition in leaves inoculated with mutant strains. Callose depositions could be identified by staining with aniline blue and observation under UV-fluorescence microscope. The leaves inoculated with either 8397 or 8396 strain showed higher level of callose deposition in the host cell wall than in leaves inoculated with wild-type strain 8004. In contrast, the leaves pretreated with xanthan failed to accumulate callose in response to inoculation with mutant strains compared with untreated leaves that accumulated callose deposition, indicating that xanthan may induce susceptibility by suppression deposition of callose on the plant cell walls. The inference is supported by the experiment conducted to assess the effect of 2-deoxy-D-Glc (2DDG), a callose synthesis inhibitor. The *Xcc* mutants 8397 and 8396 strains, following application of 2DDG, could not induce callose deposition and also they regained the ability to induce disease symptoms. The results clearly demonstrate that there is a direct correlation between the absence of callose deposition and production of disease symptom suggesting a major role for callose in pathogenesis of *Xcc* (Yun et al. 2006).

2.3 Phytoplasmal Pathogens

Phytoplasmal pathogens inducing yellowing, little leaf and virescence of floral organs as chief symptoms, are a poorly understood group of organisms, primarily because of the difficulty in culturing them in artificial media. DNA-based molecular techniques such as DNA hybridization, restriction fragment length polymorphism (RFLB) and polymerase chain reaction (PCR) have been employed to detect, identify and differentiate the phytoplasmas (Volume 1, Chapter 2). Very little information is available on the different stages of pathogenesis. All the phytoplasmas have to be introduced into the susceptible host plant cells as the first step in the process of infection, by the specific insect vector(s). In the case of human and animal mycoplsmas, the requirement of adhesion of mycoplasma to host cells for successful colonization of host tissues has been reported. This step is mediated by mycoplasma surface proteins and among them, adhesions are considered to play an important role in invasion and pathogenicty (Rottem 2002). An adhesion-like gene in *Spiroplasma kunkelii*, the causal agent of corn stunt disease has been identified. Primers and a fluorigenic probe based on the nucleotide sequence of the adhesion-like gene were prepared for developing a field-depoyable real-time PCR assay for the detection of *S. kunkelii* (Wei et al. 2006).

The interaction between the phytoplasma and plant host appears to have been studied in a few pathosystem. Plant pathogenic phytoplasmas are endocellular and lack cell walls which allow secreted phytoplasmal proteins to interact with host plant and insect vector cells directly, suggesting possible roles for these proteins. Immunodominant membrane proteins of phytoplasmas constitute a major portion of the total cellular membrane proteins in most phytoplasmas. The presence of genes encoding immuno-dominant membrane proteins has been detected in phytoplasmas such as aster yellows (AY), clover phyllody (Barbara et al. 2002), pear decline, peach yellow leaf roll and European stone fruit yellows (Morton et al. 2003). Variations in amino acid sequence and antigenicity in these proteins have been observed. They may be secreted across phytoplasmal cell membrane during protein localization and they are grouped into three types (Barbara et al. 2002). The genes of onion yellows (OY phytoplasma) that encode the SecA and SecY proteins forming essential components of the Sec protein translocation system were identified indicating the existence of a Sec system in phytoplasmas (Kakizawa et al. 2001). In a later investigation, a gene that encodes a putative secE protein, a component of the Sec protein translocation system was cloned from the OY phytoplasma. Further, a gene encoding an antigenic membrane protein (Amp) – a type of immunodominant membrane protein- of OY was cloned and sequenced. The anti-Amp antibody generated using bacterially expressed part of OY Amp, reacted specifically with an OY-infected plant extract in Western blot analysis (Fig. 2.14). A partial OY Amp protein expressed in *E. coli* was localized in the periplasm as a shorter, putatively processed from the protein. Furthermore, OY Amp protein expressed in OY and detected in OY-infected plants was also apparently processed. Because of the lack of possibility for culturing phytoplasmas at present, information on the secretion protein systems is very little and future research in this area may throw more light for better understanding of the phenomenon of phytoplasma of pathogenesis (Kakizawa et al. 2004).

Some studies directed towards the understanding of the molecular basis of the interaction between *Spiroplasma citri* and the leafhopper vector species have been taken up. Propagative nature of the relationship between *S. citri* and *Circulifer haemotoceps* involves multiplication of the pathogen in the leafhopper cells. In *S. citri*, an adhesion-related protein designated P89 was shown to be directly involved in the spiroplasma-insect cell interaction (Yu et al. 2000). Cellular recognition mediated by carbohydrates and lectins has been suggested, because the basal lamina of insect organs is highly glycosylated. Hence, carbohydrates are likely to be the targets for *S. citri* interaction (Altmann 1996).

The proteins of *C. haematoceps* were screened by putative *S. citri*-binding molecules using a spiroplasma overlay assay of protein blots (Far-Western assay). Insect proteins were separated by SDS-PAGE technique, blotted and probed with *S. citri* proteins. Spiroplasma proteins showed affinity for even leafhopper proteins. The interaction between *S. citri* proteins and insect proteins with molecular masses of 50 and 60 kDa were sugar sensitive. The insect proteins were

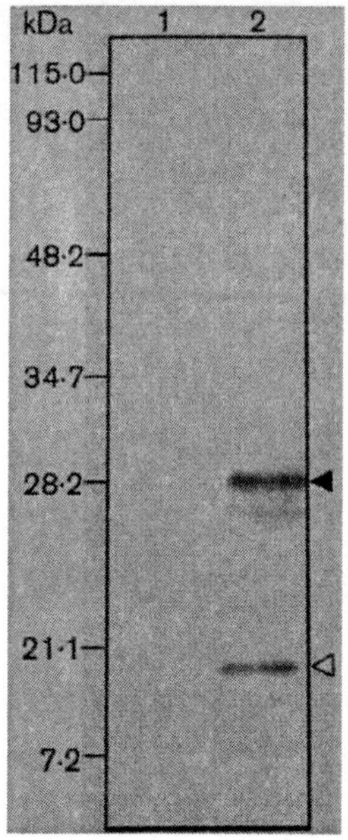

Fig. 2.14 Detection of distinct protein bands in garland chrysanthemum plants infected by onion yellows (OY) phytoplasma by Western blotting
Lane 1: Extract from healthy plant; Lane 2: Extract from an OY-infected plant showing the presence of a major band (*open arrow head*) and a minor band (*filled arrow head*)
(Courtesy of Kakizawa et al. 2004; Society of General Microbiology, Reading, UK)

identified as high mannose $N-$glycoprotein which supported the interaction of glycoprotein-lectin type with *S.citri* proteins. Lectin detection in *S. citri* showed the presence of only one protein of 24 kDa. One spiroplasma protein with a similar molecular mass of 24 kDa had an insect protein-binding capacity. This protein was the most abundant membrane protein of *S. citri* and it was designated spiralin. Experiments with purified spiralin and insect glycoprotein confirmed the bindng of spiralin to insect glycoproteins of 50 and 60 kDa. The results indicate a key role for spiralin in the transmission of *S. citri* by mediating spiroplasma adherence to epithelial cells of insect vector gut or salivary gland (Killiny et al. 2005).

The molecular mechanisms underlying phytoplasmal pathogen-vector specificity has not been clearly understood. *Candidatus Phytoplasma asteris* OY strain, line W (OY) has an abundant cell-surface membrane protein designated antigenic membrane protein (Amp) which is not homologous with any reported functional protein. By using immunofluorescence microscopy, localization of OY phytoplasma to the microfilaments of the visceral smooth muscle surrounding intestinal tract of the vector insects was visualized. Amp forms a complex with three insect proteins namely actin, myosin heavy chain and myosin light chain as revealed by the affinity column assay. The presence of Amp-microfilament complexes was detected only in all OY-transmitting leafhopper species, but not in the non-vectors of OY, suggesting that the formation of the Amp-microfilament complex may be a critical factor in deciding the transmission of phytoplasmas. Thus this study brings out the existence of host-specific interaction between a phytoplasmal surface protein and a host microfilament insect cells (Suzuki et al. 2006).

2.4 Viral Pathogens

Plant viruses constitute an entirely distinct group of obligate, intracellular, molecular pathogens as against the cellular and multicellular bacterial and fungal pathogens. As the viruses are basically different in their constitution, their strategies for successful pathogenesis are also different. The viruses have no known physiological functions and enzymatic activities and their entry into the host cell depends on the vectors or wounds created for artificial inoculation. Viral infection of susceptible plants is a complex process. The plant viruses have to accomplish four main steps for successful infection of the plant: (i) entry into plant cells, (ii) replication in the primarily infected cells, (iii) cell-to-cell movement through plasmodesmata (PD) and (iv) long-distance movement through vascular system. Animal viruses enter cells through receptor-mediated endocytosis. In contrast, the plant cell wall is a formidable barrier for the entry of virus and there is no receptor to mediate the initial transfer of the plant virus into a plant cell. The infectious virus particle has to be introduced by physically penetrating the cell wall. Shortly after the entry into the plant cell, the viral genome is released from the protein coat. The genome (+ sense) – single stranded (ss)-RNA- of the infecting virus is either translated directly or mRNAs are formed and translated to produce viral replicase as an early product as well as virus-specific proteins. The viral replicase or replication-associated proteins are involved in the synthesis of new viral genomes. The viral coat protein and movement proteins required to encapsidate the viral genome and for cell-to-cell movement of the virus are synthesized as the next step in viral replication. The coat protein and the viral genome are assembled to form new progeny virus particles which accumulate within the cell usually in the cytoplasm. The infectious virus particles have to move from the initially infected cells to adjacent cells and later to other tissues as the virus spreads systemically. The steps in virus replication may vary depending on the host-virus combination.

2.4.1 Movement of Plant Viruses

The plant viruses have to invade and infect as much of the host tissues as possible to maximize their chances of perpetuation. Two phases of virus movement within the host plant viz., cell-to-cell or short distance and long distance, have been recognized. If a virus is able to replicate in the initially infected cell, but unable to move to neighboring cells, the infection is designated subliminal infection of the plant species which shows high form of resistance (extreme hypersensitivity) (Cooper and Jones 1983). Varying extent of virus movement may occur depending on the level of susceptibility / resistance of the host plant species to the virus concerned.

2.4.1.1 Cell-to-Cell Movement

During the life cycle from virus accumulation to intracellular, local and systemic movement, viruses utilize plant proteins, normally involved in host-specific activities for their own purposes (Nelson and Citovsky 2005). They move cell-to-cell via plasmodesmata (PD) which are modified to a lesser or greater extent and to distant plant organs through the vascular system. The plasmodesmata (PD) function as cytoplasmic bridges that establish membrane [endoplasmic reticulum (ER) and plasma membrane] as well as cytoplasmic continuity between adjacent plant cells. The PDs form an important route for communication between plant cells, regulating cell-to-cell communication. As such, PDs are too small to allow free passage of virions or viral genomes which are far larger than the size exclusion limit (SEL) of PDs. The viruses have to first move from replication sites to PD at the cell periphery and then traverse the intercellular channels to enter the neighboring cells.

Movement Proteins

In the case of many plant viruses cell-to-cell transport is mediated by specific virus-encoded factors termed movement proteins (MPs), the function of which may be augmented by other viral proteins. The involvement of MPs in modifying the SEL of PD has been indicated. By using microinjection of dye-coupled dextrans, the SEL of unmodified PDs between mesophyll cell was found to be about 0.75–1.0 kDa. But the transgenic plants expressing *Tobacco mosaic virus* (TMV) MP allowed 9.4 kDa molecules to pass through PDs (Wolf et al. 1989). In some viral systems the structural or coat protein (CP), in addition to MP may be required to mediate virus movement. The constant streaming of plant cell cytoplasm may help in the random movement and viral MPs may be able to bind to PD or other peripheral target sites, when they encounter proteins they recognize. The majority of the cell-to-cell transport is presumably, provided by the host cell such as the cytoskeleton (Mclean et al. 1995). Later investigations suggested that the cytoskeletal network may not function alone in viral transport to and through PD. But it may act together with the endomembrane transport system of the host cell. Many viral MPs may be delivered to PD via the endoplasmic reticulum (ER), whereas actin / myosin filaments may regulate the flow of proteins in the ER membrane (Boevink and Oparka 2005).

The regulation of plasmodesmatal transport by the TMV-MP (P30) is dependent on phosphorylation in a host-dependent manner. P30 has been shown to be associated with the cytoskeletal structures as revealed by experiment involving antibody to P30 and using infectious TMV clones with GFP fused to P30. Colocalization of P30 with microtubules was observed in protoplasts and whole plants (Heinlein et al. 1995). In addition, association of P30 with cortical endoplasmic reticulum was also demonstrated (Heinlein et al. 1998; Reichel and Beachy 1998). Interaction of P30 with pectin methyl esterase (PME) isolated from tobacco leaf cell walls was also reported by Chen et al. (2000). Participation of microfilaments in the cell-to-cell movement of TMV and involvement of microtubules and microtubule-associated proteins in the degradation of the MP have been demonstrated (Gillespie et al. 2002; Kragler et al. 2003). Disruption of microfilaments by virus-induced gene silencing significantly reduced spread of TMV from cell to cell. However, accumulation of virus within the infected cells was not affected. In addition, another TMV factor, the 126 kDa protein was demonstrated to be involved in viral transport along with microfilaments. This protein was shown to associate with viral replication complexes, modulate their size and potentially mediate their interaction with and movement along the microfilament net work (Liu et al. 2005b).

Many viruses form replication centers enriched in endoplasmic reticulum (ER) and many viral proteins are located to ER. The role of ER translocation and plasmodesmatal targeting was explored using calreticulin, a cellular protein that localizes to PD. The N-terminal signal peptide critically influenced the ability of calreticulin to accumulate within plasmodesmata (Chen et al. 2005). Furthermore, calreticulin interacted with the TMV-MP and overexpression of calrecticulin in transgenic plants resulted in redirecting TMV from PD to microtubules and consequently compromised cell-to-cell transport of the virus (Chen et al. 2005). Although interaction between viral MPs and other cellular proteins has been suggested only protein kinases have been demonstrated to have a role in intercellular movement of TMV (Citovsky et al. 1993; Trutnyeva et al. 2005). In order to identify domains of the protein necessary for membrane association, deletion mutants of TMV-MP were produced. A membrane fraction was isolated from tobacco BY-2 protoplasts infected with wild-type and mutant TMV that produce MP carrying a 3-amino acid deletion. Deletion of two major hydrophobic regions and other hydrophic regions reduced the membrane association. In addition, elimination of phosphorylation sites also adversely affected membrane association. The results indicated that MP function requires interaction membrane although membrane association alone was insufficient for function and that TMV-MP is an integral or tightly associated membrane protein that includes two hydrophobic transmembrane domains (Fujiki et al. 2006).

The intracellular and intercellular movement of virus replication complexes (VRCs) of TMV was monitored in intact leaf tissue from 12 to 36 h post infection (hpi) by using confocal microscopy. The VRCs in primary infected cells were associated with ER from 12 hpi. High intracellular mobility by VRCs was seen at 14 hpi ($\approx$160 nm/sec) followed by progressive reduction in mobility between 14 and 16 hpi, becoming stationary at 18 hpi adjacent to PD. Then VRCs traversed the PD between 18 and 20 hpi. The process of formation and movement of VRCs in adjacent cells in 3–4 h as against 20 h taken in primary infected cells. Application of inhibitors of filamentous actin and myosin blocked the rapid movement of the VRCs and the spread to adjacent cells. Inhibitors of microtubles did not have any effect on this process. Based on the results it was hypothesized that cell-to-cell spread of TMV infection may be accomplished by subviral replication complexes that initiate TMV replication immediately after entry to adjacent cells (Kawakami et al. 2004).

Coat protein (CP), in addition to providing protection to the viral genome from inactivation, is also required for the viral movement. In cells infected by *Cowpea mosaic virus* (CPMV) tubules that penetrate the PD are formed and consequently PDs lose their characteristic desmotubules. Two overlapping genes produce peptides (58 kDa/48 kDa) which are needed along with CP for virus mobilization to the tubules. The 48 kDa protein was shown to be involved in the tubule formation. Such tubular structures are involved in cell-to-cell movement of CPMV (Kasteel et al. 1996). In the case of *Cauliflower mosaic virus* (CaMV) infected cells, tubules passing through PD are produced. CaMV is a ds-DNA virus, whose movement functions are attributed to *gene1*. The gene I protein was localized to the cell wall in infected cells (Linstead et al. 1988). The formation of tubules in CaMV-infected protoplasts has been reported by Perbal et al. (1993). The requirements of CMV for cell-to-cell movement appear to be different. Both MP and CP are necessary for the cell-to-cell movement. CMV variants lacking a CP could not move from cell-to-cell (Canto et al. 1997). Virion assembly was not found to be a prerequisite for CMV movement, since assembly-defective CMV variants efficiently spread and induced local lesions (Kaplan et al. 1998). Although tubules were formed in transfected protoplasts, they do not appear to contribute to viral movement (Canto and Palukaitis 1999).

Triple Gene Block (TGB) Movement Proteins

The cell-to-cell movement of many groups of plant viruses such as potexviruses carlaviruses, foveaviruses, pomoviruses and hordeiviruses is mediated by proteins translated from a set of three overlapping ORFs known as triple gene block (TGB). The three proteins (TGBps) encoded by TGB of potexviruses are required for the cell-to-cell movement of these viruses (Angell et al. 1996).

Two classes of triple gene blocks have been distinguished: class 1 TGB includes the hordeiviruses having a TGB1 encoding a 42–63 kDa protein and class 2 TGB contains potexviruses and carlaviruses with a TGB1 encoding a product of 25 kDa (Solovyev et al. 1996). Viral CP is also required for the movement of virus with class 2 TGB, whereas class 1 TGB in hordeiviruses mediates the movement of hordeiviruses independently of viral CP (Donald and Jackson 1996). The accumulation levels of TGBp1 of *Beet necrotic yellow vein virus* (BNYV) was adversely affected by mutations in TGBp2 and TGBp3, indicating the existence of a highly coordinated control for the interactions among the elements of the movement machineries (Lauber et al. 1998). In addition, TGBp2 and TGBp3 of *Potato mop top virus* were considered to be required for the transport of TBGp1 to and through plasmodesmata (Zamyatnin et al. 2004).

The TGBps are structurally and functionally different from the MPs. Among the three TGBps, TGBp1 seems to exist in soluble and cytoplasmic inclusion forms. The soluble form of TGBp1 has several functions: binding cooperatively to ssRNA, increasing the SEL of plasmodesmata and forming a ribonucleoprotein complex (Angell et al. 1996; Wung et al. 1999; Lough et al. 2000). TGPBp1 functions as a gene silencing suppressor (Voinnet et al. 2000) in addition to its independent movement from cell-to-cell and ability to mediate cell-to-cell movement of TGBp2 and TGBp3 (Yang et al. 2000b; Krishnamurthy et al. 2002). TGBp2 and TGBp3 have been demonstrated to be endoplasmic reticulum (ER)-targeted proteins and ER targeting is essential for virus movement (Krishnamurthy et al. 2003). Single Arg-Ala substitutions at position 11 of TGBp1 of *Bamboo mosaic virus* (BaMV) did not affect cell-to-cell movement of the virus. But mutants with Arg-16, Arg-21, or both Arg-16 and Arg-21 of TGBp1 replaced with alanine were defective in virus movement (Lin et al. 2004).

Based on sequence analysis, the protein region comprising the NTPase/helicase domain in in both potex-like and hordei-like TGBp1s was predicted to possess NT-Pase activity in vitro (Liou et al. 2000; Kalinina et al. 2002). The ATPase activity may be required for cell-to-cell movement, particularly for plasmodesmata dilation by the potex-like TGBp1 (Lough et al. 2000; Howard et al. 2004). The TGBp1 helicase activity was essential for protein / RNA translocation through plasmodesmata to adjacent cell for both types of viruses with TGB (Morozov and Solovyev 2003). The suppressing activity of RNA silencing has been demonstrated to another function of PVX-TGBp1 protein (Voinnet et al. 2000). It was also suggested that silencing suppression was essential for the TGBp1 protein to mediate viral cell-to-cell movement (Bayne et al. 2005). At least two functions of TGBp1 are essentially required: involvement in viral movement per se and ability to suppress RNA silencing. A series of deletion and point mutants in the TGBp1 proteins encoded by PVX and *Poa semilatent virus* (PSLV, genus *Hordeivirus*) were analysed to map functional regions responsible for their biochemical activities in vitro. In both PVX and PSLV, the N-terminal part of the TGBp1 NTPase/helicase domain comprising motifs I, Ia and II was sufficient for ATP hydrolysis, RNA binding and homologous protein–protein interactions (Leshchiner et al. 2006).

The transport component ribonucleoprotein (RNPs) or virions involved in the cell-to-cell movement of PVX genome include TGBp1 and CP. Both proteins are able to move through plasmodesmata (Batten et al. 2003; Howard et al. 2004). The PVX TGBp1 is able to modify plasmodesmta and move between cells. In rod-shaped viruses, TGBp2 and TGBp3 move to plasmodesmata and act in concert to transport TGBp1 to and through plasmodesmata. In PVX the 8 kDa TGBp3 protein is a membrane-embedded protein that has an N-terminal hydrophobic sequence segment and hydrophilic C-terminus. TGBp3 mutants with deletions in the C-terminal hydrophilic region retain the ablity to be targeted to cell peripheral structures and to support limited PVX cell-to-cell movement. The intracellular transport of TGBp3 from sites of its synthesis in rough ER to ER-derived peripheral bodies involves a non-conventional COPII-independent pathway (Schepetilnikov et al. 2005).

For the movement of *White clover mosaic virus* (WClM) TGBp3 is essentially required (Lough et al. 1998). On the other hand, TGBp3 may or may not be absolutely required for PVX movement suggesting that different potexviruses appear to employ different mechanisms for cell-to-cell movement functions. The requirement of BaMV TGBps for movement functions and the compatibilities with those of two potexviruses PVX and *Foxtail mosaic virus* (FoMV) were examined using a statellite RNA-mediated trans-complementation assay system. The results showed that all three TGBps were required for cell-to-cell movement of BaMV. Alterations of the ratio among TGBps by ectopic expression of individual components of TGBps from satBaMV RNA factors did not affect the cell-to-cell movement of BaMV and possibly other potexviruses (Lin et al. 2006).

In virus-infected plants, RNA silencing involves a mobile silencing signal that can move cell-to-cell and also systemically through the plants. The silencing signal may be linked to the defense activity if it moves through the plant either with or ahead of an invading virus. Several plant viral genomes (like tombusvirus) encode suppressors of RNA silencing that are also effectors of long distance virus movement through the phloem. The protein P19 from *Cymbidium ringspot tombusvirus* is a suppressor of silencing not required for virus replication in isolated cells, but is necessary for invasion of leaves systemically (Havelda et al. 2003). Likewise, the potyviral Hc-Pro and cucumoviral 2b proteins have been shown to be silencing suppressors (Anandalakshmi et al. 1998; Brigneti et al. 1998). The P25 movement protein from PVX was assessed for its ability as suppressor of silencing. This protein is one of the three TGB proteins of potexviruses required for cell-to-cell-movement of the virus. It is an RNA helicase capable of moving cell-to-cell and modifying the plasmo- desmata (Angell et al. 1996; Yang et al. 2000b). Through random mutation analysis of PVX silencing suppressor P25, evidence to show that suppression of silencing is necessary, but not sufficient, for cell-to-cell movement through PD, was obtained. All mutants that were defective for silencing suppression were also non-functional in viral cell-to-cell movement. It appeared that there must be a second P25 function that is independent of silencing but also required for cell-to-cell movement. Two classes of suppressor inactive P25 mutants were identified, one of them being functional for the accessory function. Based on the analyses of short interfering RNA accumulation, it was concluded that P25 may suppress silencing by interfering with either assembly or function of the effector complexes of RNA silencing (Bayne et al. 2005).

The major difference between the two groups of TGB-containing virus is the requirement of CP for Group 2 viruses and dispensability of CP for Group1 viruses for cell-to-cell movement. This suggests that the RNP of Group 1 viruses cannot move in the form of intact virions. The studies on the subcellular localization of TGB proteins of *Potato mop-top virus* (PMTV) when expressed as N-terminal fusions to GFP from a TMV-based vector showed that both the PMTVGFP-TGB2 and GFP-TGB3 fusions were associated with cellular endomembranes, particularly the ER network and membranes surrounding the nucleus. The GFP-TGB3 labeled opposing pairs of fluorescent spots across neighboring cell walls that suggested localization at or near plasmodesmata, were observed (Cowan et al. 2002).

The TGB1 protein interaction with RNA and its transport as a part of the RNP complex out of the cell and long distance throughout the plant is generally accepted. But the details of the process in which TGB2 and TGB3 proteins facilitate RNP intercellular transport are not clearly understood. The interaction of TGB2 and TGB3 proteins of *Potato mop-top virus* (PMTV) with components of the secretory and endocytic pathways was investigated. These proteins were expressed as N-terminal fusions to green fluorescent protein (GFP) or monomeric red fluorescent protein (mRFP). The fluorophore-labeled TGB2 and TGB3 showed an early association with ER and colocalized in mobile granules that used the ER-actin network for intracellular movement. Both proteins were able to increase the SEL of PD and TGB3 accumulated at PD in the absence of TGB2. Both fusion proteins were incorporated into vesicular structures during expression cycle. TGB2 was associated with its structures independently, while TGB3 required TGB2 for its incorporation into the vesicles. Furthermore, in addition to localization to the ER and motile granules, mRFP-TGB3 was incorporated into vesicles when expressed in PMTV-infected epidermal cells, indicating recruitment by virus-expressed TGB2. The experiments involving labeling of the TGB fusion proteins-containing vesicles with FM4-64, indicated that the endocytic pathway was involved in viral intracellular movement (Haupt et al. 2005).

Double-Gene Block Movement Proteins

Two small internal overlapping ORFs required for cell-to-cell movement are present in carmoviruses. The 5′ORF of *Carnation mottle virus* (CarMV), encoding P7 has RNA-binding properties (Marcos et al. 1999). Two small proteins, p8 and p9 mediated the cell-to-cell movement of *Turnip crinkle virus* (TCV) which did not require viral CP (Li et al. 1998). The ability of TCV p8 for virus movement was inversely correlated with the p8 RNA-binding. A single amino acid change in this protein adversely affected the movement of TCV in *Arabidopsis* (Wobbe et al. 1998). The protein p8 localized to the nucleus, whereas p9 protein was present in both cytoplasm and nucleus (Cohen et al. 2000). The GFP fusions of p8 and p9 did not form punctate patterns in the cell walls of infected cells or appear to be associated with the cytoskeleton (Cohen et al. 2000). The TCV p8 protein was found to interact with a protein from *Arabidopsis* containing two possible transmembrane helices and several potential phosphorylation sites (Lin and Heaton 2001). In the case of *Tomato bushy stunt virus* two internal overlapping ORFs encoding p22, a soluble protein and p19, a membrane-associated protein have been identified (Scholthof et al. 1995), whereas p22 is needed for cell-to-cell movement. The p19 influcences the lesion diameter suggesting an auxillary host-dependent activity in virus movement (Scholthof et al. 1995; Chu et al. 1999).

Olive latent virus 1 (OLV-1), a definitive species of the genus *Necrovirus*, family *Tombusviridae* infects olive, citrus and tulip. The proteins encoded by ORF3 and ORF4, 8 and 6 kDa respectively are involved in the cell-to-cell movement, whereas the 30 kDa CP coded by ORF5 aids in long distance translocation of OLV-1 (Pantaleo et al. 1999). The cellular localization of the 8 kDa (p8K) and 6 kDa (p6k) proteins was monitored by cell fractionation, gold immunotagging and by fusion with the green fluorescent protein (GFP) marker. The time course of accumulation of p8K was determined by using Western blot analysis of total protein extracts from *Nicotiana benthamiana* infected by OLV-1. In these extracts p8K was detectable from 2 d.p.i (days postinoculation) till 6 d.p.i, reaching the maximum at 4 d.p.i (Fig. 2.15). In situ localization of the p8K using specific antisera, very light labeling at 3 d.p.i in the nuclei was seen in infected cells exposed to p8k antiserum. A much stronger lebeling on the fibrous structures in the cytoplasm occurred. Nuclei of cells at 4, 5 and 6 d.p.i were not labeled. In contrast labeling was abundant and largely restricted to bundles of cytoplasmic filaments (Fig. 2.16). Expression in *N. benthamiana* protoplasts of p8K-GFP and p6K-GFP 18 h after transformation was studied. The fusion p8k-GFP was seen throughout the cytoplasm and the nucleus. On the other hand, p6K-GFP appeared to be associated with nuclear envelope and with structures adjacent to the plasma membrane which was tentatively identified as ER. The OLV-1 p8K association with nuclei resembled that of the p8 MP of *Turnip*

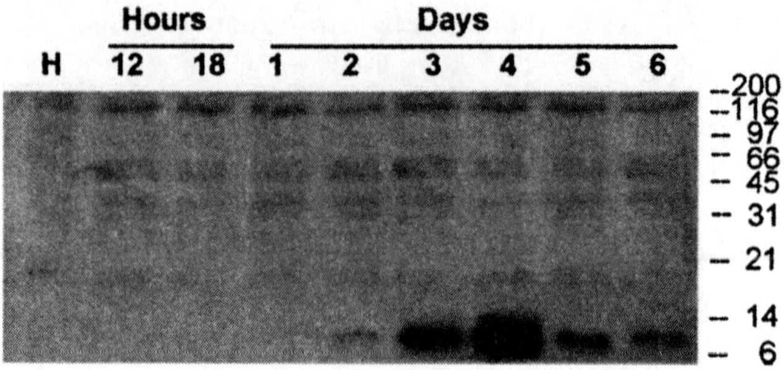

Fig. 2.15 Immunodetection of *Turnip crinkle virus* movement protein (p8K) by Western blotting from 2 dpi till 6 dpi reaching its maximum concentration at 4 dpi
Molecular weight markers (kDa) are indicated at right extremity. H: Healthy control; Sampling time in hours and days post-inoculation (dpi) are indicated at the top. (Courtesy of Castellano et al. 2005; Springer Science, Heidelberg, Germany)

crinkle virus (TCV). The presence of OLV-1 p8K in the nucleoplasm, as revealed by immunolabeling experiments, seems to be transient (Castellano et al. 2005).

Potyviral Movement Proteins

Potyviruses are the most numerous causing several devasting diseases in various economically important crops. The viral genome (about 10,000 nucleotides) encodes a polyprotein that is processed by three proteinases to synthesize ten mature viral protein. The proteinases are P1, HC-Pro and NIa and mature proteins are designated P1, HC-Pro, P3, 6K1, CI, 6K2, VPg, NIa-Pro NIb and CP. All viral proteins are involved at one or other stages of the viral cycle. Three essential HC-Pro regions responsible for different functions have been recognized. The N-terminal part controls virus transmission by aphids, virulence (symptoms severity), genome amplificaion and virus accumulation. The central region controls both long-distance movement, replication maintenance functions. The C-terminal region of HC-Pro is a cysteine-type proteinase and has a role in virus cell-to-cell movement (Syller 2006). The helper component-proteinase (HC-Pro) is a multifunctional protein involved in a number of processes of the virus life cycle (Revers et al. 1999). HC-Pro is involved in the viral genome amplification (Klein et al. 1994), short and long-distance movement (Kasschau and Carrington 2001) and aphid transmission (Huet et al. 1994). Further, HC-Pro has also been shown to be a pathogenicity determinant (Revers et al. 1999; Redondo et al. 2001). Its role in the synergism observed in mixed infection (Vance et al. 1995) and ability to suppress plant defense responses based on gene silencing (Kasschau and Carrington 2001) have also been demonstrated. The coordinate activity of HC-Pro with capisd protein (CP) for transmission of potyviruses by aphid vectors and virus movement appear to indicate a direct interaction between these proteins. By using HC-Pro-specific monoclonal antibodies in a two-site triple antibody sandwich (TAS)-ELISA assay, an interaction between HC-Pro and CP of *Lettuce mosaic virus* (LMV) was demonstrated in

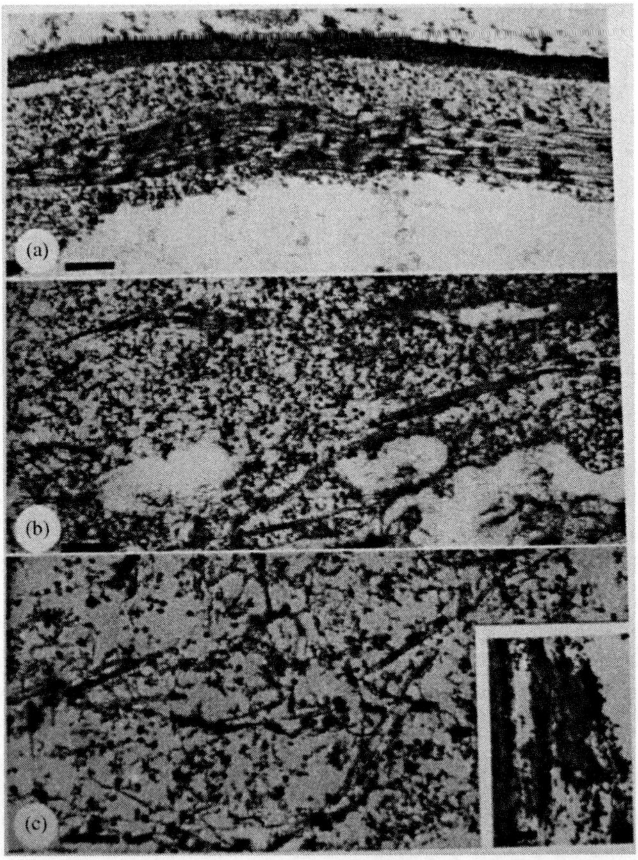

Fig. 2.16 Immunolocalization of *Olive latent virus* movement protein p6K in locally infected *Nicotiana benthamiana* cell at 4 (**a**), 5 (**b**) and 6 (**c**) days post- inoculation (dpi) by exposure to gold-labeled serum to p6K
Note that labeling by p6K is restricted mostly to the cytoplasmic fibrous structures; Inset in C shows filaments in a control cell. Bar = 200 nm. (Courtesy of Castellano et al. 2005; Springer Science, Heidelberg, Germany)

extracts of LMV-infected leaves, as well as for two other potyviruses *Plum pox virus* (PPV) and *Potato Virus Y* (PVY) (Roudet-Tavert et al. 2002).

Analysis of naturally occurring non-transmissible variants of potyviruses and site-directed mutagenesis revealed two conserved motifs in HC-Pro and one in CP. The terminal part of HC-Pro including the conserved motif KITC was apparently involved in binding to aphid mouth parts (Blanc et al. 1998). The second motif, PTK in HC-Pro was implicated in binding to CP (Peng et al. 1998). The involvement of HC-Pro in RNA silencing suppression activity has intensified several studies on the virus product. Transient expression of viral suppressors of RNA silencing, using *A. tumefaciens* has been demonstrated to be useful (Voinnet et al. 1999). The functionality of HC-Pro of *Plum Pox virus* (PPV) in standard silencing suppression assays has been shown (González-Jara et al. 2005). The structure of HC-Pro from *Tobacco etch virus* (TEV) was determined using analytical ultracentrifugation and single particle electron microscopy combined with three-dimensional reconstruction (Ruiz-Ferrer et al. 2005). A transient *Agrobacterium*-mediated expression system was used to produce PPV HC-Pro in *Nicotiana benthamiana* leaves from constructs that incorporated the 59 region of the viral genome. The expressed PPV HC-Pro assisted aphid transmission of purified virus particles in a sequential feeding assay and also complemented transmission-defective variants of PPV. The HC-Pro of another potyviruses *Tobacco etch virus* (TEV) was expressed and found to be functional for aphid transmission. The protocol developed in this study has the potential for different application to investigate different functions of HC-Pro (Goytia et al. 2006).

The cell-to-cell movement of two potyviruses *Bean common mosaic necrosis virus* (BCMNV) and *Lettuce mosaic virus* (LMV) was demonstrated to be dependent on HC-Pro. The interaction of this protein with viral CP was revealed by microinjection techniques using *E. coli* expressed proteins. Both proteins trafficked from cell-to-cell, induced an increase of plasmodesmatal SEL and facilitated cell-to-cell movement of viral RNA. As the mutations in the C-terminal part of HC-Pro abolished cell-to-cell movement of BCMNV and LMV, this part HC-Pro appears to have a role in short distance movement of potyviruses (Rojas et al. 1997). In addition to the cell-to-cell movement, the involvement of HC-Pro in different steps of the viral cycle such as aphid transmission, replication, systemic movement and as suppressor of post- transcriptional gene silencing (PTGS) has also been indicated by various studies (Plisson et al. 2003).

The cylindrical inclusion (CI) protein is the movement protein for potyviruses. The multifunctional nature of this protein with roles in virus replication, short-and long-distance movement of the virus was indicated by alanine scanning mutagenesis (Carrington et al. 1998). The CI proteins have been shown to be associated with plasmodesmata. In addition to CIs, potyviral CP is also necessary for cell-to-cell movement of *Tobacco etch virus* (TEV) has no specific protein dedicated to virus movement. The TEV-CP functions in both cell-to-cell movement and systemic transport of the virus (Dolja et al. 1995). A potential link between virus accumulation and cell-to-cell movement was identified, when the eukaryotic translation factors eIF4E and eIF (iso) 4E, which are required for potyvirus accumulation in plants (Lellis et al. 2002; Nicaise et al. 2003). These factors were also found to aid in virus cell-to-cell movement (Gao et al. 2004).

Tospoviral Movement Proteins

Tomato spotted wilt virus (TSWV) is the type member of the genus *Tospovirus* in family *Bunyaviridae* which has plant and animal-infecting members. The NSm protein has been identified as the MP and the gene encoding this protein is not present in any of the animal viruses. This protein is expressed early and transiently in infected plants and assembles into tubular structures that extend from the plasma membranes of protoplasts and insect (thrips) cells. The tubules extend through plasmodesma, possibily conducting the noneveloped nucleocapsid of TSWV to adjacent cells. The

NSm protein could be detected in cell wall and organ subcellular fractions from infected plants (Kormelink et al. 1994). The NSm protein was detected also in the midgut epithelium of *Frankliniella occidentalis* in the L2 development stage and in salivary glands and midgut muscle cells in the adult development stage (Storms et al. 1995; Storms 1998). The TSWV NSm specifically interacted with TSWV N protein and binds ssRNA in a non-sequence-specific manner (Soellick et al. 2000).

The NSm protein from TSWV was utilized as bait in yeast two-hybrid interaction trap screenings. Based on this, a protein of unknown function, At-4/1 was isolated from *A. thaliana*. The PABs against bacterially expressed At-4/1 were employed in Western blot analysis of protein extracts isolated from different plant species. The homologs of At-4/1 appeared to be encoded by many vascular plants. At-4/1 was fused to GFP and corresponding expression vectors were used in particle bombardment and agroinfiltration assays. At-4/1 assembled impuctate spots at cell periphery as visualized using confocal laser scannings. Accumulation of this protein intracellularly in only one half of a bombarded epidermal cell was characteristically seen. Further, it moved from cell-to-cell, forming a twin-structured bodies seemingly located at both orifices of plasmodesmatal pore. At-4/1 colocalized with the MP-TGBp3 known to reside in ER-derived membrane structures located in close vicinity to PD. The results demonstrated that At-4/1 belongs to a new family of plant proteins capable of directed intra-and intercellular trafficking (Paape et al. 2006).

Geminivirus Movement Proteins

The geminiviruses transmitted by whitefly vectors have bipartite genomes. The DNA A of the bipartite genome encodes genes for encapsidation and replication, whereas DNA B contains genes required for viral movement (Harrison 1985). Mutations in *BR1* or *BL1* genes in DNA B abrogates systemic moement but not the replication of *Bean dwarf mosaic virus* (BDMV) belonging to the genus *Begomovirus* (Noueiry et al. 1994). The presence of BL1 proteins in cell wall subcellular fractions from systemic hosts was detected by Von Arnim et al. (1993). BL1 potentiates its own cell-to-cell movement, modifies the SEL of plasmodesmata and moves ds-DNA from cell to cell. BR1 transports ss-and ds-DNA from nucleus (Noueiry et al. 1994). These two proteins are involved in the movement of newly formed viral DNA from nucleus to adjacent cells. BV1 is required for nuclear transport, while BL1 is necessary for movement through the plasmodesmata (Boulton et al. 1993).

Modification of Plasmodesmata (PD)

The process of increasing the size exclusion limit (SEL) of plasmodesmata is designated gating (Wolf et al. 1989) and this is the characteristic feature of movement proteins (MPs) of plant viruses. Several movement complexes (MCs) are found to be more complex than originally visualized and the different functions required for movement may be performed by separate proteins. Gating seems to be a basic requirement for cell-to-cell movement of non-tubule forming viruses. Actin is involved in deploymerizing the actin, leading to increase in SEL. In contrast, callose deposition has been shown to close PD during defense and wound responses (Roberts and Oparka 2003). The interaction between PVX TGB2 and proteins that interact with β-1,3-glucanase, a callose-degrading enzyme has been suggested to a strategy of PVX employed to gate PD to accelerate callose degradation (Fridborg et al. 2003). Myosin spokes lining the PD channel link actin to the plasma membrane. The myosin / kinesin-like protein At-4/1 found to interact with TSWV M may be involved in trafficking to the PD (von Bargen et al. 2001). In the case of TMV MP, its interaction with pectin methyl esterase (PME) may regulate the activity of PME, loosening the cell wall around PD. This may result in permitting the PD to open more easily. Alternatively the MP may recruit additional PME to PD to assist gating rather than relying on PME for targeting (Chen et al. 2005).

2.4.1.2 Long-Distance Movement of Viruses

Plant viruses can move over long distances and become systemic infecting plant organs away from the site of infection / inoculation. Majority of plant viruses use phloem for long-distance movement, whereas a few viruses such as *Lettuce necrotic yellows virus* (LYNV) and *Rice yellow mottle virus* (RYMV) are able to move through xylem (Hull 2002). Proteins and other macromolecules are actively transported in the phloem. Fisher et al. (1992) suggested that proteins are synthesized and processed in the companion cells, loaded into sieve elements of minor veins of source tissues and then loaded into companion cells in sink tissues. The movement of plant virus from cells locally infected, may be achieved due to operation of different mechanisms singly or in combination. There is a signal of silencing that spreads between cells and also to long distances from the lower to upper leaves. Long-distance spreading depends on an RNA-dependent RNA polymerase (RDR), whereas short-distance movement of the signal is not dependent on RDR (Himber et al. 2003). Silencing signal may be linked to a role in disease resistance. In contrast, the suppressor of RNA silencing encoded by plant viral genomes may lead to enhancement of susceptibility to systemic infection by viruses.

The systemic spread (movement) of plant viruses involves cell-to-cell movement into, through and out of various cell types of the vascular system, in addition to long-distance transport through sieve elements. Systemic spread may be blocked by different barriers which may be constitutive or formed as a response to infection. These barriers in some plants may allow limited or reduced rates of movement depending on the level of resistance to the viruses concerned. Reduction in the rate of cell-to-cell movement in the inoculated leaves may result in lack of systemic infection as in the case of M strain of *Cucumber mosaic virus* (M-CMV) which cannot infect zucchini squash (*Cucurbita pepo*) systemically (Shintaku and Palukaitis 1990). This block in long-distance movement has been mapped to specific sites in the viral capsid protein (CP) and it was also associated with delayed cell-to-cell movement in the inoculated leaves (Wong et al. 1999). However, coinfection of zucchini plants with M-CMV and *Zucchini yellow mosaic virus* strain A (ZYM-A) allowed M-CMV to move systemically as shown by tissue print analysis. Gel analysis of RT-PCR products obtained from nucleic acids extracted from the upper leaves of plants doubly inoculated with M-CMV and ZYMV, showed that M-CMV CP gene-specific production was absent in extracts from plants inoculated with M-CMV alone. On the other hand, the expected size band for M-CMV CP was detected in plants doubly inoculated with M-CMV and ZYMV. It is hypothesized that ZYMV-encoded HC-Pro or other proteins of ZYMV involved in systemic movement may also be able to help M-CMV to move into new uninoculated leaves (Fig. 2.17) (Choi et al. 2002).

In the case of potyviruses, the central region of HC-Pro is required for long-distance movement. The long-distance movement of *Tobacco etch virus* (TEV) was abrogated due to a mutation in this region of HC-Pro, whereas only minor defects occurred in genome amplification and cell-to-cell movement (Cronin et al. 1995). The involvement of *Plum pox virus* (PPV) HC-Pro in long-distance movement in tobacco plants was reported. PPV can systemically infect several species of *Nicotiana*. However, it replicated in tobacco leaves, but could not spread systemically in this host. The failure of PPV to move long distance could be complemented in transgenic tobacco plants expressing the 5′-terminal region of the genome of TEV that naturally infects tobacco systemically. The TEV transgene with a mutation in the HC-Pro coding sequence could not facilitiate systemic movement of PPV in tobacco. Thus the results confirmed that biologically active HC-Pro is essential for long-distance movement of PPV in tobacco. Further, the activity of HC-Pro in the systemic movement of potyvirus can be highly host-specific (Sáenz et al. 2002). The adaptive defense against RNA and DNA viruses is considered as the biological function of RNA

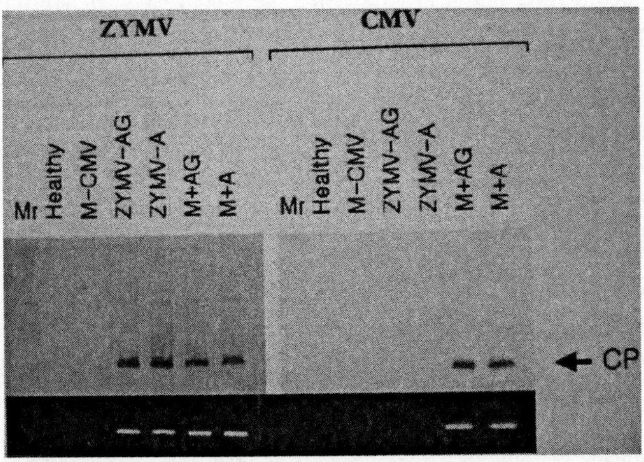

Fig. 2.17 Detection of CP protein and RNA accumulation in zucchini squash plants infected by ZYMV strains A and AG and/ or CMV strain M (M-CMV) by Western blotting (*upper panels*) and gel analysis of RT-PCR-specific products (*lower panels*)
Upper panels: The specific antisera to either CMV CP or ZYMV are indicated at the top; position of CPs of ZYMV and CMV are indicated by an arrow. Mr: Protein marker; *Lower panels*: The specificity of the primers used for RT-PCR is indicated at the top; the agarose gel of the RT-PCR products of ZYMV (*left*) and CMV (*right*) can be seen. (Courtesy of Choi et al. 2002; Society of General Microbiology, Reading, UK)

silencing which has been demonstrated first in plants (Covey et al. 1997) and later in the animal kingdom (Li et al. 2002). As a counter defense viruses encode proteins such as helper component-proteinase (HC-Pro) and *Cucumber mosaic virus* 2b that suppress RNA silencing at various steps in the pathway (Anandalakshmi et al. 1998; Kasschau and Carrington 1998).

Transcriptional gene silencing (TGS) is recognized by the absence of transcription from silenced gene and occurs where homology exists between promoter sequences of affected genes. In contrast, PTGS genes contain homologous sequence within the coding region. Although PTGS genes are actively transcribed, only low levels of steady-state mRNA can be detected. PTGS is a homology dependent RNA degradation process that may target RNA exclusively in the cytoplasm of host cells. Plant viruses replicating in the cytoplasm can function as both initiator and target of PTGS. Many plant viruses encode protein that can suppress PTGS as a counter-defensive strategy. The 2b protein of *Tomato aspermy virus* (TAV) is targeted in tobacco by an alternative host defense mechanism (Li et al. 1999). The HC-Pro encoded by potyviruses is able to "switch off" gene silencing in plants wherever it is expressed (Brigneti et al. 1998). In contrast, the 2b proteins encoded by TAV and CMV, do not have any effect in tissues where PTGS is established but are able to prevent initiation of gene silencing in newly emerging tissues (Brigneti et al. 1998; Li et al. 1999). The intracellular distribution of CMV 2b protein (Cmv2b) was examined by fusing it with GFP to understand the mechanism(s) of counter-defensive suppression effect of Cmv2b. The Cmv2b protein localized to the nuclei of tobacco suspension cells and whole plants via an arginine-rich nuclear localization signal. The nuclear targeting of 2b protein was required for the efficient suppression of PTGs, indicating that PTGS might be blocked in the nucleus. The results also indicated that the PTGS suppressor activity is important, but not sufficient, for virulence determination by Cmv 2b protein (Lucy et al. 2000).

A viral suppressor of RNA silencing that also plays a unique role in induction of viral disease has been identified. The suppressor, p69 encoded by *Turnip yellow mosaic virus* (TYMV) is essential for virus movement and it also influences viral contents and symptom severity in infected plants. However, it is dispensable for viral replication in single cells (Bozarth et al. 1992; Tsai and Dreher 1993). A virulence factor encoded by TYMV, p69 was found to suppress the small interfering RNA (siRNA) pathway, while it promoted the micro RNA (miRNA) pathway in *A. thaliana.* Plant expressing p69 exhibited disease-like symptoms in the absence of virus infection, suggesting a novel mechanism for viral virulence by promoting miRNA-guided inhibition of host gene expression (Chen et al. 2004).

The tripartite genome of *Cucumber mosaic virus* (CMV) encodes replicase function in RNA1 and RNA2, with RN3 encoding MP and CP. For local cell-to-ell movement, both MP and CP are essential and for systemic movement in squash (Canto et al. 1997; Kaplan et al. 1997). The role of CP in systemic movement was shown to be host-specific. CMV can infect cucumber systemically, whereas *Tomato aspermy virus* is not able to infect cucumber systemically (Salánki et al. 1997). Spontaneous, compensatory, single amino acid substitutions in the CP of CMV, have been demonstrated to confer the ability in engineered CP mutants to infect squash. A panel of 16 single amino acid substitution mutants was screened for the ability to sytemically infect squash. These mutants, however, could infect and move systemically in tobacco. Although all the mutants produced symptoms in squash cotyledons comparable to wild-type, 12 of the mutants were impaired in their ability to systemically infect growth chamber grown-squash plants at 19°C. Other mutants infected only low percentage of plants inoculated. This experiment indicated that single amino acid substitutions in the CP altered the ability of the virus to move systemically in squash. Five of the mutants reverted to a wild-type phenotype with >90% of the inoculated plants exhibiting systemic infection. This reversion was found to be due to compensatory mutations in the capsid protein gene conferring systemic movement (Thompson et al. 2006).

The systemic movement of *Cucumber mosaic virus* (CMV) in cucumber plants was investigated by analyzing the structures that is translocated and its putative interactions with phloem components in the phloem exudates (PE) samples. The movement of CMV as intact virus particles was shown by the rate zonal centrifugation and electron microscopy observations of PE from CMV-infected plants. The virus particles interacted with a PE protein, p48 which was found to be cucumber homolog of phloem protein 1 (PP1) from pumpkin. The PP1 from pumpkin has been shown to have PD gating property and ability to translocate in sieve tubes. Interaction between enapsidated CMV particles and p48 protein appeared to make the viral RNA more resistant to RNAase. The results indicated that the interaction with p48 protein modified the CMV particle structure and suggested that the virus particles may interact with cucumber homolog of PP1 during translocation in sieve tubes of cucumber (Requena et al. 2006).

Bamboo mosaic virus has a satellite RNA (sat BaMV) which has a single ORF for a nonstructural RNA-binding protein 20 that facilities the long-distance movement of satBaMV in *N.benthamiana*. Domain mapping indicate that the self-interacting domain of P20 protein overlaps the RNA-binding domain in the N-terminal arginine-rich motif (ARM) P20. The deletion of the ARM resulted in the loss of self-interaction ability of P20 in vitro and in vivo and impaired its intracellular targeting and efficient cell-to-cell movement in *N. benthamiana* leaves. Furthermore, RNA and protein accumulation of ARM deletion mutant of sat BaMV was significantly reduced in leaves systemically coinfected with BaMV and sat BaM (Palani et al. 2006).

Long-distance movement of viruses is essential for systemic infection of plants. Mutations within the HC-Pro coding region of *Wheat streak mosaic virus* (WSMV) were introduced by misincorporation during PCR. For nonsynomous substitutions, including the one that reverted consistently to wild-type, led to attenuation of systemic infection. Mutants bearing the GUS reporter gene were evaluated for their ability to establish primary infection foci. In vitro assays showed that HC-Pro proteinase activity was noted in the case of mutants that could not establish foci of infection. The results indicated that a complete HC-Pro deletion mutant can infect plants systemically and that certain substitutions in this dispensable gene probably prevented systemic infection by WSMV via interference (Stenger et al. 2006).

Tombusvirus P19 is a suppressor of silencing that is not required for virus replication in isolated plant cells, but is required for extensive viral invasion of leaves at different position in plants (Havelda et al. 2003). Likewise, the potyviral HC-Pro and cucumoviral 2b proteins are silencing suppressors essential for systemic infection by viruses (Ding et al. 1995; Anandalakshmi et al. 1998). The 2b protein is able to block both production and transmission of long-distance silencing signal (Guo and Ding 2002). A correlation between silencing suppression and ability to mediate long-distance movement was reported for HC-Pro protein from *Tobacco etch virus* (Kasschau and Carrington 2001).

The strategies of the host plant to defend itself against virus replication and spread is considered to be very complex. A plant host defense system targeting viral RNA with extreme sequence specificity is mediated by RNA referred to the RNA interference (RNAi). Plants have multiple RNA silencing pathways with diverse biological roles. However, viruses may modify the final outcome by their expression of proteins, suppressors of RNA silencing that defeat the host defense system depending on gene silencing mechanism (Baulcombe 2004). RNA silencing occurs via a multistep enzymatic pathway including an RNA dependent RNA polymerase (RdRP, currently termed as RDR) and several other enzymes. Specific RDRs may be needed by plants to recognize different viruses. RDR1 is required for protection of plants against tobamoviruses, whereas RDR6 protects the plants against CMV (Nelson and Citovsky 2005). During infection with PVX, RDR6 prevented systemic infection, but not local infection (Schwach et al. 2005). RDR is needed for cells to respond to a systemically moving silencing signal. Mutant TMVs expressing 126 kDa protein silencing suppressors of varying concentrations were also not changed during cell-to-cell movement of the virus (Ding et al. 2004). Delinking of RNA silencing suppressor activity from cell-to-cell movement was demonstrated for the P15 supressor from *Peanut clump peculovirus* (Dunoyer et al. 2002). Exposure of plants to abiotic stress may block viral systemic movement. In contrast, stress may help virus movement in certain pathosystems. Host heat-shock protein HSP70 and virus-encoded HSP70-related protein appear to aid viral and / host macromolecular transport through plasmodesmata (Aoki et al. 2002; Aparicio et al. 2005).

Bamboo mosaic virus (BaMV) a member of *Potexvirus* is a single-stranded positive sense RNA virus with flexuous rod shaped particles. The tertiary structure in the 3′-untranslated region (3′-UTR) of the viral RNA is involved in the minus strand RNA synthesis. Proteins present in the RNA-dependent RNA polymerase (RDR) fraction of BaMV-infected leaves interacted with the radio-labeled 3′-UTR probe in electrophoretic mobility shift assays (EMSA). Two host cellular factors p43 and p51 appeared to interact specifically with the 3′-UTR of BaMV RNA. Extracts containing p51 specifically down-regulated minus-strand RNA synthesis, when added to in vitro RDR assays. p43 was shown to be a chloroplast phosphoglycerate kinase (PGK). In virus-induced silencing (VIGS) experiment, PGK was found to be essential for efficient BaMV accumulation. Real-time PCR and Western blot analyses revealed that reduction of chloroplast PGK resulted in concomitant decrease in the accumulation of BaMV coat protein. The interaction between the chloroplast PGK and the BaMV RNA appeared to suggest that chloroplast PGK possibly facilitate the viral RNA in targeting it to the chloroplast membrane (Lin et al. 2007).

Apple chlorotic leaf spot virus (ACLSV) belonging to the genus *Trichovirus*, has ORF2 which encodes the 50 kDa protein (p50) functioning as the movement protein. The transgenic *Nicotiana occidentalis* expressing a functional ACLSV p50 showed specific resistance to *Grapevine berry inner necrosis virus* (GINV), another member of *Trichovirus*. In contrast, a transgenic plant line expressing $\Delta A'$ (deletion of the C-terminal 42 amino acids) allowed GINV to spread in inoculated leaves, but not into upper leaves, indicating that ΔA' expressed in transgenic plants interfered with the long-distance movement of GINV. Immunohistochemical analysis of GINV-inoculated leaves of $\Delta A'$ plants revealed that GINV could invade into phloem parenehyma cells through bundle sheath of minor veins. It appears that the long-distance movement of GINV is possibly blocked between the phloem cells and sieve element and/or within sieve element rather bundle sheath-phloem interfaces. Accumulation of p50 on the parietal layer of sieve elements and on sieve plates, was revealed by immunogold electron microscopy using an anti-P50 antiserum. Interference of both cell-to-cell and long-distance movements of the virus may contribute to the resistance to GINV in *N. occidentalis* (Yoshikawa et al. 2006).

Molecular basis of long-distance movement of plant viruses via vascular elements has been studied only in a few pathosystems Using a modified *Tobacco mosaic virus* (TMV) expressing green fluorescent protein (GFP), the systemic vascular invasion routes of TMV in *Nicotiana benthamiana* were investigated. TMV could enter minor, major or transport veins directly from non-vascular cells to cause systemic infection (Cheng et al. 2000). Only a few host genes have been shown to be responsible for systemic spread of viruses. Inhibition of long-distance movement of *Turnip vein-clearing virus* was attributed to a vascular specific Gly-rich protein which may probably induce callose accumulation in the phloem cell walls (Ueki and Citovsky 2002). Restriction of long-distance-movement of *Tobacco etch virus* was reported to be due to two proteins RTM1 and RTM2 functioning in the phloem of *Arabidopsis* (Chisholm et al. 2000, 2001). From viral transport suppressor screen using virus-induced gene silencing (VIGS) in *N. benthianiana*, RPN9, a regulatory subunit of the 26S proteasome was identified. Silencing *RPN9* inhibits the systemic transport of two taxonomically distinct viruses TMV and *Turnip mosaic virus* (TuMV). The 26S proteasome is a highly conserved eukaryotic protease complex controlling many fundamental biochemical processes. The inhibition of viral systemic transport, following RPN9 silencing was found to be largely due to alterations in the vascular tissue. In RPN9-silenced plants, production of extra leaf veins with increase in xylem and decrease in pheloem was in part through regulation of auxin transport and brassinosteroid signaling, which are considered to be essentially required for vascular formation (Jin et al. 2006).

2.4.2 Symptom Expression

Virus infection may remain localized, if the plant species shows hypersensitive response (HR) resulting in the formation of necrotic lesions. The movement of the virus may be limited to the primarily infected and some adjacent cells, since the virus is not able to become systemic. On the other hand, when the virus can move long distance through phloem and infect other plant organs, characteristic symptoms are expressed. But in some plant species, although the virus is able to move systemically, no visible symptom is induced, such infection is referred to as latent. Further, under certain conditions, the infected plants may exhibit severe symptoms in the early stages and the symptoms may become mild or almost absent. Such recovery from severe infection confers significant level of resistance to infection by the same virus or closely related virus. This phenomenon designated "cross-protection" has been exploited as a disease management strategy. Unlike the fungal and bacterial pathogens, viruses do not produce any enzymes or toxins that act on the cell wall or other cellular contents causing structural derangements for deriving their nutrients. In contrast, the viruses utilize host machinery for synthesizing the molecules required for the synthesis of viral genome and capsid and virus-encoded proteins. There seems to be no viral gene(s) exclusively involved in induction of symptoms in susceptible plants (Hull 2002). The symptoms of the virus disease are considered to reflect the extent of physiological derangements following infection. The host synthetic machinery is under the control of invading virus and the requirement of virus replication has to be satisfied at the expense of the needs of the host plant. In some host-virus pathosystems, evidences indicate that determinants of symptom production are located in the viral genomes.

The viral genome may be composed of one, two or more nucleic acid species. Experiments on reassortment of genomic segments have shown that specific disease symptoms may be attributed to particular genome segments. Stanley et al. (1985) reported that symptom determinants were located in the DNA1 of *Cassava latent virus*. In the case of *Cucumber mosaic virus* (CMV), RNA 2 or RNA 3 determined the reaction in some host plant species, whereas the interaction between these RNAs controlled the symptom development in other host plant species (Rao and Francki 1982). A reduced rate of cell-to-cell movement of CMV in the inoculated leaves resulted in lack of systemic infection and absence of symptoms. Such a block in long-distance movement was mapped to specific sites in the viral CP (Wong et al. 1999). The virulence of *Pea seedborne mosaic virus* (PSbMV) was shown to be associated with the 21 kDa genome-linked protein (VPg). Pea cultivars containing genes *sbm-1*, *sbm-3* and *sbm-4* closely linked on chromosome 6 showed resistance to PsbMV pathotypes 1, 3 and 4 respectively. The *sbm-1* gene may interfere with the VPg activity that may be required for the replication of PSbMV (Keller et al. 1998).

The viral CP gene has been shown to function as elicitor of host defense resulting in HR in certain host plants. The CP gene of TMV is involved in the elicitation of HR in plants with the N' gene originating from *Nicotiana sylvestris* (Saito et al. 1987). Elicitation of HR was attributed to independent amino acid substitutions (Culver et al. 1991). The elicitor function of TMV-CP may be drastically altered by a single amino acid substitution from a strong elicitor of HR in N′ gene-containing tobacco to being a weak elicitor allowing systemic spread of necrosis. Specific amino acid substitutions and deletions in the TMV-CP affected the development of chlorotic symptoms. Furthermore, the mutants with no change in the C-terminus of

CP produced more intense chlorosis in tobacco (Culver and Dawson 1989). Induction of systemic chlorosis in tobacco was associated with the CP gene of *Cucumber mosaic virus* strains. Changes in the amino acid at position 129 in the CP accounted for variations in the symptom intensity (Shintaku et al. 1992). The appearance of the 66-kDa gene product, as revealed by immunoblotting, was demonstrated to be responsible of symptoms production by *Cauliflower mosaic virus* in tobacco (Goldberg et al. 1999). The differences in the severity of symptoms induced by viruses may be attributed to the functions of viral genes. The TMV mutants LII and LIIR do not induce visible symptoms when compared to parent strain. A change from Cys to Tyr at amino acid position 348 of the p26 kDa and 183 kDa proteins was considered to be responsible for the loss of symptom induction by the mutants (Nishiguchi et al. 1985). Symptom determinants of the Holmes masked (M) and U1 strains of TMV were mapped to the 126 in 183 kDa protein (Shintaku et al. 1996). The p23 protein encoded by *Citrus tristeza virus* (CTV) was reported to be involved in expression of some symptoms in Mexican lime (Ghorbel et al. 2001).

The relationship between the virus pathogenicity determinant domain and induction of mosaic symptoms by *Pepper mild mottle virus* (PMMV) was investigated. Mutants were obtained by a base substitution in the 130 K replication protein gene causing an amino acid change or a truncation of the 3 (prime) terminal pseudoknot structure. The mutant with amino acid residue substitution at 349 position alone did not induce any visible symptom, although the mutant reached relatively high concentration. In contrary, the pseudoknot mutants induced symptoms, but exhibited low accumulation indicating that virus accumulation need not necessarily reflect the disease symptom severity. The activity to suppress posttranscriptional gene silencing (PTGS) was impaired in the mutant that failed to induce symptom. This finding suggested that the development of symptoms may be controlled by combat between host PTGS and its suppression by the 130 K replication protein rather than virus accumulation (Tsuda et al. 2007).

Mosaic is a major type of symptom induced by viruses with RNA and DNA genomes. *Cauliflower mosaic virus* (CaMV) with dsDNA genome induces chlorotic symptoms. The gene VI protein (P6) has been shown to be a virulence factor eliciting chlorosis in crucifers. The gene VI as a transgene produced virus-like symptoms in transformed plants including *Arabidopsis* (Cecchini et al. 1997). The primary role for P6 in the CaMV infection cycle is to modify the host translation machinery to facilitate the translation of the polycistronic CaMV 35S RNA. This function for P6 has been designated the translation transactivator (TAV) function. An unusual variant of P6 derived from CaMV strain D4 did not induce chlorosis upon transformation into *A. thaliana.* The level of D4 P6 produced in transgenic *Arabidopsis* line D4-2 was comparable to the amount reached in transgenic plants homozygous for W260 and CM 1841. Two versions of P6 induced strong chlorotic symptoms and stunting in *Arabidopsis.* A complementation assay proved that P6 expressed in the D4-2 line was functional, since it was able to support the systemic infection of a CM1841 mutant containing a lethal frame-shift mutation with gene VI. The TAV activity of D4 P6 was at a level comparable to W260 P6. It appears that the TAV function of

P6 may have only a minor role in the development of chlorotic symptoms in CaMV infected plants (Yu et al. 2003b).

Combination of different viral genes, noncoding regions of the viral genomes and host plant factors may function as symptom determinants. Hybrids of *Cymbidium ringspot* (CymRSV) and *Carnation Italian ringspot* (CIRV) tombusviruses were employed to identified viral symptom determinants responsible for the generalized necrosis in infected plants. The necrotic response of *N. benthamiana* was demonstrated to be due to the products of ORF1 (p35) and p19 of the hybrids. Symptoms on *N. benthamiana* infected by hybrids were entirely different. Hybrids containing chimeric ORF1 were not able to induce lethal necrosis, even if the level of viral replication was not reduced significantly. If a wild type (Wt) p33 (product of ORF1) of CymRSV was provided in trans in transgenic plants expressing p33 and its readthrough product p92, the level of necrosis characteristic to tombusvirus infection was restored. Further, the expression of p33 by a PVX vector in *N. benthamiana* caused severe chlorosis and occasionally necrosis, indicating the requirement of p33 in wild-type symptoms of tombusviruses. Production of the necrotic phenotype requires the presence of the wild-type p33 in addition to the p19 protein of tombusviruses (Burgyán et al. 2000). The geminiviruses do not have a DNA polymerase function and hence they are dependent on host factors to amplify their genome. As a consequence, an early step in the geminivirus infection process, is the induction of the required replication machinery. Geminiviral replication enhancer (REn) proteins (also designated C3, Ac3 L3 or AL3) are able to enhance viral DNA accumulation and enhance infectivity and symptom expression as in *Tomato leaf curl virus* (TLCV) (Selth et al. 2005). *Hisbiscus chlorotic ringspot virus* (HCRSV) belonging to the genus *Carmovirus* encodes p27 (27 kDa protein) and two other in frame isoforms (p25 and p22.5). Infectivity assays indicated that p27 was a determinant of symptom severity. Appearance of symptoms in systemically infected kenaf (*H. cannabinus*) was delayed in the absence of p25, due to retardation in virus systemic movement. No effect on virus movement or symptom development could be seen due to mutations disrupting expression of p22.5 (Zhou et al. 2006).

The association of satellite RNAs (sat-RNAs) with plant viruses helps them in replication. The sat-RNAs require helper virus-encoded products for replication and systemic spread. Most of the sat-RNAs have either no effect or attenuate the symptoms produced by helper viruses. However, some of them may enhance the symptom intensity. SatRNA associated with *Panicum mosaic virus* facilitates systemic spread allowing the virus to invade previously restricted tissues, in addition to enhancing the virus titre in the infected plants (Scholthof 1999). The sat C associated with *Turnip crinkle virus* (TCV) significantly increased symptom intensity from mild stunting and chlorotic symptoms to severe stunting, crinkling and dark green leaves. In contrast, sat C attenuated the symptoms of a TCV mutant that expressed low levels of defective coat protein. The TCV virion levels were substantially reduced by the presence of satC or when two amino acids were inserted at the terminus of the CP, resulting in similar enhanced symptoms. Since TCV-CP was found to be a suppressor RNA silencing, increased levels of resultant free CP could augment silencing suppression, resulting in enhanced colonization of host plant tissues (Zhang and Simon 2003).

It is well known that virus infections disrupt normal physiology of the host plant resulting in symptoms such as mosaic, leaf curl and general growth abnormalities. Underlying many of these different disease responses are changes in host gene expression. It will be possible to link specific viral and cellular processes to the development of disease response(s), if host genes that display virus-induced alterations in their expression patterns are identified. Some genes may show transient variations in expression levels that correspond with advancing virus invasion front as in the case of lipoxygenase 1 and heat shock cognate protein. The genes encoding these proteins are repressed only within pea cells displaying active *Pea seedborne mosaic virus* (PSbMV) (Aranda and Maule 1998). In contrast, significant increases in the expression of heat shock protein (HSP) 70 or polyubiquitin were recorded in cells supporting virus replication (Escaler et al. 2000). More stable expression shifts in response to CMV infection were observed in the genes encoding catalase and NADP-dependent glyceraldehyde-3-phosphate dehydrogenase, even in uninfected tissues located ahead of advancing infection front (Havelda and Maule 2000).

The gene expression profiles of *Arabidopsis thaliana* ecotype Shahdara which is as susceptible as tobacco and tomato to TMV and permits rapid accumulation of the virus in both inoculated and systemic tissues (Dardick et al. 2000). Gene expression levels in Shahdara were monitored in inoculated leaves at 4 dpi and in systemically infected leaves at 14 dpi using cDNA microarrays created by the *Arabidopsis* Functional Genomics Consortium (AFGC). The cDNA microarray technology permitted the simultaneous screening of approximately 8000–10,000 *Arabidopsis* gene for their response to TMV infection. Based on expression analysis, 68 genes that displayed significant and consistent changes in expression levels, either up or down, in either TMV-inoculated or systemically infected tissues or both were identified. At 4 dpi, all of the identified TMV-responsive genes were induced in transcription. In contrast, 35 of the 53 genes regulated at 14 dpi showed repression. This increase in number of repressed genes at 14 dpi may reflect the general repression of host genes. The identified TMV-responsible genes encode a diverse array of functional proteins that include transcription factors, antioxidants, metabolic enzymes and transporters. The TMV-responsive genes may impact a wide range of host cellular functions that may reflect the biochemical and physiological changes involved in the development of this disease syndrome (Golen and Culver 2003).

In plants, multiple viral infections resulting in synergism or antagonism are seen under natural conditions, potyvirus-associated synergistic interactions being more frequent. Increase in titres of both viruses involved in synergistic interaction was reported in the case of *Wheat streak mosaic virus* (*Potyviridae*) and *Maize chlorotic mottle virus* (*Tombusviridae*) in corn (Scheets 1998). The mechanisms and viral products involved in viral synergistic interactions are yet to be studied in detail in most cases. The potyvirus *Tobacco etch virus* genome encodes a broad-range pathogenicity enhancer that transactivates replication of heterologous viruses such as *Cucumber mosaic virus* (CMV) (*Bromoviridae*, genus *Cucumovirus*) and *Tobacco mosaic virus* (genus *Tobamovirus*). The viral RNA silencing suppressor protein P1/HC-Pro appeared to be involed in the synergistic effect, possibly by decreasing the targeting of the other virus by RNA silencing machinery (Vance 1999). Another mechanism of synergistic interaction operating between CMV and *Potato virus Y* (PVY) was suggested to be a result of complementation of a movement deficiency by one of the viruses (Ryang et al. 2004).

Tomato is infected by *Tomato spotted wilt virus* (TSWV, *Bunyaviridae*) transmitted by thrips and *Tomato chlorosis virus* (ToCV, genes *Crinivirus*) transmitted by whiteflies. Simultaneous presence of high populations thrips and whiteflies of many tropical and subtropical regions worldwide, resulting in the mixed infection of tomato plants by TSWV and ToCV. The effects of synergistic interaction between TSWV and ToCV were investigated in tomato cv. Anastasia which was susceptible to ToCV but resistant to TSWV under normal conditions and also in tomato cv. Moneymaker which was susceptible to both viruses. A dramatic increase in ToCV (a phloem-limited virus) was observed in leaf tissues of Moneymaker tomato plants co-infected with TSWV. By using in situ RNA hybridization, confinement of ToCV to phloem cells of tomato plants either singly or doubly infected by the viruses was indicated. No difference in distribution of TSWV in plant tissues could be noted between singly and doubly infected plants (Table 2.5). In the case of Anastasia tomato, all plants inoculated with ToCV showed typical symptoms of virus infection, whereas one out of five plants inoculated with TSWV showed TSWV symptoms. When ToCV preceded TSWV inoculation, resistance to TSWV was compromised in Anastasia tomato plants. Similar breakdown of TSWV resistance was not observed in plants inoculated with ToCV and TSWV simultaneously. However, breakdown of TSWV resistance was seen in other commercial cultivars such as DRW 7456 and Vinchy RZ. All the three resistant cultivars contain the TSWV resistance gene *Sw-5*. The virus infection in the inoculated plant was confirmed by performing DAS-ELISA test. Preinoculation with ToCV resulted in susceptibility to TSWV, whereas coinoculations did not alter the resistance levels (Table 2.6) (García-Cano et al. 2006).

Table 2.5 Titres of viruses as determined by DAS-ELISA test at different intervals after inoculation of tomato cv. Moneymaker

Days after inoculation	Nature of plant samples (absorbance value at 405 nm)*		
	Inoculated with ToCV + TSWV	Inoculated with TSWV	Mock-inoculated
8 days	0.416 ± 0.030 a**	0.514 ± 0.039 a	0.091 ± 0.011 b
15 days	0.735 ± 0.046 a	0.456 ± 0.024 b	0.040 ± 0.002 c
20 days	0.676 ± 0.035 a	0.0771 ± 0.044 a	0.090 ± 0.008 b
25 days	0.808 ± 0.016 a	1.124 ± 0.077 a	0.188 ± 0.004 c

*Mean ± standard error DAS-ELISA absorbance value
**Values followed by the same letter are not significantly different
Source: García-Cano et al. (2006)

Table 2.6 Differential reaction of tomato cultivars following inoculation with TSWV or ToCV + TSWV

Virus (es) inoculated		Cultivar*		
		DRWV 7456	Vinchy RZ	Anastasia
TSWV	ELISA positive	1/5**	0/5	1/5
	Symptomatic plants	1/5	3/5	1/5
ToCV+TSWV	ELISA positive	3/10	1/10	2/10
	Symptomatic plants	3/10	9/10	10/10
Mock inoculated	ELISA positive	0/5	0/5	0/5
	Symptomatic plants	0/5	0/5	0/5

*Detection by DAS-ELISA test
**Number of plants showing symptom or presence of virus(es)
Source: García-Cano et al. (2006)

2.5 Viroids

Viroids constitute a unique class of infectious agents capable of causing diseases of plants. They are composed entirely of a nonprotein-coding small, single-stranded ciruclar RNA capable of replicating autonomously in certain plant-species which may or may not express visible symptom in response to infection. The viroids can be differentiated from the statellite RNAs by their ability to complete their life cycle without a helper virus. In contrast, satellite RNAs require a helper virus for their replication. The infected plants exhibit symptoms of infection by viroids which replicate and accumulate in such host plant species (Diener 2001; Flores et al. 2005a). The viroids (30) have been classified into two families *Popsiviroidae* [type species *Potato spindle tuber viroid* (PSTVd)] and *Avsunviroidae* [type species *Avacado sunblotch viroid* (ASBVd)] each containing several genes (Flores et al. 2005b).

In all multicellular organisms, RNA and protein traffic is considered as a basic biological function. Diverse RNAs are transported in the phloem which delivers them to the proper locations. *Potato spindle tuber viroid* (PSTVd) is considered as a model system to study RNA traffic in plants. PSTVd replicates in the nucleus of cells of infected plants and its systemic movement includes nuclear transport, cell-to-cell transport and phloem transport (Zhao et al. 2001; Zhu et al. 2001). PSTVd was transported, following mechanical inoculation on leaves of *N. benthamiana* to sepals but not to other floral organs-petals, stamens and ovary of developing flowers. Although PSTVd could replicate in all floral organs, the absence of PSTVd in petals, ovary and stamens may possibly be attributed to restricted PSTVd movement into these organs and not to suppression of replication. The results indicated that PSTVd RNA movement within sieve tubes does not simply follow mass flow from source to sink organs but it is transported into selective sink organs. In addition, two PSTVd mutants could enter the phloem to spread systemically but could not exit the phloem in the leaves of tobacco. It seems that the viroids have evolved most likely, structural motifs that may mimic endogenous plant RNA motifs, so that they may be recognized by cellular factors involved in RNA trafficking (Zhu et al. 2002).

Hop stunt viroid (HSVd) infects cucumber (*Cucumis sativus* L. cv. Suyo) causing severe systemic symptoms. The phloem exudates from *Cucurbita* spp. contained a complex mixture of over 100 compounds, including the phloem filament protein PP1 (96-kDa) and phloem lectin protein PP2 (49 kDa) which were present abundantly. The genes encoding PP2 are expressed experimentally in phloem companion cells. PP2 in addition to many phloem proteins are able to move from cell-to-cell through plasmodesmata. Furthermore, PP1 and PP2 can move toward sink tissues within the assimilate stream. The phloem exudates collected from uninfected cucumbers was screened for the presence of protein(s) that can bind HSVd such as putative viroid movement proteins. Evidence showing that PP2 could interact with a variety of RNA molecules like viroid RNAs and polyadenylated mRNAs was obtained. Incubation of HSVd with cucumber phloem exudate in vitro resulted in a dramatic decrease in the electrophoretic mobility of HSVd. Binding of RNA by PP2 in in vitro appeared to be nonspecific. As the ability of PP2 to move from cell-to-cell via plasmodesmata as well as long distance in the phloem is known, the results suggested that PP2 is most likely involved in the systemic movement of viroids and also other RNAs in vivo (Owens et al. 2001).

The movement pattern of *Peach latent mosaic viroid* (PLMVd) was analyzed by in situ hybridization. In contrast to earlier report that PSTVd was absent from the shoot apex (Zhu et al. 2001), it was found that PLMVd could move into the shoot apex indicating a demarcating difference in the movement patterns of the two representative members of both viroid families. PLMVd is well distributed in shoots, leaves, bark, roots and fruits. Young expanding fruits seem to be the best source of PLMVd (Flores et al. 1992; Delgado et al. 2005). The molecular mechanisms through which viroids may induce their characteristic effects (symptoms) are not clearly understood. Viroids may modify host gene expression interfering with normal developmental pathways. The mature RNA or its replicative intermediates are likely to be the primary pathogenic effectors interacting with a host protein or RNA, as suggested by Diener (2001). But in the recent years the small interfering RNAs (siRNAs) resulting from the host RNA silencing, defensive response triggered by double-or quasi-double-stranded RNAs, are considered to induce the symptoms of viroid diseases. In support this concept, viroid-specific siRNAs of both polarities corresponding to different genomic regions, the hallmarks of RNA silencing have been detected in plants infected by PSTVd (Itaya et al. 2001), PLMVd (Martínez de Alba et al. 2002) and ASBVd (Markarian et al. 2004). Tomato plants expressing hairpin RNA derived from PSTVd, developed symptoms similar to those of PSTVd infection, suggesting viroids can cause disease symptoms by directing RNA silencing against physiologically important host genes. Furthermore, the results indicate that viroid RNAs are significantly resistant to RNA silencing-mediated degradation (Wang et al. 2004c).

The PLMVd-specific RNAs of different sizes have been detected in vivo (Landry et al. 2004). Viroids, as in the case of PLMVd, can induce an RNA silencing response in their hosts. By loading the second key RNase of the RNA silencing pathway, forming part of RNA-induced silencing complex (RISC), the viroid-specific siRNAs may be able to guide this complex against their cognate genomic RNAs. Further, some of the viroid-specific siRNAs could by chance guide RISC against host messenger RNAs. This may down-regulate the expression of specific host genes and eventually result in production of disease symptoms. A link between certain macroscopic symptoms such as the albinophenotype induced by PMLVd variants and the underlying cytopathic and molecular alterations like malformed proplastids and impairment of formation of chloroplasts, has been indicated by Flores et al. (2006).

The viroids, the smallest of the microbial pathogens, do not encode any protein and are not encapsidated. Yet they can replicate autonomously and spread systemically inducing characteristic symptoms. Distinctly different symptoms are induced by strains of a viroid with subtle differences in nucleotide sequences. A comprehensive analysis of the differential gene expression patterns of tomato plants at various stages of infection by a mild and severe strain of PSTVd was performed. Marcoarrays containing 1156 clones from the subtracted cDNA library to monitor their expression patterns throughout the infection process were constructed. Changes in tomato gene expression following infection by PSTVd and TMV were compared. The PSTVd strains altered expression of both common and unique tomato genes. These genes encode products involved in defense/stress response, cell wall structure, chloroplast function, protein metabolism and other diverse functions. The expression of some, but not all of those genes altered by PSTVd, was also altered by TMV infection. Despite the simple structure, viroids are capable of triggering complex host responses (Itaya et al. 2002).

Posttranscriptional gene silencing (PTGS), assigned with an important role in antiviral defense in plants, is triggered by the presence of ds RNAs that are cleaved by an RNase known as Dicer (Dillin 2003, Voinnet 2003). Following the cleavage, ss-RNAs, 21–25 nt in length designated small interfering RNAs (siRNAs) are formed. The siRNAs are incorporated into a multisubunit protein complex, the RNA-induced silencing complex and act as a guide to direct this RNA degradation machinery to its target RNAs. The members of the two families of viroids have been shown to induce PTGS, based on the detection of siRNA representing different regions of the viroid genome. The incubation of PLMVd transcripts of either plus or minus polarity with DCL enzyme (s) resulted in the detection of siRNA. Regardless of the polarity of the transcripts, only these regions including the long p11 harpin resulted in the formation of siRNA. The hairpin known to be implicated in the replication of PLMVd has the ability to trigger the Dicer enzyme(s). It appears unlikely that the P11 hairpin evolved specifically to be a Dicer substrate, since the viroid can evade PTGs-mediated destruction by replicating in the chloroplasts (Landry and Perreault 2005).

Appendix 1: Detection of Components of the Extracellular Matrix of Germinating Spores of *Stagonospora nodorum* (Zelinger et al. 2004)

A. Plate-Trapped Antigen (PTA) – ELISA

i. Place the spores of test fungal pathogen (*Stagonospora nodorum*) in microtitre wells for germination overnight at 18°C or in wells coated with PBS-surface washings of a plate culture of the pathogen and incubate at 4°C overnight.
ii. Wash the antigen coated wells four times with PBS; air-dry at 24°C in a laminar flow hood and store the plates in sealed polyethylene bags at 4°C or use directly.
iii. Incubate wells successively with undiluted supernatant of appropriate hybridoma-goat anti-moust polyvalent (IgG, IgA and IgM) peroxide conjugate (Sigma, A-0412 diluted in 1 in 1000 in PBST) and the substrate teramethyl benzidine, with four washings with PBST (PBS $+ 0.025 - 0.05\%$ W/V Tween-20) between incubations.
iv. Stop the color development with H_2SO_4 and record the absorbance values at 405 nm using ELISA automated microplate reader (Dynatech MRX).

B. Immunofluorescence and Confocal Microscopy

i. Fix infected leaf samples between desired time interval 0–20 h post inoculation (hpi) in 3% paraformaldehyde in pH7.2, for 3 or 1 h respectively at room temperature.
ii. Rinse the samples three times in PBST, 10 min each; block in 1% bovine serum albumin (BSA) in PBST for 1 h, wash and incubate for 1 h with appropriate MABs (SN-MG11, SN-CH9) or negative controls (IgG specific for nonpathogens)
iii. Wash the specimens (preparations), incubate with secondary goat antimouse antibody FITC-conjugate (Sigma F-9006) and dilute (1:30 v/v) in blocking solution.
iv. Wash as done before; place the specimens in a specimen holder slide (with a well created to fit the size and depth of specimen) and mount in Citiflour (Citiflour AF1, Agar Scientific Ltd, UKR-1320).
v. Follow the steps mentioned above for double labeling, introducing a blocking step (in blocking solution for one hour) after the first set of antibody labeling; wash and apply a second set of antibody labeling; apply the first MAB-SN-MG11 followed by goat anti-mouse IgG (Molecular Probes, T-862) TRITC conjugate as the reporter antibody; apply the second MAB-SN-CH9 followed by a goat antimouse μ-chain specific-FITC conjugate (Sigma, F-9256) as the reporter antibody.
vi. Label the germlings with a panel of biotinylated lectins at 4 and 24 h pi on multiwell glass slides; follow the same procedure except the substitution of the MABs with the lectin and an avidin-TRITC secondary conjugate.

vii. Observe under a Zeiss LSM 410 laser scanning confocal microscope using a 488 nm argon laser for FITC or a 543 nm helium / neon laser for TRITC.

C. Immunogold Staining

i. Allow the conidia to germinate on multi-well glass/polystyrene slides or on 7-day-old wheat leaves for six h; fix and stock as done for immunofluorescence labeling, before staining will Protogold (anionic gold particles, British BioCell International, UK) for one h at room temperature with gentle agitation.
ii. Treat the germinated conidia with immunogold labeled with MAB SN-CH9 as in immunofluorescence labeling exept the substitution of the secondary antibody with 10 nm gold anti-mouse.
iii. Rinse the stained germlings with distilled water three times 5 min for each rinse; enhance silver stain as per manufacturers' instructions (British Biocell International)
iv. Terminate the process after 3–5 min by rinsing slides or by immersion in distilled water three times, 5 min for each immersion.
v. Observe under bright field or epi-polarisation microscopy with Nomarski optics.

Appendix 2: Separation of the Fungal Chromosomal DNA Containing Toxin Gene(s) of *Alternaria alternata* by Pulsed Field Gel Electrophoresis (Masunaka et al. 2005)

A. DNA Manipulations

i. Isolate the genomic DNA of the fungal pathogen by following standard procedure.
ii. Prepare the DNA fragment with desired gene (s) (ACTT-specific genes ACTT1 and ACTT2 present only in ACT-toxin-producing isolates). The specific DNA fragment present only in ACR-toxin producing isolates is designated RP10-2.

B. Preparation of Protoplasts for Pulsed Field Gel Electrophoresis (PFGE)

i. Collect the fungi grown on V8-juice agar plates in 7 ml of sterile distilled water; add mycelia and spores to 50 ml of potato dextrose broth (PDB) in 200 ml flasks and incubate for 24 h at 28°C with reciprocating shaking (125 rpm).
ii. Harvest the fungal material by centrifugation at 3000 rpm for 5 min; wash the pellet twice with 30 ml of sterile distilled water and once with 30 ml of osmotic (OM) buffer (containing 1.2 M $MgSO_4$ and 10 mM Na_2HOP_4, pH 5.8)

by centrifugation for 5 min at 4000 rpm and incubate in 5 ml of sterile enzyme solution [20 mg of Lysing enzymes per ml (Sigma, St. Louis) and 3 mg of Kitalase per ml (Wako, Tokyo) in OM buffer] at 28°C for 6 h with shaking at 200 rpm.

iii. Transfer the digested culture into 15 ml centrifuge tubes (Corning, Acton, MA) and overlain with 2 ml of ST buffer (consisting of 0.6 M sorbitol and 100 mM Tris–HCl, pH 8.0) and centrifuge at 3000 rpm for 5 min.
iv. Transfer the protoplasts banded at the interface of OM buffer and ST buffer to another centrifuge (Corning); mix the protoplast suspension with 5 ml of STC buffer (1.0 M sorbitol, 50 mM CaCl2 and 50 mM Tris–HCl, pH 8.0) and pellet the protoplasts by centrifugation at 3000 rpm for 5 min.
v. Wash the protoplasts twice with 5 ml of SE buffer containing 1.0 M sorbitol and 50 mM EDTA at 3000 rpm for 5 min; resuspend the protoplasts in SE buffer and adjust to a final concentration of 5.0×10^8 protoplasts per ml; mix the protoplast solution with an equal volume of 1% low melting point agarose (Invitrogen, Carlsbad, CA) in SE buffer.
vi. Pipette 100 μl of protoplast-agarose mixture into plug molds (Bio-Rad, Hercules, CA); incubate plugs in ET buffer consisting of 0.5 M EDTA and 10 mM Tris–HCl, pH8.0 containing protease K (0.5 mg/ml, Wako) at 50°C for 48 h and wash the plugs three times for 1 h in 50 m EDTA and stored at 4°C until use.

C. Pulsed Field Gel Electrophoresis (PFGE)

i. Use CHEF-DRII unit (Bio-Rad) to separate chromosomal DNA by DNA by PFGE technique.
ii. Prepare agarose gels [0.8% 0.5× Tris-borate-EDTA (TBE)] with Seakem Gold agarose (BMA, Rockland, ME) and run in 0.5× TBE buffer at 8°C with following running conditions (to separate chromosomes of less than 6 Mb):

    1.5 V/cm with pulse intervals of 3600–1800 s for 115 h,
    1.5 V/cm with pulse intervals of 1800–1300 s for 24 h,
    1.8 V/cm with pulse intervals of 1300–800 s for 28 h, and
    2.4 V/cm with pulse intervals of 800–600 s for 28 h,
    use chromosomal size standards from Bio-Rad.

iii. For isolating the 1.5 Mb chromosome from isolate HC1 of the rough lemon pathotype, use running conditions as detailed below:

    5.4 V/cm with pulse intervals of 120 s for 13 h
    5.4 V/cm with pulse intervals of 180 s for 13 h
    use chromosome standards (*Saccharomyces cerevisiae* and *Schizosaccharomyces pombe*)

iv. Visualize DNA bands by ethidium bromide staining.
v. Excise the 1.5 Mb chromosome from the gel and purify using Nucleo Trap spin columns (Macheery-Nagel, Easton, PA) as per manufacturer's instructions.
vi. Use the purified 1.5 Mb chromosome from the rough lemon pathotype HC1 as the template for PCR with random primers.

References

Abramovitch RB, Anderson JC, Martin GB (2006a) Bacterial elicitation and evasion of plant innate immunity. Nature Rev 7:601–611

Abramovitch RB, Janjuseivic R, Stebbins CE, Martin GB (2006b) Type III effector Avr PtoB requires intrinsic E3 ubiquitin ligase activity to suppress plant cell death and immunity. Proc Natl Acad Sci USA 103:2851–2856

Abramovitch RB, Kim YJ, Chen S, Dickman MB, Martin GB (2003) *Pseudomonas* type III effector AvrPtoB induces plant disease susceptibility by inhibition of host programmed cell death. EMBO J 22:60–69

Abramovitch RB, Martin GB (2004) Strategies used by bacterial pathogens to suppress plant defenses. Curr Opin Plant Biol 7:356–364

Aharoni A, Vorst O (2001) DNA microarrays for functional plant genomics. Plant Mol Biol 48:99–118

Ahn JH, Walton JD (1997) A fatty acid synthase gene in *Cochliobolus carbonum* required for production of HC toxin cyclo(D-Prolyl-L-alanyl-D-alanyl-L-2-amino-9, 10-epoxi-8-oxodecanoyl). Mol Plant Microbe Interact 19:207–214

Aizawa SI (1996) Flagellar assembly in *Salmonella typhimurium*. Mol Microbiol 19:1–5

Akamatsu H, Taga M, Kodama M, Johnson RD, Otani H, Kohmotoa K (1999) Molecular karyotypes for *Alternaria* plant pathogens known to produce host-specific toxins. Curr Genet 35:647–656

Akimitsu K, Peever TL, Timmer LW (2003) Molecular, ecological and evolutionary approaches to understanding *Alternaria* diseases of citrus. Mol Plant Pathol 4:435–446

Alebeek G-JWM, van Scherpenzeel K, Beldman G, Schols HA, Voragen AGJ (2003) Partially esterified oligogalacturonides are the preferred substrates for pectin methylesterase of *Aspergillus niger*. Biochem J 372:211–218

Alegria MC, Docena C, Khater L, Ramos CHI, da Silva ACR, Farah CS (2004) New protein–protein interactions identified for the regulatory and structural components and substrates of the type III secretion system of the phytopathogens *Xanthomonas axonopodis* pv. *citri*. Bacteriol 186:6186–6197

Alfano JR, Charkowski AO, Deng WL, Badel JL, Petnicki-Ocwieja T, van Dijk K, Collmer A (2000) The *Pseudomonas syringae* Hrp pathogenicity island has a tripartite mosaic structure composed of a cluster of type III secretion genes bounded by exchangeable effector and conserved effector loci that contribute to parasitic fitness and pathogenicity in plants. Proc Natl Acad USA 97:4856–4861

Alfano JR, Collmer A (2004) Type III secretion system effector proteins: double agents in bacterial disease and plant defense. Annu Rev Phyopathol 42:385–414

Aldrige P, Hughes T (2002) Regulation of flagellar assembly. Curr Opin Microbiol 5:160–165

Allen C, Gay J, Simon-Buela L (1997) A regulatory locus *pehSR* controls polygalacturonase production and other virulence functions in *Ralstonia solanacearum*. Mol Plant Microbe Interact 10:1054–1064

Altmann F (1996) *N*-glycosylation in insects revisited. Trend Glycosci Glycotechnol 8:101–114

Amrein H, Makart S, Granado J, Shakya R, Schneider-Pokorny J, Dudler R (2004) Functional analysis of genes involved in the synthesis of syringolin A by *Pseudomonas syringae* pv. *syringae*. Mol Plant Microbe Interact 17:90–97

Anandalakshmi R, Pruss GJ, Ge X, Marathe R, Smith TH, Vance CB (1998) A viral suppressor of gene silencing in plants. Proc Natl Acad Sci USA 95:13079–13084

Anderson JC, Pascuzzi PE, Xiao F, Sessa G and Martin GB (2006) Host-mediated phosphorylation of type III effector AvrPto promotes *Pseudomonas* virulence and avirulence in tomato. Plant Cell 18:502–514

Andrie RM, Martinez JP, Ciuffetti LM (2005) Development of *ToxA* and *ToxB* promoter driven fluorescent protein expression vectors for use in filamentous ascomycetes. Mycologia 97:1152–1161

Angell SM, Davies C, Baulcombe DC (1996) Cell-to-cell movement of *Potato virus X* is associated with a change in the size exclusion limit of plasmodesmata in trichome cells of *Nicotiana clevelandii.* Virology 216:197–201

Angot A, Peeters N, Lechner E, Vailleau F, Band C, Gentzbittel L, Sartorel E, Genschik P, Boucher C, Genin S (2006) *Ralstonia solanacearum* requires F-box-like domain-containing type III effectors to promote disease on several host plants. Proc Natl Acad Sci USA 103:14620–14625

Aoki K, Kragler F, Xoconstle-Cazares B, Lucas WJ (2002) A subclass of plant heat shock cognate 70 chaperones carries a motif that facilitates trafficking through plasmodesmata. Proc Natl Acad Sci USA 99:16342–16347

Aparicio F, Thomas CL, Lederer C, Niu Y, Wang D, Maule AJ (2005) Virus induction of heat-shock protein 70 reflects a general response to protein accumulation in plant cytosol. Plant Physiol 138:529–536

Aranda MA, Maule AJ (1998) Virus-induced host gene shut-off in animals and plants. Virology 243:261–267

Arnold DL, Pitman A, Jackson RW (2003) Pathogenicity and other genomic islands in plant pathogenic bacteria. Mol Plant Pathol 4:407–420

Arora SK, Ritchings BH, Almira EC, Lory S, Ramphal R (1997) A transcriptional activator, Fle Q regulates mucin adhesion and flagellar gene expression in *Pseudomonas aeruginosa* in a cascade manner. J Bacteriol 179:5574–5581

Arrebola E, Cazorla FM, Romero D, Pérez-García A, de Vincete A (2007) A nonribosomal peptide synthetase gene (*mgoA*) of *Pseudomonas syringae* pv. *syringae* is involved in mangotoxin biosynthesis and is required for full virulence. Mol Plant Microbe Interact 20:500–509

Asai T, Tena G, Plotnikova J, Willmann MR, Chiu WL, Gomez-Gomez L, Boller T, Ausubel FM, Sheen J. 2002. MAP kinase signaling cascade in *Arabidopsis* innate immunity. Nature 415:977–983

Asakura M, Okuno T, Takano Y (2006) Multiple contributions of peroxisomal metabolic function to fungal pathogenicity in *Colletrotichum lagenarium.* Appl Environ Microbiol 72:6345–6354

Atmakuri K, Cascales E, Christie PJ (2004) Energetic components Vir D4, VirB11 and VirB4 mediate early DNA transfer reactions required for bacterial type IV secretion. Mol Microbiol 54:1199–1211

Ausubel FM (2005) Are innate immune signaling pathways in plants and animals conserved? Nature Rev Immunol 6:973–979

Axtell MJ, Staskawicz BJ 2003. Initiation of RPS2-specified disease resistance in Arabidopsis is coupled to the AvrRpt2 directed elimination of RIN4. Cell 112:369–377

Azpiroz-Leehan R, Feldmann KA (1997) T-DNA insertion mutagenesis in *Arabidopsis* going back and forth. Trends Genet 13:152–156

Badel JL, Charkowski AO, Deng WL, Collmer A (2002) A gene in the *Pseudomonas syringae* pv. *tomato* Hrp pathogenicity island conserved effector locus, *hopPtoA1*, contributes to efficient formation of bacterial colonies in planta and is duplicated elsewhere in the genome. Mol Plant Mol Interact 15:1014–1024

Badel JL, Nomura K, Bandyopadhyay SRS, Collmer A, He SY (2003) *Pseudomonas syringae* pv. *tomato* DC 3000 HopPtoM (CEL ORF3) is important for lesion formation but not growth in tomato and is secreted and translated by the Hrp type III secretion system in a chaperone-dependent manner. Mol Microbiol 49:1239–1251

Badel JL, Shimizu R, Oh HS, Collmer A (2006) *Pseudomonas syringae* pv. *tomato avrE1/hopM1* mutant is severely reduced in growth and lesion formation in tomato. Mol Plant Microbe Interact 19:99–111

Bae H, Bowers JH, Tooley PW, Bailey BA (2005) *NEP1* orthologs encoding necrosis and ethylene inducing proteins exist as a multigene family in *Phytophthora megakarya,* causal agent of black pod disease on cocao. Mycol Res 109:1373–1385

Bae H, Kim MS, Sicher RC, Bae HJ, Bailey BA (2006) Necrosis-and ethylene-inducing peptide from *Fusarium oxysporum* induces a complex cascade of transcripts associated with signal transduction and cell death in *Arabidopsis*. Plant Physiol 141:1056–1067

Bai J, Choi SH, Ponciano G, Leung H, Leach JE (2000) *Xanthomonas oryzae* pv. *oryzae* avirulence genes contribute differently and specifically to pathogen aggressiveness. Mol Plant Microbe Interact 13:1322–1329

Bai GH, Desjardins AE, Plattner RD (2002) Deoxynivalenol-non-producing *Fusarium graminearum* causes initial infection but does not cause disease spread in wheat spikes. Mycopathologia 153:91–98

Bailey BA, Jennings JC, Anderson JD (1997) The 24-kDa protein from *Fusarium oxysporum* f.sp. *erythroxyli*: occurrence in related fungi and the effect of growth medium on its production. Canad J Microbiol 43:45–55

Bailey A, Mueller E, Bowyer P (2000) Ornithine decarboxylase of *Stagonospora* (*Septoria*) *nodorum* is required for virulence toward wheat. J Biol Chem 275:14242–14247

Baker SE, Kroken S, Inderbitizin P, Asvarak T, Li BY, Shi L, Yoder OC, Turgeon BG (2006) Two polyketide synthase-encoding genes are required for biosynthesis of the polyketide virulence factor, T-toxin by *Cochliobolus heterostrophus*. Mol Plant Microbe Interact 19:139–149

Baldé JA, Francisco R, Queiroz A, Reglalado AP, Ricardo CP, Veloso MM (2006) Immunolocalization of a class III chitinase in two muskmelon cultivars reacting differently *Fusarium oxysporum* f.sp *melonis*. J Plant Physiol 163:19–25

Balhadere PV, Foster AJ, Talbot NJ (1999) Identification of pathogenicity mutants of rice blast fungus *Magnaporthe grisea* by insertional mutagenesis. Mol Plant Microbe Interact 12: 129–143

Ballas N, Citovsky V (1997) Nuclear localization signal building protein from *Arabidopsis* mediates nuclear import of *Agrobacterium* Vir D2 protein. Proc Natl Acad Sci USA 94: 10723–10728

Banar DW, Bogdanove AJ, Beer SV, Collmer A (1994) *Erwinia chrysanthemi hrp* genes and their involvement in soft rot pathogenesis and elicitation of the hypersensitive response. Mol Plant Microbe Interact 7:573–581

Barbara DJ, Morton A, Clark MF, Davies DL (2002) Immuno-dominant membrane proteins from two phytoplasmas in the aster yellows clade (chloroante aster yellows and clover phyllody) are highly divergent in the major hydrophilic region. Microbiol 148:157–167

Barber CE, Tang JL, Feng JX, Pan MQ, Wilson TJG, Slater H (1997) A novel regulatory system required for pathogenicity of *Xanthomonas campestris* is mediated by a small diffusible signal molecule. Mol Microbiol 24:555–566

Barny MA, Gaudriault S, Brisset MN, Paulin JP (1999) Hrp-secreted proteins: their role in pathogenicity and HR-elicitation. Acta Horti 489:353–358

Barras G, van Gijsegem F, Chatterjee AK (1994) Extracellular enzymes and pathogenesis of soft-rot *Erwinia*. Annu Rev Phytopathol 32:201–234

Barta TM, Kinscherf TG, Willis DK (1992) Regulation of tabtoxin production by the *lemA* gene in *Pseudomonas syringae*. J Bacteriol 174:3021–3029

Batten JS, Yoshinari S, Hemenway C (2003) *Potato virus X*: a model system for virus replication, movement and gene expression. Mol Plant Pathol 4:125–131

Baulcombe D (2004) RNA silencing in plants. Nature 431:356–363

Bayne EH, Rakitina DV, Morozov SY, Baulcombe DC (2005) Cell-to-cell movement of *Potato potex virus X* is dependent on suppression of RNA silencing. Plant J 44:471–482

Beaulieu C, Boccara M, van Gijsegem F (1993) Pathogenic behaviour of pectinase-defective *Erwinia chrysanthemi* mutants on different plants. Mol Plant Microbe Interact 6:197–202

Bender CL (2000) Regulatory interactions between the Hrp type III protein secretion system and coronatine biosynthesis in *Pseudomonas syringae* pv. *tomato* DC 3000. Microbiology (Reading) 146:2447–2456

Bender CL, Alarcon-Chaidez F, Gross DC (1999) *Pseudomonas syringae* phytotoxins: mode of action, regulation and biosynthesis by peptide and polyketide synthetases. Microbiol Mol Biol Rev 63:266–272

Benedetti CE, Arruda P (2002) Altering the expression of the chlorophyllase gene *ATHCOR1* in transgenic *Arabidopsis* caused changes in the chlorophyll-to-chlorophyllide ratio. Plant Physiol 128:1255–1263

Benito EP, ten Have A, van'T Klooster JW, Kan JAL (1998) Fungal and plant gene expression during synchronized infection of tomato leaves by *Botrytis cinerea*. Eur J Plant Pathol 104:207–220

Berks BC, Palmer T, Sargent F (2003) The Tat protein translocation pathway and its role in microbial physiology. Adv Microbial Physiol 47:187–254

Berks BC, Sargent F, Palmer T (2000) The Tat protein export pathway. Mol Microbiol 35:260–274

Bermpohl A, Dreier J, Bahro R, Eichenlaub R (1996) Exopolysaccharides in the pathogenic interaction of *Clavibacter michiganensis* subsp *michiganensis* with tomato plants. Microbiol Res 151:391–399

Bernhard F, Coplin DL, Geider K (1993) A gene cluster for amylovoran synthesis in *Erwinia amylovora*: characterization and relationship to *cps* genes in *Erwinia stewartii*. Mol Gen Genet 239:158–168

Bhambra KG, Wang ZY, Soanes DM, Wakley GE, Talbot NJ (2006) Peroxisomal carnitine acetyl transferase is required for elaboration of penetration hyphae during plant infection by *Magnaporthe grisea*. Mol Microbiol 61:46–60

Bhatnagar D, Ehrlich KC, Cleveland TE (2003) Molecular genetic analysis and regulation of aflatoxin biosynthesis. Appl Microbiol Biotechnol 61:83–93

Bingle LEH, Simpson TJ, Lazarus (1999) Ketosynthase domain probes identify two subclasses of fungal polyketide synthase genes. Fungal Genet Biol 26:209–223

Bins AN, Costantino P (1998) The Agrobacterium oncogenes. In: Kondorsi HP, Hooykaas PJJ (eds) The Rhizobiaceae (Spainik). Kluwer Academic Publishers, Dordrecht, The Netherlands, pp 252–256

Birch PRJ, Rehmany AP, Pritchard L, Kamoun S, Beynon JL (2003) Trafficking arms: oomycete effectors enter host plant cells. Trends Microbiol 14:8–11

Bird CB, Malone B, Rice LG, Ross PF, Eppley R, Abouzied MM (2002) Determination of total fumonisins in corn by competitive enzyme-linked immunosorbent assay-collaborative study. J AOAC Int 85:404–410

Blair DF (1995) How bacteria sense and swim. Annu Rev Microbiol 49:489–520

Blair DF, Berg HC (1991) Mutations in the MotA protein of *Escherichia coli* reveal domains critical for proton conduction. J Mol Biol 221:1433–1442

Blanc S, Ammar ED, Garcia-Lam Pasona S, Dolja VV, Llave C, Baker J, Pirone TP (1998) Mutations in the potyvirus helper component protein: effects on interactions with virions and aphid stylets. J Gen Virol 79:3119–3122

Blanco FA, Judelson HS (2005) A bZIP transcription factor from *Phytophthora* interacts with a protein kinase and is required for zoospore motility and plant infection. Mol Microbiol 56: 638–648

Boevink P, Oparka KJ (2005) Virus–host interactions during movement process. Plant Physiol 138:4–6

Bogdanove AJ, Kim JF, Wei ZM, Kolchinsky P, Charkowski AO, Collmer A, Beer SV (1998) Homology and functional similarity of an hrp-linked pathogenicity locus *dspEF* of *Erwinia amylovora*. Proc Natl Acad Sci USA 95:1325–1330

Boller T (1991) Ethylene in pathogenesis and disease resistance. In: Mattoo AK, Suttle JC (eds) The Plant Hormone Ethylene. CRC Press, Boca Raton, FL, USA

Bolton GW, Nester EW, Gordon MP (1986) Plant phenolic compounds induce expression of the *Agrobacterium tumefaciens* loci needed for virulence. Science 232:983–985

Bonas V, van den Ackerveken G, Buttner D, Hahn K, Marios E, Nennstiel D, Noel L, Rossier D, Szurek B (2000) How the bacterial plant pathogen *Xanthomonas campestris* pv. *vesicatoria* can conquer the host? Mol Plant Pathol 1:73–76

Bonshtien A, Lev A, Gibly A, Debbie P, Avni A, Sessa G (2005) Molecular properties of the *Xanthomonas* AvrRxv effector and global transcriptional changes determined by its expression in resistant tomato plants. Mol Plant Microbe Interact 18:300–310

Both M, Eckert SE, Csukai M, Miller E, Dimopoulos G, Spanu PD (2005) Transcript profiles of *Blumeria graminis* development during infection reveal a cluster of genes that are potential virulence determinants. Mol Plant Microbe Interact 18:125–133

Bottini AJ, Bowen JR, Gilchrist DG (1981) Phytotoxins II characterization of a phytotoxic fraction from *Alternaria alternata* f.sp *lycopersici*. Tetrahedron Lett 22:2723–2726

Boulton MI, Pallaghy CK, Chatani M, MacFarlane S, Davies JW (1993) Replication of *Maize streak virus* mutants in maize protoplasts: evidence for a movement protein. Virology 192: 85–93

Boureau T, ElMaarouf-Bouteau H, Garnier A, Brisset MN, Perino C, Pucheu I, Barny MA (2006) Dsp A/E, a type III effector essential for *Erwinia amylovora* pathogenicity and growth in planta induces cell death in host apple and nonhost tobacco plants. Mol Plant Microbe Interact 19: 16–24

Bozarth CS, Weiland JJ, Dreher TW (1992) Expression of ORF-69 of *Turnip yellow mosaic virus* is necessary for viral spread in plants. Virology 187:124–130

Brandwegt BF, Mesbah LA, Takken FLW, Laurent PL, Kneppers TJA, Hille J, Nijkamp HJJ (2000) A longevity assurance gene homology of tomato mediates resistance to *Alternaria alteranta* f.sp. *lycopersici* toxins and fumonisin B1. Proc Natl Acad Sci USA 97:4961–4966

Briggs DE (1978) Barley. John Wiley, New York, USA

Brigneti G, Voinnet O, Li WX, Ji LH, Ding SW, Baulcombe DC (1998) Viral pathogenicity determinants are suppressors of transgene silencing in *Nicotiana benthamiana*. EMBO J 17: 6739–6746

Brito N, Espino JJ, González C (2006) The endo-beta-4-xylanase *Xyn11A* is required for virulence in *Botrytis cincerea*. Mol Plant Microbe Interact 19:25–32

Brito B, Marenda M, Berberis P, Boucher C, Genin S (1999) *PrhJ* and *hrpG*, two new components of the plant signal-dependent regulatory cascade controlled by *PrhA* in *Ralstonia solanacearum*. Mol Microbiol 31:237–251

Bronstein PA, Marrichi M, Cartinhour S, Schneider DJ, DeLisa MP (2005) Identification of a twin-arginine translocation system in *Pseudomonas syringae* pv. *tomato* DC 3000 and its contribution to pathogenicity and fitness. J Bacteriol 187:8350–8461

Brown GE, Burns JK (1998) Enhanced activity of abscission enzymes predisposes oranges to invasion by *Diplodia natalensis* during ethylene degreening. Postharvest Biol Technol 14: 217–227

Brown DW, Dyer RB, MCCormick SP, Kendra DF, Plattner RD (2004) Functional demarcation of the *Fusarium* core trichothecene gene cluster. Fungal Genet Biol 41:454–462

Brown IR, Mansfield JW, Taira S, Roine E, Romantschuk M (2001) Immunocytochemical localization of HrpA and HrpZ supports a role for the Hrp pilus in the transfer of effector proteins from *Pseudomonas syringae* pv. *tomato* across the host plant cell wall. Mol Plant Microbe Interact 14:394–404

Buddenhagen I, Kelman A (1964) Biological and physiological aspects of bacterial wilt caused by *Pseudomonas solanacearum*. Annu Rev Phytopathol 2:203–230

Burdman S, Shen Y, Lee SW, Xue Q, Ronald P (2004) Rax H / Rax R: a two-component regulatory system in *Xanthomonas oryzae* pv *oryzae* required for AvrXa21 activity. Mol Plant Microbe Interact 17:602–612

Burgyán J, Hornyik C, Szittya G, Silhavy D, Bisztray G. (2000) The ORF1 products of tombusviruses play a crucial role in lethal necrosis of virus-infected plants. J Virol 74: 10873–10881

Burr TJ, Bazzi C, Süle S, Ottent L (1998) Crown gall of grape: biology of *Agrobacterium vitis* and the development of disease control strategies. Plant Dis 82:1288–1297

Büttner D, Bonas U (2003) Port of entry the type III secretion translocon. Trends Microbiol 10:186–192

Büttner D, Gurlebeck D, Noel D, Bonas U (2004) Hpa B from *Xanthomonas campestris* pv. *vesicatoria* acts as an exit control protein in type III-dependent protein secretion. Mol Microbiol 54:755–768

Büttner D, Nennstiel D, Kluesner B, Bonas U (2002) Functional analysis of HrpF, a putative type III translocon protein from *Xanthomonas campestris* pv. *vesicatoria*. Bacteriol 184: 2389–2398

Caldelari I, Mann S, Crooks C, Palmer T (2006) The Tat pathway of the plant pathogen *Pseudomonas syringae* is required for optimal virulence. Mol Plant Microbe Interact 19: 200–212

Callahan TM, Rose MS, Meade MJ, Ehrenshaft M, Upchurch RG (1999) *CFP* the putative cercosporin transporter of *Cercospora kikuchii* is required for wild type cercosporin production, resistance and virulence on soybean. Mol Plant Microbe Interact 12:901–910

Calvo AM, Bok J, Brooks W, Keller NP (2004) *veA* is required for toxin and sclerotial production in *Aspergillus parasiticus*. Appl Environ Microbiol 70:4733–4739

Cangelosi GA, Ankenbauer RG, Nester EW (1990) Sugars induce the *Agrobacterium* virulence genes through a periplasmic binding protein and a transmembrane signal protein. Proc Natl Acad Sci USA 87:6708–6712

Cangelosi GA, Martineti G, Leigh JA, Lee CC, Theiness C, Nester EW (1989) *Agrobacterium tumefaciens* chvA protein in export of β-1,2-glucan. J Bacteriol 171:1609–1615

Canto T, Palukaitis P (1999) Are tubules generated by the 3a protein for *Cucumber mosaic virus movement?* Mol Plant Microbe Interact 12:985–993

Canto T, Prior DAM, Hellwald KH, Oparka KJ, Palukaitis P (1997) Characterization of *Cucumber mosaic virus* IV. Movement protein and coat protein are both essential for cell-to-cell movement of *Cucumer mosaic virus*. Virology 237:237–248

Cao T, Duncan RA, McKenry MV, Shackel KA, DeJong TM, Kirkpatrick BC (2005) Interaction between nitrogen-fertilized peach trees and expression of *syrB*, a gene involved in syringomycin production in *Pseudomonas syringae* pv. *syringae*. Phytopathology 95:581–586

Carrington JC, Jensen PE, Schaad MC (1998) Genetic evidence for an essential role for potyvirus CI protein in cell-to-cell movement. Plant J 14:393–400

Carter JP, Spink J, Cannon PF, Daniels MJ, Osbourn AE (1999) Isolation, characterization and avenacin sensitivity of a diverse collection of cereal root-colonizing fungi. Appl Environ Microbiol 65:3364–3372

Carzaniga R, Bowyer P, O'Connell RJ (2001) Production of extracellular matrices during development of infection structures by the downy mildew *Peronospora parasitica*. New Phytol 149:83–93

Cascales E, Christie PJ (2004a). *Agrobacterium* VirB10, an ATP energy sensor reqired for type IV secretion. Proc Natl Acad Sci USA 101:17228–17233

Cascales E, Christie PJ (2004b) Definition of a bacterial type IV secretion pathway for a DNA substrate. Science 304:1170–1173

Castellano MA, Loconsole G, Grieco F, Di Sansebastiano GP, Martelli GP (2005) Subcellulr localization and immuno-detection of movement proteins of *Olive latent virus*. Arch Virol: DOI 10.1007/s00705-004-0472-y

Catanzariti AM, Dodds PN, Lawrence GJ, Ayliffe MA, Ellis JG (2006) Haustorially expressed secreted proteins from flax rust are highly enriched for avirulence factors. Plant Cell 18: 243–256

Cecchini E, Gong Z, Geri C, Covey SN Milner JJ (1997) Transgenic *Arabidopsis* lines expressing gene VI protein from *Cauliflower mosaic virus* variants exhibit a range of symptom-like phenotype and accumulate inclusion bodies. Mol Plant Microbe Interact 10:1094–1101

Centis S, Guillas D, Séjalon N, Esquerre-Tugaye MT, Dumas B (1997) Endopolygalacturonase genes from *Colletotrichum lindemuthianum*: cloning of *CLPG2* and comparison of its expression to that of *CLPG1* during saprophytic and parasitic growth of the fungus. Mol Plant Microbe Interact 10:769–775

Chagué JC, Danit LV, Siewers V, Gronover CS, Tudzynski P, Tudzynski B, Sharon A (2006) Ethylene sensing and gene activation in *Botrytis cinerea*: a missing link in ethylene regulation of fungus plant interactions? Mol Plant Microbe Interact 19:33–42

Chalfie M, Tu Y, Euskirchen G, Ward WW, Prasher DC (1994) Green fluorescent protein as a marker for gene expression. Science 263:803–805

Champoiseau P, Daugrios JH, Girard JC, Royer M, Rott PC (2006) Variation in albicidin biosynthesis genes and in pathogenicity of *Xanthomonas albilineans*, the sugarcane leaf scald pathogen. Phytopathology 96:33–45

Chang JH, Rathjen JP, Bernal AJ, Staskawicz BJ, Michelmore RW (2000) AvrPto enhances growth and necrosis caused by *Pseudomonas syringae* pv. *tomato* lines lacking either *Pto* or *Prf*. Mol Plant Microbe Interact 13:568–571

Chang C, Stewart RC (1998) The two-component system. Plant Physiol 117:723–731

Che FS, Nakajima Y, Tanaka N, Iwano M, Yoshida T, Takayama S (2000) Flagellin from an incompatible strain of *Pseudomonas avenae* induces a resistance respsonse in cultured rice cells. J Biol Chem 275:32347–32356

Chen HQ, Cao L, Dekkers KL, Rollins JA, Ko NJ, Timmer LW, Chung KR (2005) A gene with domains related to transcripts regulation is required for pathogenicity in *Colletotrichum acutatum* causing Key lime anthracnose. Mol Plant Pathol 6:513–525

Chen C, Harel A, Gorovoits R, Yarden O, Dickman MB (2004) MAPK regulation of sclerotial development in *Sclerotinia sclerotiorum* is linked with pH and cAMP sensing. Mol Plant Microbe Interact 17:404–413

Chen Z, Kloek AP, Cuzick A, Moeder W, Tang D, Innes RW, Klessig DF, McDowell JM, Kunkel BM (2004) The *Pseudomonas syringae* type III effector AvrRpt2 functions downstream or independently of SA to promote virulence on *Arabidopsis thaliana*. Plant J 37:494–504

Chen H, Lee MH, Daub ME, Chung KR (2007) Molecular analysis of the cercosporin biosynthetic gene cluster in *Cerocospora nicotianae*. Mol Microbiol 64:755–770

Chen J, Li WX, Xie D, Peng JR, Ding SW (2004) Viral virulence protein suppresses RNA silencing-mediated defense but upregulates the role of microRNA in host gene expression. Plant Cell 16:1302–1313

Chen JH, Rathijen JP, Bernal AJ, Staskawicz BJ, Michelmore RW (2000) AvrPto enhances growth and necrosis caused by *Pseudomonas syringae* pv. *tomato* in tomato leaves lacking either *Pto* or *Prf*. Mol Plant Microbe Interact 13:568–571

Chen MH, Sheng J, Hind G, Handa A, Citovsky V (2000) Interaction between the *Tobacco mosaic virus* movement protein and host cell pectin methylesterases is required for viral cell-to-cell movement. EMBO J 19:913–920

Chen MH, Tian GW, Gafni Y, Citovsky (2005) Effects of calrecticulin on viral cell-to-cell movement. Plant Physiol 138:1866–1876

Cheng YQ, Ahn JH, Waltson JD (1999) A putative branched-amino acid transaminase gene required for HC-toxin biosynthesis and pathogenicity in *Cochliobolus carbonum*. Microbiology (Reading) 145:3539–3546

Cheng NH, Su CL, Carter SA, Nelson RS (2000) Vascular invasion routes and systemic accumulation patterns of *Tobacco mosaic virus* in *Nicotiana benthamiana*. Plant J 23:349–362

Chisholm ST, Coaker G, Day B, Staskawicz BJ (2006) Host–microbe interactions: shaping the evolution of the plant immune response. Cell 124:803–814

Chisholm ST, Mahajan SK, Whitham SA, Yamamoto ML, Carrington JC (2000) Cloning of the Arabidopsis *RTM1* gene, which controls restriction of long-distance movement of *Tobacco etch virus*. Proc Natl Acad Sci USA 97:489–494

Chisholm ST, Parra MA, Anderberg RJ, Carrington JC (2001) Arabidopsis *RTM1* and *RTM2* genes function in phloem to restrict long-distance movement of *Tobacco etch virus*. Plant Physiol 127:1667–1675

Chiu WL, Niwa Y, Zeng W, Hirano T, Kobayashi H, Sheen J (1996) Engineered GFP as a vital reporter in plants. Curr Biol 6:325–330

Choi WB, Kang SH, Lee YW, Lee YH (1998) Cycli-AMP restores appressorium formation inhibited by polyamines in *Magnaporthe grisea*. Phytopathology 88:58–62

Choi SK, Yoon JY, Ryu KH, Choi JK, Palukaitis P, Park WM (2002) Systemic movement of a movement-deficient strain of *Cucumber mosaic virus* in Zucchini squash is facilitated by a cucurbit-infecting potyvirus. J Gen Virol 83:3173–3178

Choquer M, Dekkers KL, Chen HQ, Cao L, Ueng PP, Daub ME, Chung KR (2005) The *CTB1* gene encoding a fungal polyketide synthase is required for cercosporin biosynthesis and fungal virulence of *Cercospora nicotianae*. Mol Plant Microbe Interact 18:468–476

Choquer M, Lee MH, Bau HJ, Chung KR (2007) Deletion of a MFS transporter-like gene in *Cerocospora nicotianae* reduces cercosporin toxin accumulation and fungal virulence. FEBS Lett 481:489–494

Christensen AB, Cho BH, Naesby M, Gregersen PL, Brandt J, Madriz-Ordeñana K, Collinge DB, Thordal-Christensen H (2002) The molecular characterization of two barley proteins establishes the novel PR-17 family of pathogenesis-related proteins. Mol Plant Pathol 3:135–144

Christie PJ (2004) Type IV secretion: the *Agrobacterium* VirB1D4 and related conjugation systems. Biochem Biophys Acta 1694:219–234

Chu ZQ, Jia JW, Zhou XJ, Chen XY (1999) Isolation of glycoproteins from *Verticillium dahliae* and their phytotoxicity. Acta Bot Sin 41:972–976

Chu M, Park JW, Scholthof HB (1999) Separate regions on the *Tomato bushy stunt virus* p22 protein mediate cell-to-cell movement versus elicitation of effective resistance responses. Mol Plant Microbe Interact 12:285–292

Chun SY, Parikson JS (1988) Bacterial motility: membrane topology of the *Escherichia coli* motB protein. Science 239:276–278

Ciesiolka LD, Hwin T, Geralds JD, Minsavage GV, Saenz R, Bravo M (1999) Regulation of expression of avirulence gene of interaction factors by sequence analysis of avrBsT. Mol Plant Microbe Interact 12:35–44

Citovsky V, De Vos G, Zambryski P (1988) Single-stranded DNA binding protein encoded by the *virE* locus of *Agrobacterium tumefaciens*. Science 240:501–504

Citovsky V, Guralnick B, Simon MN, Wall JS (1997) Nuclear import of *Agrobacterium* VirD2 and VirE2 proteins in maize and tobacco. Proc Natl Acad Sci USA 91:3210–3214

Citovsky V, Kapelnikov A, Oliel S, Zakai N, Rojas MR, Gilbertson RL (2004) Protein interactions involved in nuclear import of *Agrobacterium* VirE2 protein in vitro and in vivo. J Biol Chem 279:29528–29533

Citovsky V, Mc Lean BG, Zupan J, Zambryski PC (1993) Phosphorylation of *Tobacco mosaic virus* cell-to-cell movement protein by a developmentally regulated plant cell wall associated protein kinase. Genes Develop 7:901–910

Ciuffetti LM, Francl LJ, Ballance GM, Bockus WW, Lamari L, Meinhardt SW, Rasmussen JB (1998) Standardization of toxin nomenclature in the *Pyrenophora tritici-repentis* / wheat interaction. Canad J Plant Pathol 20:421–424

Ciuffetti LM, Tuori RP, Gaventa JM (1999) A single gene encodes a selective toxin causal to the development of tan spot of wheat. Plant cell 9:135–144

Clement JA, Mastin SG, Porter R, Butt TM, Beckett A (1993) Germination and the role of extracellular matrix in adhesion of urediniospores of *Uromyces viciae-fabae* to synthetic surfaces. Mycol Res 97:585–593

Clendennen SK, May GD (1997) Different gene expression in ripening banana fruit. Plant Physiol 115:463–469

Coaker G, Falick A, Staskawicz B (2005) Activation of a phytopathogenic bacterial effector protein by a eukaryotic cyclophilin. Science 308:548–550

Cohen Y, Qu F, Gisel A, Morris TJ, Zambryski PC (2000) Nuclear localization of *Turnip crinkle virus* movement protein p8. Virology 273:276–285

Cohn J, Martin GB (2003) Pathogen recognition and signal transduction in plant immunity. In: Ezekowitch RABH, Hoffman JA (eds) Infection, disease: innate immunity. Humana Press Inc., Totowa, NJ, USA, pp 3–27

Cohn JR, Martin GB (2005) *Pseudomonas syringae* pv. *tomato* type III effectors AvrPto and AvrPtoB promote ethylene-dependent cell death in tomato. Plant J 44:139–154

Collmer A, Badel JL, Charkowski AO, Deng WL, Fouts DW, Ramos AR, Rehm AH, Anderson DM, Schneewind O, van Dijk K, Alfano JR (2000) *Pseudomonas syringae* pv. *syringae* Hrp type III secretion system and effector proteins. Proc Natl Acad Sci USA 97:8770–8777

Collmer A, Lindeberg M, Petnicki-Ocwieja T, Schneider DJ, Alfano R (2002) Genomic mining type III secretion system effectors in *Pseudomonas syringae* yields new picks for all TTSS prospectors. Trends Microbiol 10:462–469

Commènil P, Belingheri L, Dehorter B (1998) Antilipase antibodies prevent infection of tomato leaves by *Botrytis cinerea*. Physiol Mol Plant Pathol 52:1–14

Condemine G, Castillo A, Passeri F, Enard C (1999) The PecT repressor coregulates synthesis of exopolysaccharides and virulence factors in *Erwinia chrysanthemi*. Mol Plant Microbe Interact 12:45–52

Cooper JI, Jones AT (1983) Responses of plants to viruses: proposals for the use of terms. Phytopathology 73:127–128

Cooper RM, Kemp B, Day R, Gomez-Vasquez R, Beeching JR (2001) Pathogenicity and resistance in Xanthomonas blight of cassva. In: De Boer SH (ed), Plant pathogenic bacteria. Kluwer, Dordrecht

Coquoz JL, Buchala AJ, Meuwly P, Metraux JP (1995) Arachidonic acid induces local but not systemic synthesis of salicylic acid and confers systemic resistance in potato plants to *Phytophthora infestans* and *Alternaria solani*. Phytopathology 85:1219–1224

Corbett M, Virtue S, Bell K, Birch P, Burr T, Hyman L, Lilley K, Poock S, Toth I, Salmond G (2005) Identification of a new quorum sensing controlled virulence factor in *Erwinia carotovora* subsp. *atroseptica* secreted via type II targeting pathway. Mol Plant Microbe Interact 18:334–342

Cornelis GR, van Gijsegem F (2000) Assembly and function of type III secretory systems. Annu Rev Microbiol 54:735–774

Cousin A, Mehrabi R, Guilleroux M, Dufresne M, van der Lee T, Waalwijk C, Langin T, Kema GHJ (2006) The MAP kinase-encoding gene *MgFus3* of the non-appressorium phytopathogen *Mycosphaerella graminicola* is required for penetration and in vitro pycnidia formation. Mol Plant Pathol 7:269–278

Covey SN, Al-Kaff NS, Langara A, Turner DS (1997) Plants combat infection by gene silencing. Nature 385:781–782

Cowan GH, Loliopoulou F, Ziegler A, Torrance L (2002) Subcellular localization, protein interactions and RNA binding of *Potato moptop virus* triple gene block proteins. Virology 298:106–115

Cristobal SJ, deGier JW, Nielsen H, von Heijne G (1999) Competition between Sec- and TAT-dependent protein translocation in *Escherichia coli*. EMBO J 18:2981–2990

Cronin S, Verchot J, Haldeman-Cahill R, Schaad MC, Carrington JC (1995) Long-distance movement factor: a transport function of the potyvirus helper component proteinase. Plant cell 7:549–559

Crowhurst RN, Binnie SJ, Bowen JK, Hawthorne BT, Plummer KM, Rees-George J, Rikkerink EHA, Templeton MD (1997) Effect of disruption of a cutinase gene (*cut A*) on virulence and tissue specificity of *Fusarium solani* f.sp. *Cucurbita* race 2 toward *cucurbita maxcha* and *Cucurbita moschata*. Mol Plant Microbe Interact 10:355–368

Cui Y, Chatterjee A, Hasegawa H, Chatterjee AK (2006) *Erwinia carotovora* subspecies produce duplicate variants of ExpR LucR homologs that activate *r8mA* transcription but differ in their interactions with *N*-acylhomoserine lactone signals. J Bacteriol 188:4715–4726

Culver JN, Dawson WO (1989) Point mutations in the coat protein gene of *Tobacco mosaic virus* induce hypersensitivity in *Nicotiana sylvestris*. Mol Plant Microbe Interact 2:209–213

Culver JN, Lindebeck AGC, Dawson WO (1991) Virus–host interactions: induction of chlorotic and necrotic responses in plants by tobamoviruses. Annu Rev Phytopathol 29:193–217

Cumagan CJR, Bowden RL, Jurgenson JE, Leslie JF, Miedaner T (2004) Genetic mapping of pathogenicity and aggressiveness of *Gibberella zeae* (*Fusarium graminearum*) toward wheat. Phytopathology 94:520–526

Cunnac S, Occhialini A, Barberis P, Boucher C, Genin S (2004) Inventory and functional analysis of the large Hrp regulon in *Ralstonia solanacearum*: identification of novel effector proteins translocated to plant host cells through the type III secretion system. Mol Microbiol 53:115–128

Curtis MJ, Wolpert TJ (2002) The oat mitochondrial permeability transition and its implication in victorin binding and induced cell death. Plant J 29:295–312

Czymmek KJ, Bourett TM, Howard RJ (2005) Fluorescent protein probes in fungi. In: Savidge T, Pothoulakis (eds) Methods in microbiology – microbial imaging, vol 34. Elsevier, Amsterdam, pp 27–62

Daniels R, Vanderleyden J, Michels J (2004) Quorum sensing and swarming migration in bacteria. FEMS Microbiol Rev 28:261–289

Dardick CD, Golen S, Culver JN (2000) Susceptibility and symptom development in *Arabidopsis thaliana* to *Tobacco mosaic virus* is influenced by virus cell-to-cell movement. Mol Plant Microbe Interact 13:1139–1144 da Silva AC, Ferro JA, Reinach FC Farah CS, Furlan LR, Quaggio RB (2002) Comparison of the genomes of two *Xanthomonas* pathogens with differeing host specificities. Nature 417: 459–463

DebRoy S, Thilmony R, Kwack YB, Nomura K, He SY (2004) A family of conserved bacterial effectors inhibits salicylic acid-mediated basal immunity and promotes disease necrosis. Proc Natl Acad USA 101:9927–9932

Deising H, Nicholson RL, Haug M (1992) Adhesion pad formation and the involvement of cutinase and esterases in the attachment of uredospores to the host cuticle. Plant cell 4:1101–1011

De Kievit TR, Iglweski BH (2000) Bacterial quorum-sensing in pathogenic relationships. Infection Immunol 68:4839–4849

Dekkers LA, You BJ, Gowda VS, Liao HL, Lee MH, Bau HJ et al (2007) The *Cercospora nicotianae* gene encoding dual o-methyltransferase and FAD-dependent monooxygenase domains mediates cercosporin toxin biosynthesis. Fungal Genet Biol 44:444–454

Delgado S, Martinez de Alba E, Hernández C, Flores R (2005) A short double-stranded RNA motif of *Peach latent mosaic viroid* contains initiation and the self-cleavage sites of both polarity strands. J Virol 79:12934–12943

Dellagi A, Brisset MN, Paulin JP, Expert D (1998) Dual role of desferrioxamine in *Erwinia amylovora* pathogenicity. Mol Plant Microbe Interact 11:743–742

Deng W, Chen L, Wood DW, Metcalfe T, Liang X, Gordon MP, Comai L, Nester EW (1998) *Agrobacterium* VirD2 protein interacts with plant host cyclophilins. Proc Natl Acad Sci USA 95:7040–7045

Denny TP (1995) Involvement of bacterial polysaccharides in plant pathogenesis. Annu Rev Phytopathol 33:173–197

Denny TP, Schell MA (1994) Virulence and pathogenicity of *Pseudomonas solanacearum*: genetic and biochemical perspectives. In: Bills D, Kung SD (eds), Bacterial pathogenesis and disease resistance. World Science, River Edge, NJ, USA

Desjardins AE, Munkvold GP, Plattner RD, Proctor RH (2002) *FUM1*-a gene required for fumonisin biosynthesis but not for maize ear rot and ear infection by *Gibberella moniliformis* in field tests. Mol Plant Microbe Interact 15:1157–1164

Dowell FE, Pearson T, Maghirang EB, Xie F, Wicklow DT (2002) Reflectance and transmittance spectroscopy applied to detecting fumonisin in single corn kernels infected with *Fusarium verticillioides*. Cereal Chem. 79:222–226

DeZwaan TM, Carroll AM, Valent B, Sweigard JA (1999) *Magnaporthe grisea* Pth11p is a novel plasma membrane protein that mediates appressorium differentiation in response to inductive substrate cues. Plant Cell 11:2013–2030

Dickman MB, Ha YS, Yang Z, Adams B, Huang C (2003) A protein kinase from *Colletotrichum trifolii* is induced by plant cutin and is required for appressorium formation. Mol Plant Microbe Interact 16:411–421

Dickman MB, Park YK, Oltersdorf T, Li W, Clemente T, French R (2001) Abrogation of disease development in plants expressing animal antiapoptotic genes. Proc Natl Acad Sci USA 98:6957–6962

Dickman MB, Podila GK, Kolattukudy PE (1989) Insertion of cutinase gene into wound pathogen enables it to infect intact host. Nature (London) 342:446–448

Dickman MB, Yarden O (1999) Serine/threonine protein kinase and phosphates in filamentous fungi. Fungal Genet Biol 26:99–117

Diener TO (2001) The viroid: biological oddity or evolutionary fossil? Adv Virus Res 57:137–184

Dillin A (2003) The specifics of small-interfering RNA specificity. Proc Natl Acad Sci USA 100:6289–6291

Ding H, Balsiger S, Guggenbuhl C, Hohl HR (1994) A putative IgG-binding 65-kDa adhesin involved in adhesion and infection of soybeans by *Phytophthora megasperma* f.sp. *glycinea*. Physiol Mol Plant Pathol 44:363–378

Ding SW, Li WX, Symons RH (1995) A novel naturally occurring hybrid gene encoded by a plant RNA virus facilitates long-distance virus movement. EMBO J 14:5762–5772

Ding XS, Liu J, Cheng NH, Folimonov A, Hou YM, Bao Y, Katagi C, Carter SA, Nelson RS (2004) The *Tobacco mosaic virus* 126-kDa protein associated with virus replication and movement suppresses RNA silencing. Mol Plant Microbe Interact 17:583–592

Di Pietro A, Garcia-MacEira FI, Meglecz E, Roncero MI (2001) A MAP kinase of the vascular wilt fungus *Fusarium oxysporum* is essential for root penetration and pathogenesis. Mol Microbiol 39:1140–1152

Doehlemann G, Berndt P, Hahn M (2006) Different signaling pathways involving a Gα protein, cAMP and a MAP kinase control germiantion of *Botrytis cinerea* conidia. Mol Microbiol 59:821–835

Dolja VV, Haldeman-Cahill R, Montgomery AE, vanden Bosch KA, Carrington JC (1995) Capsid protein determinants involved in cell-to-cell and long-distance movement of tobacco etch potyvirus. Virology 297:1007–1016

Dombek P, Ream W (1997) Functional domains of *Agrobacterium tumefaciens* single-stranded DNA-binding protein Vir E2. J Bacteriol 179:1165–1173

Donald RGK, Jackson AO (1996) RNA-binding activities of *Barley stripe mosaic virus* gamma b fusion proteins. J Gen Virol 77:879–888

Dorman CJ (2004) H-NS: a universal regulator for a dynamic genome. Nature Rev Microbiol 2:391–400

Dorman CJ, Deighan P (2003) Regulation of gene expression by histone-like proteins in bacteria. Curr Opin Genet Dev 13:179–184

Dow JM, Crossman L, Findlay K, He YQ, Feng JX, Tang JL (2003) Biofilm dispersal in *Xanthomonas campestris* is controlled by cell–cell signaling and is required for full virulence to plants. Proc Natl Acad Sci USA 100:10995–11000

Dow JM, Newman MA, von Roepenack E (2000) The induction and modulation of plant defense responses by bacterial lipopolysaccharides. Annu Rev Phytopathol 38:241–261

Dowd C, Wilson IW, McFadden H (2004) Gene expression profile changes in cotton root and hypocotyl tissues in response to infection with *Fusarium oxysporum* f.sp. *vasinfectum*. Mol Plant Microbe Interact 17:654–667

Dreier J, Meletzus D, Eichenlaub R (1997) Characterization of the plasmid encoded virulence region *pat-1* of phytopathogenic *Clavibacter michiganensis* subsp. *michiganensis*. Mol Plant Microbe Interact 10:195–206

Duan YP, Castañeda A, Zhao G, Erdos G, Gabriel DW (1999) Expression of a single, host-specific bacterial pathogenicity gene in plant cells elicits divisions, enlargement and cell death. Mol Plant Microbe Interact 12:556–560

Duban ME, Lee K, Lynn DG (1993) Strategies in pathogenesis: mechanistic specificity in the detection of genic signals. Mol Microbiol 7:637–645

Duckely M, Hohn B (2003) The VirE2 protein of *Agrobacterium tumefaciens*: the Yin and Yang of T-DNA transfer. FEMS Microbiol Lett 223:1–6

Duckely M, Oomen C, Axthelm F, Van Gelder P, Waksman G, Engel A (2005) The VirE1 Vir E2 complex of *Agrobacterium tumefaciens* interacts with single-stranded DNA and forms channels. Mol Microbiol 58:1130–1142

Dufrense M, Osbourn AE (2001) Definition of tissue-specific and general requirement for plant infection in a phytopathogenic fungus. Mol Plant Microbe Interact 14:300–307

Dufrense M, Perfect S, Pellier AL, Bailey JA and Langin T (2000) AGAL4-like protein is involved in the switch between biotrophic and necrotrophic phases of infection process of *Colletrotichum lindemuthianum* on common bean. Plant cell 12:1579–1589

Dumas F, Duckely M, Pelczar P, Van Gelder P, Hohn B (2001) An *Agrobacterium* VirE2 channel for transferred-DNA transport into plant cells. Proc Natl Acad Sci USA 98: 485–490

Dunger G, Arabolaze AL, Gottig N, Orellano EG, Ottado J (2005) Participation of *Xanthomonas axonopodis* pv. *citri hrp* cluster in citrus canker and nonhost plant responses. Plant Pathol 54:781–788

Dunny GM, Winans SC (1999) Cell–cell signaling in bacteria. ASM Press, Washington, DC

Dunoyer P, Pfeffer S, Fritsch C, Hemmer O, Voinnet O, Richards KE (2002) Identification, subcellular localization and some properties of a cysteine-rich suppressor of gene silencing encoded by *Peanut clump virus*. Plant J 29:555–567

Dye F, Berthelot K, Griffon B, Delay D, Delmotte FM (1997) Alkylsyringamides, new inducers *Agrobacterium tumefaciens* virulence genes. Biochimie 79:3–6

Dye BT, Schell K, Miller DJ, Ahlquist P (2005) Detecting protein–protein interaction in live yeast by flow cytometry. Cytometry A63A:77–86

Ebbole DJ, Jin Y, Thon M, Pan H, Bhattarai E, Thomas T, Dean R (2004) Gene discovery and gene expression in rice blast fungus, *Magnaporthe grisea:* analysis of expressed sequence tags. Mol Plant Microbe Interact 17:1337–1347

Ehrlich KC, Chang PK, Yu J, Cotty PJ (2004) Aflatoxin biosynthesis cluster gene *cypA* is required for G aflatoxin formation. Appl Environ Microbiol 70:6518–6524

Ehrlich KC, Cotty PJ (2004) An isolate of *Aspergillus flavus* used to reduce aflatoxin contamination in cotton seed has a defective polyketide synthase gene. Appl Mcirobiol Biotechnol 65: 473–478

Eisenbrandt R, Kalkum M, Lai EM, Lurz R, Kado CI, Lunka E (1999) Conjugative pili of IncP plasmids and the Ti plasmid T pilus are composed of cyclic subunits. J Biol Chem 274: 22548–22555

Ellenberg J, Lippincott-Schwartz J, Presley JF (1998) Two color green fluorescent protein time-lapse imaging. Biotechniques 25:838–846

Engerbrecht J, Nealson KL, Silverman M (1983) Bacterial bioluminescence: isolation and genetic analysis of the functions from *Vibrio fischeri*. Cell 32:773–781

Engerbrecht J, Silverman M (1987) Nucleotide sequence of the regulatory locus controlling expression of bacterial genes for bioluminescence. Nucleic Acids Res 15:10455–10467

Engstrom P, Zambryski P, Van Montagu M, Stachel S (1987) Characterization of *Agrobacterium tumefaciens* virulence proteins induced by the plant factor acetosyringone. J Mol Biol 197: 635–645

Eriksson RR, Anderson RA, Pirhonen M, Palva ET (1998) Two component regulators involved in the global control of virulence of *Erwinia carotovora* subsp. *carotovora.* Mol Plant Microbe Interact 11:743–752

Escaler M, Aranda MA, Thomas CL, Maule AJ (2000) Pea-embryonic tissues show common responses to the replication of a wide range of viruses. Virology 267:318–325

Espino JJ, Brito N, Noda J, González C (2005) *Botrytis cinerea* endo-β-1,4-glucanase Cel5A is expressed during infection but is not required for pathogenesis. Physiol Mol Plant Pathol 66:213–221

Espinosa A, Alfano JR (2004) Disabling surveillance: bacterial type III secretion system effectors that suppress innate immunity. Cell Microbiol 6:1027–1040

Espinosa A, Guo M, Tam VC, Fu ZO, Alfano JR (2003) The *Pseudomonas syringae* type III-secreted protein HopPtoD2 possesses protein tyrosine phosphatase activity and suppresses programmed cell death in plants. Mol Microbiol 49:377–387

Eudes F, Comeau A, Rioux S, Collin J (2001) Impact of trichothecenes on *Fusarium* head blight (*Fusarium graminearum*) development in spring wheat (*Triticum aestivum*). Canad Plant Pathol 23:318–322

Fagan H (1979) Postbloom fruit drop of citrus associated with a form of *Colletotrichum gloeosporioides*. Ann Appl Biol 91:13–20

Fang EGC, Dean RA (2000) Site-directed mutagenesis of the *magB* gene affects growth and development in *Magnaporthe grisea*. Mol Plant Microbe Interact 13:1214–1217

Farsing JW, Auffarth K, Basse CW (2005) Identification of cis-active elements in *Ustilago maydis* mig2 promoters conferring high-level activity during pathogenic growth in maize. Mol Plant Microbe Interact 18:75–87

Favaron F, Sella L, D'Ovidio R (2004) Relationships among endopolygalacturonase, oxalate, pH and plant polygalacturonase-inhibiting protein (PGIP) in the interaction between *Sclerotinia sclerotiorum* and soybean. Mol Plant Microbe Interact 17:1402–1409

Feldman MF, Cornelis GR (2003) The multitalented type III chaperones, all you can do with 15-kDa. FEMS Microbiol Lett 219:151–158

Felix G, Boller T (2003) Molecular sensing of bacteria in plants. The highly conserved RNA-binding motif RNP-1 of bacterial cold shock proteins is recognized as an elicitor signal in tobacco. J Biol Chem 278:6201–6208

Fellbrich G, Romanski A, Varet A, Blume B, Brunner F, Engelhardt S, Felix G, Kemmerling B, Krzymowska M, Nürnberger T (2002) NPP1, a *Phytophthora*-associated trigger of plant defense in parsley and *Arabidopsis*. Plant J 32:375–390

Ferrari S, Galletti R, Vairo D, Cervone F, de Lorenzo G (2006) Antisense expression of the *Arabidopsis thaliana AtPGIP1* gene reduces polygalacturonase-inhibiting protein accumulation and enhances susceptibility to *Botrytis cinerea*. Mol Plant Microbe Interact 19:931–936

Fields PI, Groisman EA, Heffron F (1989) A *Salmonella* locus that controls resistance to microbial proteins from phagocytic cells. Science 243:1059–1062

Fisher DB, Wu Y, Ku MSB (1992) Turnover of soluble proteins in the wheat sieve tube. Plant Physiol 100:1433–1441

Flego D, Marits R, Eriksson ARB, Kõiv V, Karlsson MB, Heikinheimo R, Palva ET (2000) A two-component regulatory system, pehR-pehS, controls endopoly-galacturonase production and virulence in plant pathogen *Erwinia caratovora* subsp. *carotovora*. Mol Plant Microbe Interact 13:447–455

Flores R, Delago S, Rodio M, Ambrós S, Hernandez C, Di Serio F (2006) *Peach latent mosaic viroid*: not so latent. Mol Plant Pathol 7:209–221

Flores R, Fernandez C, Martinez de Alba E, Doro's JA, Di Serio F (2005a) Viroids and viroid–host interactions. Annu Rev Phytopathol 43:117–139

Flores R, Hernández C, Avinent L, Hermoso A, Llácer G, Juárez J, Arregni JM, Navarro L, Desvignes JC (1992) Studies on the detection, transmission and distribution of *Peach latent mosaic viroid* in peach trees. Acta Horti 309:325–350

Flores R, Randles JW, Owens RA, Bar-Joseph M, Diener TO (2005b) Viroids. In: Fauquet CM, Mayo MA, Maniloff I, Desselberger U, Ball AL (eds) Virus taxonomy – eigth report of the International Committee on Taxonomy of Viruses, Elsevier Academic Press, London, pp1145–1159

Frandsen RJN, Nielsen N, Maolanon N, Sørensen JC, Olsson S, Nielsen J, Giese H (2006) The biosynthetic pathway for aurofusarin in *Fusarium graminearum* reveals a close link between the naphthquinones and napthapyrones. Mol Microbiol 61:1069–1080

Franza T, Sauvage C, Expert D (1999) Iron regulation and pathogenicity in *Erwinia chrysanthemi* 3937: role of the Fur repressor protein. Mol Plant Microbe Interact 12:119–128

Frederick RD, Ahmad M, Majerczak DR, Arroyo-Rodriquez AS, Manulis S, Coplin DL (2001) Genetic organization of the *Pantoea stewartii* subsp *stewartii hrp* gene cluster and sequence analysis of the *hrpA, hrpC, hrpN* and *WtsE* operons. Mol Plant Microbe Interact 14: 1213–1222

Fridborg I, Grainger J, Page A, Coleman M, Findlay K, Angell S (2003) TIP, a novel host factor linking callose degradation with the cell-to-cell movement of *Potato virus X*. Mol Plant Microbe Interact 16:132–140

Fujiki M, Kawakami S, Kim RW, Beachy RN (2006) Domains of *Tobacco mosaic virus* movement protein essential for its membrane association. J Gen Virol 87:2699–2707

Fujiwara A, Takamura H, Majumder P, Yoshida H, Kojima M (1998) Functional analysis of the protein encoded by the chromosomal virulence gene (*acvB*) of *Agrobacterium tumefaciens*. Ann Phytopathol Soc Jpn 64:191–193

Fullner KJ, Lara JC, Nester EW (1996) Pilus assembly of *Agrobacterium* T-DNA transfer genes. Science 273:1107–1109

Fuqua WC, Winans SC, Greenberg EP (1994) Quorum sensing in bacteria: the LuxR-*LuxI* family of cell-density-responsible transcriptional regulators. J Bacteriol 176:269–275

Gachon C, Saindrenan P (2004) Real-time PCR monitoring of fungal development in *Arabidopsis thaliana* infected by *Alternaria brassicola* and *Botrytis cinerea*. Plant Physiol Biochem 42: 367–371

Gaffoor I, Brown DW, Plattner R, Proctor RH, Qi W, Trail F (2005) Functional analysis of the polyketide synthase genes in the filamentous fungus *Gibberella zeae* (Anamorph – *Fusarium graminearum*). Eukary Cell 4:1926–1933

Galan JE, Collmer A (1999) Type III secretion machines: bacterial devices for protein delivery into host cells. Science 284:1322–1328

García-Cano E, Resende RO, Fernández-Muñoz R, Moriones E (2006) Syngergistic interaction between *Tomato chlorosis virus* and *Tomato spotted wilt virus* results in breakdown of resistance in tomato. Phytopathology 96:1253–1263

García-Rodríguez FM, Schrammeijer B, Hooykaas PJJ (2006) The *Agrobacterium* VirE3 effector protein: a potential plant transcriptional activator. Nucleic Acids Res 34:6496–6504

Gardner JM, Kono Y, Tatum JH, Suzuki Y, Takeuchi S (1985) Structure of major component of ACRL toxins, host-specific phytotoxic compound produced by *Alternaria citri*. Agri Biol Chem 49:1235–1238

Gao Z, Johansen S, Eyers S, Thomas CL, Ellis THN, Maule AJ (2004) The potyvirus recessive resistance gene, *sbm1*, identifies a novel role for translation initiation factor eIF4E in cell-to-cell trafficking. Plant J 40:376–385

Gaudriault S, Paulin JP, Barny MA (2002) The DspB/F protein *Erwinia amylovora* is a type III secretion chaperone ensuring efficient intrabacterial production of the Hrp-secreted DspA/E pathogenicity factor. Mol Plant Pathol 3:313–320

Geider K, Zhang Y, Ullrich H, Langlotz (1999) Expression of virulence factors secreted by *Erwinia amylovora*. Acta Horti 489:347–351

Gelvin SB (2003) *Agrobacterium*-mediated plant transformation: the biology behind the 'gene-jockeying' tool. Microbiol Mol Biol Rev 67:16–37

Gheysen G, Villarroel R, Van Montagu M (1991) Illegitimate recombination in plants: a model for T-DNA integration. Genes Develop 5:287–297

Ghorbel R, López, C, Fagoaga C, Moreno P, Navarro L, Flores R, Peña L (2001) Transgenic citrus plants expressing the citrus tristeza p23 protein exhibit viral-like symptoms. Mol Plant Pathol 2:27–36

Giesbert S, Lepping HB, Tenberge KB, Tudzynski P (1998) The xylanolytic system of *Claviceps purpurea*: cytological evidence of secretion of xylanases in infected rye tissue and molecular characterization of two xylanase genes. Phytopathology 88:1020–1030

Gietl C, Koukolikova-Nicola Z, Hohn B (1987) Mobilization of T-DNA from *Agrobacterium* to plant cells involves a protein that binds single-stranded DNA. Proc Natl Acad Sci USA 84:9006–9010

Gilchrist DG (1998) Programmed cell death in plant disease. The purpose and promise of cellular suicide. Annu Rev Phytopathol 24:104–111

Gillespie T, Boevink P, Haupt S, Roberts AG, Toth R, Valentine T, Chapman S, Oparka KJ (2002) Movement protein reveals that microtubules are dispensable for the cell-to-cell movement of *Tobacco mosaic virus*. Plant Cell 14:1207–1222

Gloux K, Touze T, Pagot Y, Jouan A, Blanco C (2005) Mutation of *ousA* alter the virulence of *Erwinia chrysanthemi*. Mol Plant Microbe Interact 18:150–157

Goicoechea N, Aguirreolea J, Cenoz S, Garcia-Mina JM (2000) *Verticillium dahliae* modifies the concentration of proline, soluble sugars, starch, soluble protein and abscisic acid in pepper plants. Eur J Plant Pathol 106:19–25

Goldberg KB, Kiernan J, Shepherd RJ (1999) A disease syndrome associated with expression of gene VI of caulimoviruses may be a nonhost reaction. Mol Plant Microbe Interact 4: 182–189

Golen S, Culver JN (2003) *Tobacco mosaic virus* induced alterations in the gene expression profile of *Arabidopsis thaliana*. Mol Plant Microbe Interact 16:681–688

Gomez-Gomez L, Boller T (2002) Flagellin perception: a paradigm for innate immunity. Trends Plant Sci 7:251–256

González-Jara P, Atencio FA, Martinez-Garcia B, Barajas D, Tenllado F, Diaz-Ruiz JK (2005) A single amino acid mutation in the *Plum pox virus* helper component-proteinase gene abolishes both synergistic and RNA silencing suppression activities. Phytopathology 95:894–901

Gopalan S, He SY (1996) Bacterial genes involved in the elicitation of hypersensitive response and pathogenesis. Plant Disease 80:604–610

Görnhardt B, Rouhara I, Schmelzer E (2000) Cyst germination proteins of the potato pathogen *Phytophthora infestans* share homology with human mucins. Mol Plant Microbe Interact 13: 32–42

Göttfert M, Grob P, Hennecke H (1990) Proposed regulatory pathway encoded by the *nodV* and *nodW* genes, determinants of host specificity in *Bradyrhizobium japonicum*. Proc Natl Acad Sci USA 87:2680–2684

Goytia E, Fernández-Calvino L, Martínez-García B, López-Abe Ila D, López-Moya JJ (2006) Production of *Plum pox virus* HC-Pro functionally active for aphid transmission in a transient-expression system. J Gen Virol 87:3413–3423

Greenberg JT, Yao N (2004) The role and regulation of programmed cell death in plant–pathogen interactions. Cell Microbiol 6:201–211

Grell MN, Holm KB, Giese H (2005) Two novel *Blumeria graminis* f.sp. *hordei* genes are induced in planta and up-regulated in *mlo* virulent isolates. Physiol Mol Plant Pathol 66:79–89

Grogan RG, Kimble KA, Misagi I (1975) A stem canker of tomato caused by *Alternaria alternata* f.sp. *lycopersici*. *Phytopathology* 65:880–886

Gu K, Yang B, Tian D, Wu L, Wang D, Sreekala C, Yang F, Chu Z, Wang GL, White FF (2005) R gene expression induced by a type III effector triggers disease resistance in rice. Nature 435:1122–1125

Gunawardena U, Hawes MC (2002) Tissue specific localization of root infection by fungal pathogens: role of root border cells. Mol Plant Microbe Interact 15:1128–1136

Gunawardena U, Rodriquez M, Straney D, Romeo JT, Van Etten HD, Hawes MC (2005) Tissue-specific localization of pea root infection by *Nectria haematococca* – Mechanisms and consequences. Plant Physiol 137:1363–1374

Guo M, Chancey ST, Tian F, Ge Z, Jamir Y, Alfano JR (2005) *Pseudomonas syringae* type III chaperones Shc 01, ShcS1, and Shc S2 facilitate translocation of their cognate effectors and can substitute for each other in the secretion of Hop01-1. J Bacteriol 187:4257–4269

Guo HS, Ding SW (2002) A viral protein inhibits the long range signaling activity of the gene silencing signal. EMBO J 21:298–407

Gürlebeck D, Thiene F, Bonas U (2006) Type III effector proteins from the plant pathogen *Xanthomonas* and their role in the interaction with the host plant. J Plant Physiol 163:233–255

Gustin MC, Albertyn J, Alexander M, Davenport K (1998) MAP kinase pathways in the yeast *Saccharomyces cerevisiae*. Microbiol Mol Biol Rev 62:1264–1300

Guttman DS, Vinatzer BA, Sarkar SF, Ranall MV, Kettler JT and Greenberg JT (2002) A functional screen for the type III (Hrp) secretome of the plant pathogen *Pseudomonas syringae*. Science 295:1722–1726

Hahn M, Mendgen K (1997) Characterization of in planta induced rust genes isolated from a haustorium specific cDNA library. Mol Plant Microbe Interact 10:427–434

Hahn M, Neef U, Struck C, Göttfert M, Mendgen K (1997) A putative amino acid transporter is specifically expressed in haustoria of the rust fungus *Uromyces fabae*. Mol Plant Microbe Interact 10:435–438

Hamer JE, Howard RJ, Chunley FG, Valent B (1988) A mechanism for surface attachment in spores of a fungal plant pathogen. Science 239:288–290

Harel A, Bercovich S, Yarden O (2006) Calcineurin is required for sclerotial development and pathogenicity of *Sclerotinia sclerotiorum* in an oxalic acid-independent manner. Mol Plant Microbe Interact 19:682–693

Harel A, Gorovits R, Yarden O (2005) Changes in protein kinase A activity accompany sclerotial development in *Sclerotinia sclerotiorum*. Phytopathology 95:397–404

Hao G, Zhang H, Zheng D, Burr TJ (2005) *LuxR* homolog *avhR* in *Agrobacterium vitis* affects the development of a grape-specific necrosis and a tobacco hypersensitive response. J Bacteriol 187:185–192

Harrison BD (1985) Advances in geminivirus research. Annu Rev Phytopathol 23:55–82

Hasdai M, Elmaci C, Goldschmidt EE, Droby S, Porat R (2004) Isolation of a thioredoxin h cDNA from grapefruit peel tissue that is induced upon infection by *Penicillium digitatum* and elicitation of pathogen resistance. Physiol Mol Plant Pathol 65:277–283

Haseloff BI, Freeman TL, Valmeekam V, Melkus MW, Oner F, Valachovic MS, San Francisco MJD (1998) The *exuT* gene of *Erwinia chrysanthemi* EC 16: nucleotide sequence, expression, localization and relevance of the gene product. Mol Plant Microbe Interact 11:270–276

Hastings JW, Greenberg EP (1999) Quorum sensing: the explanation of a curious phenomenon reveals a common characteristic of bacteria. J Bacteriol 181:2667–2668

Hatta R, Ito K, Hosaki Y, Tanaka T, Tanaka A, Yamomoto M, Akimitsu K, Tsuge T (2002) A conditionally dispensable chromosome controls, host-specific pathogencity in the fungal plant pathogen *Alternaria alternata*. Genetics 161:59–70

Hatta R, Shinjo A, Ruswandi S, Kitani K, Yamamoto M, Akimitsu K, Tsuge T (2006) DNA transposon fossils present on the conditionally dispensable chromosome controlling AF-toxin biosynthesis and pathogenicity of *Alternaria alternata*. J Gen Plant Pathol 72: 210–219

Hauck P, Thilmony R, He SY (2003) A *Pseudomonas syringae* type III effector suppresses cell wall based extracellular defense in susceptible *Arabidopsis* plants. Proc Natl Acad Sci USA 100:8577–8582

Haupt S, Cowan GH, Ziegler A, Roberts AG, Oparka KJ, Torrance L (2005) Two plant-viral movement proteins traffic in endocytic recycling pathway. Plant Cell 17:164–181

Havelda Z, Hornyik C, Crescenzi A, Burgyan J (2003) In situ characterization of *Cymbidium ringspot tombusvirus* infection-induced posttranscritptional gene silencing in *Nicotiana benthamiana*. J Virol 77:6082

Havelda Z, Maule AJ (2000) Complex Spatial responses to *Cucumber mosaic virus* infection in susceptible *Cucurbita pepo* cotyledons. Plant cell 12:1975–1985

Hawes MC, Smith LY, Howarth AJ (1988) *Agrobacterium tumefaciens* mutants deficient in chemotaxis to root exudates. Mol Plant Microbe Interact 1:182–186

Hauck P, Thilmony R, He SY (2003) A *Pseudomonas syringae* type III effector suppresses cell wall-based extracellular defense in susceptible *Arabidopsis* plants. Proc Natl Acad Sci USA 100:8577–8582

He SY (1998) Type III protein secretion systems in plant and animal pathogenic bacteria. Annu Rev Phytopathol 36:363–392

He P, Shan L, Lin NC, Martin GB, Kemmerling B, Nürnberger T, Sheen J (2006) Specific bacterial suppressors of MAMP signaling upstream of MAPKKK in *Arabidopsis* innate immunity. Cell 125:563–573

He YW, Xu M, Lin K, Ng YJA, Wen CM, Wang LH, Liu ZD, Zhang HB, Dong YH, Dow JM, Zhang LH (2006) Genome scale analysis of diffusible signal factor regulon in *Xanthomonas campestris* pv. *campestris*: identification of novel cell-cell communication-dependent genes and functions. Mol Microbiol 59:610–622

Hein R, Tsien RY (1996) Engineering green fluorescent protein for improved brightness, longer wavelengths and fluorescence resonance energy transfer. Curr Biol 6:178–182

Heinlein M, Epel BL, Padgett HS, Beachy RN (1995) Interaction of tobamovirus movement proteins with the plant cytoskeleton. Science 270:1983–1985

Heinlein M, Padgett HS, Gens JS, Pickard BG, Casper SJ, Epel BL, Beachy RN (1998) Changing patterns of localization of *Tobacco mosaic virus* movement protein and replicase to the endoplasmic reticulum and microtubules during infection. Plant cell 10:1107–1120

Herbers K, Conrads-Strauch J, Bonas U (1992) Race-specificity of plant resistance to bacterial spot disease determined by repetitive motifs in a bacterial avirulence protein. Nature 356:172–174

Herlache TC, Zhang HS, Reid CL, Carle S, Zheng D, Basaran P, Thaker MA, Burr AT, Burr TJ (2001) Mutations that affect *Agrobacterium vitis*-induced grape necrosis also alter its ability to cause a hypersensitive response on tobacco. Phytopathology 91:966–972

Herrera-Estrella A, Van Montagu M, Wang K (1990) A bacterial peptide acting as a plant nuclear targeting signal: the amino-terminal portion of *Agrobacterium* Vir D2 protein directs a β-galactosidase fusion protein into tobacco nuclei. Proc Natl Acad Sci USA 87:9534–9537

Hildebrand M, Aldridge P, Geider K (2006) Characterization of *hns* genes from *Erwinia amylovora*. Mol Gen Genomics 275: 310–319

Hillen W, Berens C (1994) Mechanisms underlying expression of Tn10 encoded tetracycline resistance. Annu Rev Microbiol 48:345–369

Himber C, Dunoyer P, Moissiard G, Ritzenthaler C, Voinnet O (2003) Transitivity-dependent and-independent cell-to-cell movement of RNA silencing. EMBO J 22:4523–4533

Hoat TX, Nakayashiki H, Tosa Y, Mayama S (2006) Specific cleavage of ribosomal RNA during victorin-induced apoptotic cell death in oat. Plant J 46:922–933

Hoi JWS, Herbert C, Bacha N, O'Connell R, Lafitte C, Borderies G, Rossignol M, Rouge P, Dumas B (2007) Regulation and role of a STE 12-like transcription factor from the plant pathogen *Colletotrichum lindemuthianum*. Mol Microbiol 64:68–82

Hooykaas PJJ, Beijersbergen AGM (1994) The virulence system of *Agrobacterium tumefaciens*. Annu Rev Phytopathol 32:157–179

Hooykaas PJJ, den Dult-Ras H, Schilperont RA (1988) The *Agrobacterium tumefaciens* T-DNA gene 6*b* is an *onc* gene. Plant Mol Biol 11:791–794

Hossain MM, Shibata S, Aizawa SI, Tsuyumu S (2005) Motility is an important determinant for pathogenesis of *Erwinia carotovora* subsp. *carotovora*. Physiol Mol Plant Pathol 66:134–143

Hoston A, Chosed R, Shu H, Orth K, Mudgett MB (2003) *Xanthomonas* type III effector XopD targets SUMO-conjugated proteins in planta. Mol Microbiol 50:377–389

Hossain MM, Tsuyumu S (2006) Flagella-mediated motility is required for biofilm formation by *Erwinia carotovora* subsp. *carotovora*. J Gen Plant Pathol 72:34–39

Hou ZM, Xue CY, Peng YL, Katan T, Kistler HC, Xu JR (2002) A mitogen-activated protein kinase gene (*MGV1*) in *Fusarium graminearum* is required for female fertility, heterokaryon formation and plant infection. Mol Plant Microbe Interact 15:1119–1127

Howard RJ, Ferrari MA, Roach DH, Money NP (1991) Penetration of hard substrates by a fungus employing enormous turgor pressures. Proc Natl Acad Sci USA 88:11281–11284

Howard AR, Heppler ML, Ju HJ, Krishnamurthy K, Payton ME, Verchot-Lubicz J (2004) *Potato virus X* TGBp1 induces plasmodesmata gating and moves between cells in several host species whereas CP moves only in *N. benthamiana* leaves. Virology 328:185–197

Hu W, Yuan J, Jin QL, Hart P, He SY (2001) Immunogold labeling of Hrp pili of *Pseudomonas syringae* pv. *tomato* assembled in minimal medium and in planta. Mol Plant Microbe Interact 14:234–241

Huet H, Gal-On A, Meir E, Locoq H, Raccah B (1994) Mutations in the helper component protease gene of *Zucchini yellow mosaic virus* affect its ability to mediate aphid transmissibility. J Gen Virol 75:275–279

Hughes HB, Carzaniga R, Rawlings SL, Green JR, O'Connell RJ (1999) Spore surface glycoproteins of *Colletotrichum lindemuthianum* are recognized by a monoclonal antibody which inhibits adhesions to polystyrene. Microbiology 145:1927–1936

Huguet E, Hahn K, Wengelnik K, Bonas U (1998) *hpaA* mutants of *Xanthomonas campestris* pv. *vesicatoria* are affected in pathogenicity but retain the ability to induce host specific hypersensitive reaction. Mol Microbiol 29:1379–1390

Hull R (2002) Matthew's plant virology, 4th edn. Academic Press, San Diego, CA, USA

Hyytiäinen H, Montesano M, Palva ET (2001) Global regulators ExpA (GacA) and KdgR modulate extracellular enzyme gene expression through the Rsm-rsmB system in *Erwinia carotovora* subsp. *carotovora*. Mol Plant Microbe Interact 14:931–938

Ishiga Y, Funato A, Tachiki T, Toyoda K, Shiraishi T, Yamada T, Ichinose Y (2002) Expression of the 12-oxophytodienoic acid 10,11-reductase gene in the compatible interaction between pea and fungal pathogen. Plant Cell Physiol 43:1210–1220

Isshiki A, Akimitsu K, Yamamoto M, Yamamoto H (2001) Endopolygalacturonase is essential for citrus black rot caused by *Alternaria citri* but not brown spot caused by *Alternaria alternata.* Mol Plant Microbe Interact 14:749–757

Isshiki A, Ohtani K, Kyo M, Yamamoto H, Akimitsu K (2003) Green fluorescent detection of fungal colonization and endopolygalacturonase gene expression in interaction of *Alternaria citri* with citrus. Phytopathology 93:768–773

Itaya A, Folimonov A, Matsuda Y, Nelson RS, Ding B (2001) Potato spindle viroid as inducer of RNA silencing in infected tomato. Mol Plant Microbe Interact 14:1332–1334

Itaya A, Matsuda Y, Gonzales RA, Nelson RS, Ding B (2002) *Potato spindle tuber viroid* strains of different pathogenicity induces and suppresses expression of common and unique genes in infected tomato. Mol Plant Microbe Interact 15:990–999

Ito S, Kawaguchi T, Nagata A, Tamura H, Matsushita H, Takahara H, Tanaka S, Ikeda T (2004b) Distribution of the *FoToml* gene encoding tomatinase in formae speciales of *Fusarium oxysporum* and identification of a novel tomatinase from *F. oxysporum* f.sp. *radicis-lycopersici*, the causal agent of Fusarium crown and root rot of tomato. J Gen Plant Pathol 70:195–201

Ito S, Nagata A, Kai T, Takahara H, Tanaka S (2005) Symptomless infection of tomato plants by tomatinase producing *Fusarium oxysporum* formae speciales nonpathogenic on tomato plants. Physiol Mol Plant Pathol 66:183–191

Ito K, Tanaka T, Hatta R, Yamamoto M, Akimitsu K, Tsuge T (2004a) Dissection of the host range of the fungal plant pathogen *Alternaria alternata* by modification of secondary metabolism. Mol Microbiol 52:399–411

Ito S, Takahara H, Kawaguchi T, Tanaka S, Kameya-Iwaki M (2002) Post-transcriptional silencing of the tomatinase gene in *Fusarium oxysporum* f.sp. *lycopersici.* J Phytopahtol 150:474–480

Jackson RW, Athanassopoulos E, Tsiamis G, Mansfield JW, Sesma A, Arnold DL,Gibbon MJ, Murillo J, Taylor JD, Vivian K (1999) Identification of a pathogenicity island which contains genes for virulence and avirulence on a large native plasmid in the bean pathogen *Pseudomonas syringae* pv. *phaseolicola*. Proc Natl Acad Sci USA 96:10875–10880

Jacques MA, Josi K, Darrasse A, Samson R (2005) *Xanthomonas axonopodis* pv. *phaseoli* var. *fuscans* is aggregated in stable biofilm population size in the phyllosphere of field-grown beans. Appl Environ Microbiol 71:2008–2015

Jafra S, Figura I, Hugouvieux-Cotte-Pattat N, Lojkowska E (1999) Expression of *Erwinia chrysanthemi* pectinase genes *pelI, pelL* and *pelZ* during infection of potato tubers. Mol Plant Microbe Interact 12:845–851

Jakubowski SJ, Cascales E, Krishnamoorthy V, Christie PJ (2005) *Agrobacterium tumefaciens* VirB9, an outer-membrane-associated component of a type IV secretion system, regulates substrate selection and T-pilus biogenesis. J Bacteriol 187:3486–3495

Jamir Y, Guo M, Oh HS, Petnicki-Ocwieja T, Chen S, Tang X, Dickman MB, Collmer AR, Alfano JR (2004) Identification of *Pseudomonas syringae* type III effectors that can suppress programmed cell death in plants and yeast. Plant J 37:554–565

Janjusevic R, Abramovitch RB, Martin GB, Stebbins CE (2006) A bacterial inhibitor of host programmed cell death defenses is an E3 ubiquitin ligase. Science 311:222–226

Jansen C, Wettstein D, von Schäfer W, Kogel KH, Felk A, Maier FJ (2005) Infection patterns in barley and wheat spikes inoculated with wild-type and trichodiene synthase gene-dirupted *Fusarium graminearum.* Proc Natl Acad Sci USA 102:16892–16897

Jencmionka NJ, Maier FJ, Losch AP, Schafer W (2003) Mating, conidiation and pathogencitiy of *Fusarium graminearum*, the main causal agent of the head blight disease of wheat are regulated by the MAP kinase *gpmk1*. Curr Genet 43:87–95

Jensen DF, Schulz A (2004) Exploitation of GFP-technology with filamentous fungi. Mycol ser: Handbook Fungal Biotechnol, 20:441–451

Jin Q, He SY (2001) The Hrp pilus: a conduit for type III secretion in *Pseudomonas syringae.* Science 294:2556–2558

Jin Q, Hu W, Brown I, McGhee G, Hart P, Jones AL, He SY (2001) Visualization of secreted Hrp and Avr proteins along the Hrp pilus during type III secretion in *Erwinia amylovora* and *Pseudomonas syringae.* Mol Microbiol 40:1129–1139

Jin H, Li S, Villegas A, Jr (2006) Down-regulation of the 26S proteasome subunit RPN9 inhibits viral systemic transport and alters plant vascular development. Plant Physiol 142:651–661

Jin S, Prusti RK, Roitsch T, Ankenbauer RG, Nester EW (1990a) Phosphorylation of the VirG protein of *Agrobacterium tumefaciens* by the autophosphorylated VirA protein: essential role in biological activity of VirG. J Bacteriol 172:4945–4950

Jin S, Roitsch T, Ankenbauer RG, Gordon MP, Nester EW (1990b) The VirA protein of *Agrobacterium tumefaciens* is autophorylated and is essential for *vir* gene regulation. J Bacteriol 172:525–530

Jin Q, Thilmony R, Zwiesler-Vollick J, He SY (2003) Type III protein secretion in *Pseudomonas syringae*. Microbes Infect 5:301–310

Jones DA, Takemoto D (2004) Plant innate immunity-direct and indirect recognition of general and specific pathogen-associated molecules. Curr Opin Immunol 16:48–62

Joshi M, Rong X, Moll S, Kers J, Franco C, Loria R (2007) *Streptomyces turgidiscabies* secretes a novel virulence protein Nec1, which facilitates infection. Mol Plant Microbe Interact 20: 599–608

Joubert DA, Kars I, Wagemakers L, Bergmann C, Kemp G, Vivier MA, van Kan JAL (2007) A polygalacturonase-inhibiting protein from grapevine reduces the symptoms of the endopolygalacturonase BcPG2 from *Botrytis cinerea* in *Nicotiana benthamiana* leaves without any evidence for in vitro interaction. Mol Plant Microbe Interact 20:392–402

Judelson HS, Roberts S (2002) Novel protein kinase induced during sporangial cleavage in the oomycete *Phytophthora infestans*. Eukary Cell 1:687–695

Kadotani N, Nakayashiki H, Tosa Y, Mayama S (2003) RNA silencing in phytopathogenic fungus *Magnaporthe oryzae*. Mol Plant Microbe Interact 16:769–776

Kakizawa S, Oshima K, Kuboyama T (2001) Cloning and expression analysis of phytoplasma translocation genes. Mol Plant Microbe Interact 14:1043–1050

Kakizawa S, Oshima K, Nishigawa H, Jung HY, Wei W, Suzuki S, Tanaka M, Miyata S, Ugaki M, Namba S (2004) Secretion of immunodominant membrane protein from onion yellows phytoplasma through Sec protein-translocation system in *Escherichia coli*. Microbiology 150: 135–142

Kalinina NO, Rakitina DV, Solovyev Ag, Schiemann J, Morozov SY (2002) RNA helicase activity of the plant virus movement proteins encoded by the first gene of the triple gene block. Virology 296:321–329

Kamdar HV, Rowley KB, Clements D, Patil SS (1991) *Pseudomonas syringae* pv *phaseolicola* genomic clones harboring heterologous DNA sequences suppress the same phaseolotoxin deficient mutants. J Bacteriol 173:1073–1079

Kamoun S (2003) Molecular genetics of pathogenic oomycetes. Eukary Cell 2:197–199

Kamoun S (2006) A catalogue of the effector secretome of plant pathogenic Oomycetes. Annu Rev Phytopathol 44:41–60

Kamoun S, van West P, Vleeshowers VG, de Groot KE, Govers F (1998) Resistance of *Nicotiana benthamiana* to *Phytophthora infestans* is mediated by the recognition of the elicitor protein INF1. Plant Cell 10:1413–1426

Kanda A, Yasukohchi M, Ohnishi K, Kiba A, Okuno T, Hikichi Y (2003) Ectopic expression of *Ralstonia solanacearum* effector protein popA early in invasion results in loss of virulence. Mol Plant Microbe Interact 16:447–455

Kang Z, Huang L, Buchenauer H (2000) Cytochemistry of cell wall component alteration in wheat roots infected by *Gauemannomyces graminis* var. *tritici* Z Pflanzenkrank Pflanzensch 107: 337–351

Kang L, Li J, Zhao T, Xiao F, Tang X, Thilmony R, He S, Zhou JM (2003) Interplay of the *Arabidopsis* nonhost resistance gene *NHO1* with bacterial virulence. Proc Natl Acad Sci USA 100:3519–3524

Kang L, Tang X, Mysore KS (2004) *Pseudomonas* type III effector AvrPto suppresses the programmed cell death induced by two nonhost pathogens in *Nicotiana benthamiana* and tomato. Mol Plant Microbe Interact 17:1328–1336

Kaplan IB, Gal-On A, Palukaitis P (1997) Characterization of *Cucumber mosaic* virus. III localization of sequences in the movement protein controlling systemic infection in cucurbits. Virology 230:343–349

Kaplan JM, Zhang L, Palukaitis P (1998) Characterization of *Cucumber mosaic virus*. V. Cell-to-cell movement requires capsid protein but not virions. Virology 246:221–231

Kariola T, Brader G, Li J, Palva ET (2005) Chlorophyllase 1, a damage control enzyme, affects the balance between defense pathways in plants. Plant Cell 17:282–294

Karjalainen R, Lounatmaa K (1986) Ultrastructure of penetration and colonization of wheat leaves by *Septoria nodorum*. Physiol Mol Plant Pathol 29:263–270

Kars I, Krooshof GH, Wagemakers L, Joosten R, Benen JAE, van Kan JAL (2005a) Necrotizing activity of five *Botrytis cinerea* endopolygalacturonases produced in *Pichia pastoris*. Plant J 43:213–225

Kars I, McCalman M, Wagemakers L, van Kan JAL (2005b) Functional analysis of *Botrytis cinerea* pectin methyl esterase genes by PCR-based targeted mutagenesis: *Bcpme1* and *Bcpme2* are dispensable for virulence of strain BO5.10. Mol Plant Pathol 6:641–652

Kars I, van Kan JAL (2004) Extracellular enzymes and metabolites involved in pathogenesis of *Botrytis*. In: Elad Y, Williamson B, Tudzynski P, Nelen N (eds) Botrytis: biology, pathology and control. Kluwer Academic Publishers, Dordrecht, The Netherlands, pp 99–118

Kasschau KD, Carrington JC (1998) A counter defensive strategy of plant viruses: suppression of posttranscriptional gene silencing. Cell 95:461–470

Kasschau KD, Carrington JC (2001) Long-distance movement and replication maintenance functions correlate with silencing suppression activity of potyviral HC.Pro. Virology 285: 71–81

Kasteel DTJ, Perbal CM, Boyer JC, Wellink J, Goldbach RW, Maule AJ, van Lent JWM (1996) The movement proteins of *Cowpea mosaic virus* and *Cauliflower mosaic virus* induce tubular structures in plant and insect cells. J Gen Virol 77:2857–2864

Katoh H, Issihiki A, Masunaka A, Yamamoto H, Akimitsu K (2006) A virulence-reducing mutation in the postharvest citrus pathogen *Alternaria citri*. Phytopathology 96:934–940

Katoh H, Ohtani K, Yamamoto H, Akimitsu K (2007) Overexpression of a gene encoding a catabolite repression element in *Alternaria citri* causes severe symptoms of black rot in citrus fruit. Phytopathology 97:557–563

Kato H, Wada M, Muraya K, Malik K, Shiraishi T, Yamada T (1995) Characterization of nuclear factors for elicitor-mediated activation of the promoter of the pea phenylalanine ammonia-lyase gene 1. Plant Physiol 108:129–139

Kawakami S, Watanabe Y, Beachy RN (2004) *Tobacco mosaic virus* infection spreads cell to cell as intact replication complexes. Proc Natl Acad Sci USA 101:6291–6296

Kawamura C, Moriwaki J, Kimura N, Fujita Y, Fuji S, Hirano T (1997) The melanin biosynthesis genes of *Alternaria alternata* can restore pathogenicity of the melanin-deficient mutants of *Magnaporthe grisea*. Mol Plant Microbe Interact 10:446–453

Keller KE, Johansen E, Martin RR, Hampton RO (1998) Potyvirus genome-linked protein (VPg) determines *Pea seedborne mosaic virus* pathotype-specific virulence in *Pisum sativum*. Mol Plant Microbe Interact 11:124–130

Kemen E, Kemen AC, Rafiqi M, Hempel U, Mendgen K, Hahn M, Voegele RT (2005) Identification of a protein from rust fungi transferred from haustoria into infected plant cells. Mol Plant Microbe Interact 18:1130–1139

Kemp BP, Horne J, Bryant A, Cooper (2004a) *Xanthomonas axonopodis* pv. *manihatis gumD* gene is essential for EPS production and pathogenicity and enhances epiphytic survival on cassava (*Manihot esculenta*). Physiol Mol Plant Pathol 64:209–218

Kemp G, Stanton L, Bergmann CW, Clay RP, Albersheim P, Darvill A (2004b) Polygalacturonase-inhibiting proteins can function as activators of polygalacturonase. Mol Plant Microbe Interact 17:888–894

Keshavarzi M, Soylu S, Brown I, Bonas U, Nicole M, Rossiter J, Mansfield J (2004) Basal defenses induced in pepper by lipopolysaccharides are suppressed by *Xanthomonas campestris* pv. *vesicatoria*. Mol Plant Microbe Interact 17:805–815

Killiny N, Catroviejo M, Saillard C (2005) *Spiroplasma citri* spiralin acts in vitro as a lectin binding to glycoproteins from its insect vector *Circulifer haematoceps*. Phytopathology 95:541–548

Kim JF, Alfano JR (2002) Pathogenicity islands and virulence plasmids of bacterial plant pathogens. Curr Topics Microbiol Immunol 264:127–147

Kim MG, Cunha L, da, McFall AJ, Belkhadir Y, DebRoy S, Dangl JL, Mackey D (2005c) Two *Pseudomonas syringae* type III effectors inhibit RIN4-regulated basal defense in *Arabidopsis*. Cell 121:749–759

Kim HS, Desveaux D, Singer AU, Patel P, Sondek J, Dangl JL (2005a) The *Pseudomonas syringae* effector AvrRpt 2 cleaves its C-terminally acylated target, RIN4, from *Arabidopsis* membranes to block RPM1 activation. Proc Natl Acad Sci USA 102:6496–6501

Kim JE, Han KH, Jin J, Kim H, Kim JC, Yun SH, Lee YW (2005b) Putative polyketide synthase and laccase genes for biosynthesis of aurofusarin in *Gibberella zeae.* Appl Environ Microbiol 71:1701–1708

Kim JG, Jeon E, Oh J, Moon JS, Hwang I (2004a) Mutational analysis of *Xanthomonas* harpin HpaG identifies a key functional region that elicits the hypersensitive response in nonhost plants. J Bacteriol 186:6239–6247

Kim J, Kang Y, Choi O, Jeong Y, Jeong JE, Jim JY, Kim M, Moon JS, Suga H, Hwang I (2007) Regulation of polar flagellum genes is mediated by quorum sensing and Flh DC in *Burkholderia glumae.* Mol Microbiol 64:165–179

Kim J, Kim J-G, Kang Y, Jang JY, Jog GJ, Lim JY, Kim S, Suga H, Nagamatsu T, Hwang I (2004b) Quorum sensing and the LysR-type transcriptional activator ToxR regulate toxoflavin biosynthesis and transport in *Burkholderia glumae*. Mol Microbiol 54:921–931

Kim YS, Lee HH, Ko MK, Song CE, Bae CY, Lee YH, Oh BJ (2001) Inhibition of fungal appressorium formation by pepper (*Capsicum annuum*) esterase. Mol Plant Microbe Interact 14:80–85

Kim YJ, Lin NC, Martin GB (2002b) Two distinct *Pseudomonas* effector proteins interact with the Pto kinase proteins interact with the Pto kinase and activate plant immunity. Cell 109:589–598

Kim MC, Panstruga R, Elliott C, Muller J, Devoto A, Yoon HW, Park HC, Cho MJ, Schulze-Lefert P (2002a) Calmodulin interacts with MLO protein regulate defense against mildew in barley. Nature 416:447–450

Kim JG, Park BK, Yoo CH, Jeon E, Oh J, Hwang I (2003) Characterization of the *Xanthomonas axonopodis* pv. *glycines* Hrp pathogenicity island. J Bacteriol 185:3155–3166

Kimura A, Takano Y, Furusawa I, Okuno T (2001) Peroxisomal metabolic function is required for appressoirum mediated plant infection by *Colletotrichum lagenarium.* Plant Cell 13:1945–1957

Klein PG, Klein RR, Rodriguez-Cerezo E, Hunt AG, Shaw JG (1994) Mutational analysis of the *Tobacco vein mottling virus* genome. Virology 204:759–769

Klimpel A, Schulze Gronover C, Williamson B, Stewert JA, Tudzinsky B (2002) The adenylate cyclase (BAC) in *Botrytis cinerea* is required for full pathogenicity. Mol Plant Microbe Interact 3:439–450

Koebnik R (2001) The role of bacterial pili in protein and DNA translocation. Trends Microbiol 9:586–590

Kloek AP, Brooks DM, Kunkel BN (2000) A *dsbA* mutant of *Pseudomonas syringae* exhibits reduced virulence and partial impairment of type III secretion. Mol Plant Pathol 1:139–150

Kohmoto K, Akimitsu K, Otani H (1991) Correlation of resistance and susceptibility of citrus to *Alternaria alternata* with sensitivity to host-specific toxins. Phytopathology 81:719–722

Kohmoto K, Itoh Y, Shimomura N, Kondoh Y, Otani M, Kodama M, Nishimura S, Nakatsuka S (1993) Isolation and biological activities of two host-specific toxins from the tangerine pathotype of *Alternaria alternata*. Phytopathology 83:445–450

Kohmoto K, Otani H, Tsuge T (1995) *Alternaria alternata* pathogens. In: Kohmoto K, Singh US, Sing RP (eds) Pathogenesis and host specificity in plant diseases: histopathological, biochemical, genetic and molecular bases. Eukaryotes, vol 2. Pergamon, Oxford, UK, pp 51–63

Kojima K, Kikuchi T, Takano Y, Oshiro E, Okuno T (2002) The mitogen-activated protein kinase gene MAF1 is essential for the early differentiation phase of appressorium formation in *Colletotrichum lagenarium.* Mol Plant Microbe Interact 15:1268–1276

Kormelink R, Storms M, Van Lent JWM, Peters D, Goldbach R (1994) Expression and subcellular location of the NSM protein of *Tomato spotted wilt virus* (TSWV), a putative viral movement protein. Virology 200:535–548

Koutsoudis MD, Tsaltas D, Minogue TD, von Bodman SB (2006) Quorum-sensing regulation governs bacterial adhesion, biofilm development and host colonization in *Pantoea stewartii* subsp *stewartii.* Proc Natl Acad Sci USA 103:5983–5988

Kragler F, Curin M, Trutneyva K, Gansch A, Waigmann E (2003) MPB2C, a microtubule associated plant protein binds to and interferes with cell-to-cell transport of *Tobacco mosaic virus* movement protein. Plant Physiol 132:1870–1883

Krishnamurthy K, Heppler M, Mitra R, Blancaflor E, Payton M, Nelson RS, Verchot-Lubicz J (2003) The *Potato virus X* TGBp3 protein associates with the ER network for virus cell-to-cell movement. Virology 309:135–151

Krishnamurthy K, Mitra R, Payton M, Verchot-Lubicz J (2002) Cell-to-cell movement of PVX 12K, 8K or coat proteins may depend on the host, leaf development stage, and the PVX 25k protein. Virology 300:269–281

Kubo Y, Nakamura H, Kobayashi K, Okuno T, Furasawa I (1991) Cloning of melanin biosynthetic gene essential for appressorial penetration of *Colletotrichum lagenarium*. Mol Plant Microbe Interact 4:440–445

Kumar RB, Xie YH, Ananth Das (2000) Subcellular localization of the *Agrobacterium tumefaciens* T-DNA transport pore proteins: VirB8 is essential for the assembly of the transport pore. Mol Microbiol 36:608–617

Kunze G, Zipfel C, Robatzek S, Nienhaus K, Boller T, Felix G (2004) The N terminus of bacterial elongation factor Tu elicits innate immunity in *Arabidopsis* plants. Plant Cell 16:3496–3507

Kwon YH, Epstein L (1993) A 90-kDa glycoprotein associated with adhesion of *Nectria hematococca* macroconidia to substrate. Mol Plant Microbe Interact 6:481–487

Lacroix B, Vaidya M, Tzfira T, Citovsky V (2005) The VirE3 of *Agrobacterium* mimics a host cell function required for plant genetic transformation. EMBO J 24:428–437

Lahey KA, Yuan R, Burns JK, Ueng PP, Timmer LW, Chung KR (2004) Induction of phytohormones and differential gene expression in citrus flowers infected by fungus *Colletotrichum acutatum*. Mol Plant Microbe Interact 17:1394–1401

Lai EM, Eisenbrandt R, Kalkum M, Lanka E, Kado CI (2002) Biogenesis of T pili in *Agrobacterium tumefaciens* requires precise VirB2 propilin cleavage and cyclization. J Bacteriol 184:327–330

Lairini K, Pérez-Espinosa A, Pineda M, Ruiz-Rubio M (1996) Purification and characterization of tomatinase from *Fusarium oxysporum* f.sp. *lycopersici*. Appl Environ Microbiol 62:1604–1609

Lan L, Deng X, Zhou J, Tang X (2006) Genome-wide gene expression analysis of *Pseudomonas syringae* pv. *tomato* DC 3000 reveals overlapping and distinct pathways regulated by *hrpL* and *hrpRS*. Mol Plant Microbe Interact 19:976–987

Landry P, Perreault J (2005) Identification of a *Peach latent mosaic viroid* hairpin able to act as a dicer-like substrate. J Virol 79:6540–6545

Landry P, Thompson D, Perreault JP (2004) The role of viroids in gene silencing: the model case of *Peach latent mosaic virus*. Canad J Plant Pathol 26:31–38

Lauber E, Bleykasten-Grosshans C, Erhardt M, Bouzoubaa S, Jonard G, Richards KE, Guilley H (1998) Cell-to-cell movement of *Beet necrotic yellow mosaic virus*. I. Heterologous complementation experiments provide evidence for specific interaction among the triple gene block proteins. Mol Plant Microbe Interact 11:618–625

Lazazzera BA, Grossman AD (1988) The ins and outs of peptide signaling. Trends Microbiol 7:288–294

Lederberg I (1992) Encyclopedia of microbiology. Academic Press, New York

Lee BN, Kroken S, Chou DYT, Robbertse B, Yoder OC (2005) Functional analysis of all nonribosomal peptide synthetases in *Cochliobolus heterostrophus* reveals a factor NSP6, involved in virulence and resistance to oxidative stress. Eukary Cell 4:545–555

Leister RT, Katagiri F (2000) A resistance gene product of the nucleotide binding site / leucine rich repeats class can form a complex with bacterial avirulence proteins in vivo. Plant J 22:345–354

Lellis AD, Kasschau KD, Whitham SA, Carrington JC (2002) Loss-of-susceptibility mutants of *Arabidopsis thaliana* reveal an essential role for eIF (iso) 4E during potyvirus infection. Curr Biol 12:1046–1051

Leshchiner AD, Solovyev AG, Morozov SY, Kalinina NO (2006) A minimal region in the NTPase/helicase domain of the TGBp1 plant virus movement protein is responsible for ATPase activity and cooperative RNA binding. J Gen Virol 87: Doi 10.1099/vir.0.8187-0

Levin DE (2005) Cell wall integrity signaling in *Saccharomyces cerevisiae*. Microbiol Mol Biol Rev 69:262–291

Li D, Ashby AM, Johnstone K (2003) Molecular evidence that the extracellular cutinase Pbc1 is required for pathogenicity for *Pyrenopeziza brassicae* on oilseed rape. Mol Plant Microbe Interact 16:545–552

Li CM, Brown I, Mansfield J, Stevens C, Boureau T (2002) The Hrp pilus of *Pseudomonas syringae* elongates from its tip and acts as a conduit for translocation of the effector protein HrpZ. EMBO J 21:1909–1915

Li L, Ding SL, Sharon A, Orbach M, Xu JR (2007) Mir1 is highly upregulated and localized to nuclei during infectious hyphal growth in the rice blast fungus. Mol Plant Microbe Interact 20:448–458

Li HW, Li WX, Ding SW (2002) Induction and suppression of RNA silencing by an animal virus. Science 296:1319–1321

Li X, Lin H, Zhang W, Zou Y, Zhang J, Tang X, Zhou JM (2005) Flagellin induces innate immunity in nonhost interaction that is suppressed by *Pseudomonas syringae* effectors. Proc Natl Acad Sci USA 102:12990–12995

Li HW, Lucy AP, Guo HS, Li WX, Ji LH, Wong SM, Ding SW (1999) Strong host resistance targeted against a viral suppressor of the plant gene silencing defense mechanism. EMBO J 18:2683–2691

Li WZ, Qu F, Morris T (1998) Cell-to-cell movement of *Turnip crinckle virus* is controlled by two small open reading frames that function in trans. Virology 244:405–418

Li L, Xue C, Bruno K, Nishimura M, Xu JR (2004) Two PAK kinase genes, CHM1 and MST20, have distinct functions in *Magnaporthe grisea*. Mol Plant Microbe Interact 17:547–556

Li W, Yuan R, Burns JK, Timmer LW, Chung KR (2003) Genes for hormone biosynthesis and regulation are highly expressed in citrus flowers infected with the fungus *Colletotrichum acutatum* causal agent of postbloom fruit drop. J Amer Soc Hort Sci 128:578–583

Liljeroth E, Santen K, Bryngelsson J (2001) PR protein accumulation in seminal roots of barley and wheat in response to fungal infection the importance of cortex senescence. J Phytopathology 149:447–456

Lim MTS, Kunkel BN (2004a) The *Pseudomonas syringae* type III effector AvrRpt2 promotes virulence indepdently of RIN4, a predicted virulence target in *Arabidopsis thaliana*. Plant J 40:790–798

Lim MTS, Kunkel BN (2004b) Mutations in the *Pseudomonas syringae avrRpt2* gene that dissociate its virulence and avirulence activities lead to decreased efficiency in AvrRpt2 induced disappearance of RIN 4. Mol Plant Microbe Interact 17:313–321

Lim MTS, Kunkel BN (2005) The *Pseudomonas syringae avrRpt2* gene contributes to virulence on tomato. Mol Plant Microbe Interact 18:626–633

Lin MK, Chang BY, Liao JT, Lin NS, Hsu YH (2004) Arg-16 and Arg-21 in the N-terminal region of the triple-gene-block protein 1 of *Bamboo mosaic virus* are essential for virus movement. J Gen Virol 85:251–259

Lin JW, Ding MP, Hsu YH, Tsai CH (2007) Chloroplast phosphoglycerate kinase, a gluconeogenetic enzyme is required for efficient accumulation of *Bamboo mosaic virus*. Nucleic Acids Res 35:424–432

Lin B, Heaton LA (2001) An *Arabidopsis thaliana* protein interacts with a movement protein of *Turnip crinkle virus* in yeast cells and in vitro. Gen Virol 82:1245–1251

Lin MK, Hu CC, Lin NS, Chang BY, Hsu YH (2006) Movement of potexviruses requires species-specific interactions among the cognate triple gene block proteins, as revealed by a trans-complementation assay based on the *Bamboo mosaic virus* satellite RNA-mediated expression system. J Gen Virol 87:1357–1367

Lin NC, Martin GB (2005) An avrPto/avrPtoB mutant of *Pseudomonas syringae* pv. *tomato* DC 3000 does not elicit Pto-mediated resistance and is less virulent on tomato. Mol Plant Microbe Interact 18:43–51

Lindberg M, Collmer A (1992) Analysis of eight *out* genes in a cluster required for pectic enzyme secretion by *Erwinia chrysanthemi*: sequence comparison with secretion genes from other gram negative bacteria. J Bacteriol 174:7385–7397

Lindgren PB (1997) The role of *hrp* genes during plant–bacterial interactions. Annu Rev Phytopathology 35:129–152

Linstead PJ, Hills GJ, Plaskitt KA, Wilson JG, Harker CL, Maule AJ (1988) The subcellular location of the gene I product of *Cauliflower mosaic virus* is consistent with a function associated with virus spread. J Gen Virol 69:1089–1098

Liou DY, Hsu YH, Wung CH, Wang WH, Lin NS, Chang BY (2000) Functional analyses and identification residues essential for the ATP-utilizing activity of the triple gene block protein 1 of *Bamboo mosaic potexvirus*. Virology 277:336–344

Liu JZ, Blancaflor EB, Nelson RS (2005b) The *Tobacco mosaic virus* 126-kilodalton protein, a constituent of the virus replication complex, alone or within the complex aligns with and traffics along microfilaments. Plant Physiol 138:1853–1865

Liu S, Dean RA (1997) G protein α-subunit genes control growth, development and pathogenicity of *Magnaporthe grisea*. Mol Plant Microbe Interact 10:1075–1086

Liu H, Zhang S, Schell MA, Denny TP (2005a) Pyramiding unmarked deletions in *Ralstonia solanacearum* shows that secreted proteins in addition to plant cell wall-degrading enzymes contribute to virulence. Mol Plant Microbe Interact 18:1296–1305

Llama-Palacios A, Lopéz-Solanilla E, Poza-Carrion C, Solanilla E, Poza-Carrion C, García-Ólmedo F, Rodríguez-Palenzuela P (2003) The *Erwinia chrysanthemi phoP- phoQ* operon plays an important role in growth at low pH, virulence and bacterial survival in plant tissue. Mol Microbiol 49:347–357

Llama-Palacios A, Lopéz-Solanilla E, Rodríguez-Palanzuela P (2005) Role of the PhoP-PhoQ system in the virulence of *Erwinia chrysanthemi* strain 3937: involvement in sensitivity to plant antimicrobial peptides, survival at acid, pH, and regulation of pectolytic enzymes. J Bacteriol 187:2157–2162

López-Solanilla E, Bronstein PA, Schneider AR, Collmer A (2004) HopPtoN is a *Pseudomonas syringae* TTSS effector and cysteine protease that suppresses pathogen-induced necrosis associated with both compatible and incompatible plant interactions. Mol Microbiol 54:353–365

López-Solanilla E, Llama-Palacios A, Collmer A, García-Olmedo F, Rodríguez-Palenzuela P (2001) Relative effects on virulence of mutations in the *sap*, *pel* and *hrp* loci of *Erwinia chrysanthemi*. Mol Plant Microbe Interact 14:386–393

Lorang JM, Tuori RP, Martinez JP, Sawyer TL, Redman RS, Rollins JA, Wolpert TJ, Johnson KB, Rodiguez RJ, Dickman MB, Ciufetti LM (2001) Green fluorescent protein is lighting up fungal biology. Appl Environ Microbiol 67:1987–1994

Losada L, Susan T, Pak K, Zeyad S, Rozenbaum I, Hutcheson SW (2004) Identification of a novel *Pseudomonas syringae* Psy61 effector with virulence and avirulence functions by a Hrp1-dependent promoter-trap assay. Mol Plant Microbe Interact 17:254–262

Lough TJ, Netzler NE, Emerson SJ, Sutherland P, Carr F, Beck DL, Lucas WJ, Forster RLS (2000) Cell-to-cell movement of potex viruses: evidence for a ribonucleoprotein complex involving the coat protein and the first triple gene block protein. Mol Plant Microbe Interact 13:962–974

Lough TJ, Shash K, Xoconostle-Cazares B, Hofstra KR, Beck DL, Balmori E, Forster RLS, Lucas WJ (1998) Molecular dissection of the mechanism by which potexvirus triple gene block proteins mediate cell-to-cell transport of infectious RNA. Mol Plant Microbe Interact 11:801–814

Lu SW, Kroken S, Lee BN, Robbertse B, Churchill ACL, Yoder OC, Turgeon BG (2003) A novel class of gene controlling virulence in plant pathogenic ascomycete fungi. Proc Natl Acad Sci USA 100:5980–5985

Lu M, Tang X, Zhou JM (2001) Arabidopsis NH01 is required for general resistance against *Pseudomonas* bacteria Plant Cell 13:437–447

Lu SE, Wang N, Wang J, Chen ZJ, Gross DC (2005) Oligonucleotide microarray analysis of the Sal A regulon controlling phytotoxin production by *Pseudomonas syringae* pv. *syringae*. Mol Plant Microbe Interact 18:324–333

Lucy PA, Guo HS, Li WX, Ding SW (2000) Suppression of posttranscriptional gene silencing by a plant viral protein localizaed in the nucleus. EMBO J 19:1672–1680

Luke HH, Wheeler HE (1955) Toxin production of *Helminthosporium victoriae*. Phytopathology 45:453–458

Lund ST, Stall RE, Klee HJ (1988) Ethylene regulates the susceptible response to pathogen infection in tomato. Plant Cell 10:371–382

Luo M, Dang P, Bausher MG, Holbrook CC, Lee RD, Lynch RE, Guo BZ (2005) Identification of transcripts involved in resistance responses to leaf spot disease caused by *Cercosporidium personatum* in peanut (*Arachis hypogaea*). Phytopathology 15:381–387

Lyngkjaer MF and Carver TLW (2000) Conditioning of cellular defense responses to powdery mildew in cereal leaves by prior attack. Mol Plant Pathol 1:41–49

Lysøe E, Klemsdal SS, Bone KR, Frandsen RJN, Johnsen T, Thrane U, Giese H (2006) The *PKS4* gene of *Fusarium graminearum* is essential for zearalenone production. Appl Environ Microbiol 72:3924–3932

Mackey D, Belkhadir Y, Alonso JM, Ecker JR, Dangl JL (2003) *Arabidopsis* RIN4 is a target of the type III virulence effector Avr Rpt2 and modulates *RPS2* mediated resistance. Cell 112: 379–389

Mackey D, Holt BF, Wiig A, Dangl JL (2002) RIN4 interacts with *Pseudomonas syringae* type III effector molecules and is required for RPM-1 mediated resistance in *Arabidopsis*. Cell 108:743–754

Maekawa N, Yamamoto M, Nishimura S, Kohmoto K, Kuwata M, Watanabe Y (1984) Studies on host-specific AF-toxins produced by *Alternaria alternata* strawberry pathotype causing Alternaria black spot of strawberry 1. Production of host-specific toxins and their biological activity. Ann Phytopathol Soc Jpn 50:600–609

Makino S, Sugio A, White F, Bogdanove AJ (2006) Inhibition of resistance gene mediated defense in rice by *Xanthomonas oryzae* pv. *oryzicola*. Mol Plant Microbe Interact 19:240–249

Malz S, Grell MN, Thrane C, Maier FJ, Rosager P, Felk A, Albertsen KS, Salmon S, Bohn L, Schäfer-W and Giesse H (2005) Identification of a gene cluster responsible for the biosynthesis of aurofusarin in the *Fusarium graminearum* species complex. Fungal Genet Biol 42:420–433

Manning VA, Ciuffetti LM (2005) Localization of Ptr ToxA produced by *Pyrenophora tritici-repentis* reveals protein import into wheat mesophyll cells. Plant Cell 17:3203–3212

Marcos JF, Vilar M, Pérez-Payá E, Pallás V (1999) In vivo detection, RNA-binding properties and characterization of the RNA-binding domain of the p7 putative movement protein from *Carnation mottle carmovirus* (CarMV). Virology 255:354–365

Markarian N, Li HW, Ding SW, Semanick JS (2004) RNA silencing as related to viroid induced symptom expression. Arch Virol 149:397–406

Marks AR (1996) Cellular functions of immunophilins. Physiol Rev 76:631–649

Marois E, van den Ackerveken G, Bonas U (2002) The *Xanthomonas* type III effector protein AvrBs3 modulates plant gene expression and induces cell hypertrophy in the susceptible host. Mol Plant Microbe Interact 15:637–646

Marques CN, Salisbury VC, Greenman J, Bowker KE, Nelson SM (2005) Discrepancy between viable counts and light output as viability measurements following cirprofloxacin challenge of self-bioluminescent *Pseudomonas aeroginosa* biofilms. J Antimicro Chemo 56: Advanced Access published online

Martin GB, Bogdanov AJ, Sessa G (2003) Understanding the functions of plant disease resistance proteins. Annu Rev Plant Biol 54:23–61

Martínez de Alba AE, Flores R, Hernàndez C (2002) Two chloroplastic viroids induce the accumulation of the small RNAs associated with posttranscriptional gene silencing. J Virol 76: 3094–3096

Martin-Hernandez AM, Dufresne M, Hugouvieux V, Melton R, Osbourn AE (2000) Effects of targeted replacement of the tomatinase gene on the interaction of *Septoria lycopersici* with tomato plants. Mol Plant Microbe Interact 13:1301–1311

Martinez JP, Oesch NW, Ciuffetti LM (2004) Characterization of the multiple-copy host selective toxin gene, *ToxB* in pathogenic and nonpathogenic isolates of *Pyrenophora tritici-repentis*. Mol Plant Microbe Interact 17:467–474

Masunaka A, Ohtani K, Peever TL, Timmer LW, Tsuge T, Yamamoto M, Yamamoto H, Akimitsu K (2005) An isolate of *Alternaria alternata* that is pathogenic to both tangerines and rough lemon and produces two selective toxins, ACT-and ACR-toxins. Phytopathology 95:241–247

Masunaka A, Tanaka A, Tsuge T, Peever TL, Timmer LW, Yamamoto M, Yamamoto H, Akimitsu K (2000) Distribution and characterization of AKT homologs in the tangerine pathotype of *Alternaria alternata*. Phytopathology 90:762–768

Matsuda Y, Sameshima T, Moriura N, Inoue, K, Nonomura T, Kakutani K, Hiroaki N, Kusakari S, Takamatsu S, Toyoda H (2005) Identification of individual powdery mildew fungi infecting leaves and direct detection of gene expression by single conidium polymerase chain reaction. Phytopathology 95:1137–1143

Matsuda Y, Toyoda H, Kurita A, Ouchi S (1997) In situ PCR technique based on priciking microinfection for cDNA cloning in single cells of barley coleoptile and powdery mildew pathogen. Plant Cell Reprod 16:612–618

Matsumoto H, Jitarcerat P, Baba Y, Tsuyumu S (2003a) Comparative study of regulatory mechanisms of pectinase production by *Erwinia carotovora* subsp. *carotovora* and *E. chrysanthemi*. Mol Plant Microbe Interact 16:226–237

Matsumoto H, Muroi H, Umehara M, Yoshitake Y, Tsuyumu S (2003b) Peh Production, flagellum synthesis and reduced virulence in *Erwinia carotovora* subsp. *carotovora*. Mol Plant Microbe Interact 16:389–397

Matthysee AG (1983) Role of bacterial cellulose fibrils in *Agrobacterium tumefaciens* infection. J Bacteriol 154:906–915

Matthysee AG (1994) Conditioned medium promotes the attachment of *Agrobacterium tumefaciens* strains *NT1* to carrot cells. Protoplasma 183:131–136

Matthysee AG, McMahan S (1998) Root colonization by *Agrobacterium tumefaciens* is reduced by *Cel, altB, altD* and *altR* mutants. Appl Environ Microbiol 64:2341–2345

Matthysee AG, Yarnall HA, Young N (1996) Requirement for genes with homology to ABC transport systems for attachment and virulence of *Agrobacterium tumefaciens*. J Bacteriol 178:5302–5308

Mayama S, Tani T, Midland SL, Sims JJ, Keen NT (1986) The purification of victorin and its phytoalexin elicitors activity in oat leaves. Physiol Mol Plant Pathol 29:1–18

McCormick SP, Harris LJ, Alexander NJ, Ouellet T, Saparno A, Allard S, Desjardins AE (2004) *Tri1* in *Fusarium graminearum* encodes a P450 oxygenase. Appl Environ Microbiol 70: 2044–2051

Mc Evoy JL, Park E, Conway WS (2003) Virulence factors involved in post-harvest apple decay by fungal pathogens. Phytopathology 93: S105(Abst.)

Mehrabi R, van der Lee T, Waalwijk C, Kema GHJ (2006) *MgSlt2,* a cellular integrity MAP kinase gene of the fungal wheat pathogen *Mycosphaerella* is dispensable for penetration but essential for invasive growth. Mol Plant Microbe Interact 19:389–398

Mehrabi R, Waalwijk C, van der Lee T, Ware S, Kema GHJ (2004) A MAP kinase gene, *MgMK1* is essential for pathogenicity in *Mycosphaerella graminicola*. In: Proceedings of 7th European Conference on Fungal Genetics page 75, Copenhan, Denmark

McKenny D, Brown KE, Allison DG (1995) Influence of *Pseudomonas aeruginosa* exproducts on virulence of *Burkholderia cepacia*: evidence for interspecies communication. J Bacteriol 177:6989–6992

McLean BG, Zupan J, Zambryski PC (1995) *Tobacco mosaic virus* movement protein associates with the cytoskeleton in tobacco cells. Plant Cell 7:2101–2114

Medina-Suarez R, Manning K, Fletcher J, Aked J, Bird CR, Seymour GB (1997) Gene expression in the pulp of ripening bananas. Plant Physiol 115:453–461

Medzhitov R, Janeway CA Jr (2002) Decoding the patterns of self and nonself by the innate immune system. Science 296:298–300

Meehan F, Murphy HC (1946) A new *Helminthosporium* blight of oats. Science 104:413–414

Melchers LS, Maroney MJ, den Dulk Ras A, Thompson DV, van Vuuren HAJ (1990) Octopine and nopaline strains of *Agrobacterium tumefaciens* differ in virulence: molecular characterization of the *virF* locus. Plant Mol Biol 14:249–259

Melton RE, Flegg LM, Brown JKM, Oliver RP, Daniels MJ, Osburn AE (1998) Heterologous expression of *Septoria lycopersici* tomatinase in *Cladosporium fulvum*: effects of compatible and incompatible interactions with tomato seedlings. Mol Plant Microbe Interact 11: 228–236

Mendgen K, Hahn M (2002) Plant infection and the establishment of fungal biotrophy. Trends Plant Sci 7:352–356

Meng X, Bonasera JM, Kim JF, Nissimen RM, Beer SV (2006) Apple proteins that interact with Dsp A/E, a pathogenicity effector of *Erwinia amylovora*, the fire blight pathogen. Mol Plant Microbe Interact 19:53–61

Mercure EW, Kunoh H, Nicholson RL (1994a) Adhesion of *Colletotrichum graminicola* to corn leaves: a requirement for disease development. Physiol Mol Plant Pathol 45:421–440

Mercure EW, Leite B, Nicholson RL (1994b) Adhesion of ungerminated conidia of *Colletotrichum graminicola* to artificial hydrophobic surfaces. Physiol Mol Plant Pathol 45:421–440

Merighi M, Majerczak, DR, Coplin DL (2005) A novel transcriptional autoregulatory loop enhances expression of the *Pantoea stewartii* subsp. *stewartii* Hrp type III secretion system. FEMS Microbiol Lett 243:479–487

Merighi M, Majerczak DR, Zianni M, Tessanne K, Coplin DL (2006) Molecular characterization of *Pantoea stewartii* subsp. *stewartii* HrpY, a conserved response regulator of the Hrp type III secretion system, and its interaction with the *hrpS* promoter. J Bacteriol 188:5089–5100

Metz M, Dahlbeck D, Morales CQ, Al Sady B, Clark ET, Staskawicz BJ (2005) The conserved *Xanthomonas campestris* pv. *vesicatoria* effector protein XopX is a virulence factor and suppresses host defense in *Nicotiana benthamiana*. Plant J 41:801–814

Meyer D, Cunnac S, Guéneron M, Declerq C, Gijsegem F, van, Lauber E, Boucher C, Arlat M (2006) PopF1 and PopF2, two proteins secreted by the type III protein secretion system of *Ralstonia solanacearum* are translocators belonging to the HrpF/NopX family. J Bacteriol 188:4903–4917

Mikes V, Milat ML, Ponchat M, Ricci P, Blein JP (1997) The fungal elicitor cryptogein is a sterol carrier protein. FEBS Lett 416:190–192

Mikes V, Milat ML, Ponchet M, Panabieres F, Ricci P, Blein JP (1998) Elicitins, proteinaceous elicitors of plant defense, are a new class of sterol carrier proteins. Biochemi Biophys Res Commun 245:133–139

Meyer R, Slater V, Dubery IA (1994) A phytotoxic protein-lipopolysaccharide complex produced by *Verticillium dahliae*. Phytochemistry 35:1449–1453

Moissiard G, Voinnet O (2004) Viral suppression of RNA silencing in plants. Mol Plant Pathol 5:71–82

Molina L, Rezzonico F, Défago G, Duffy B (2005) Autoinduction in *Erwinia amylovora* evidence of an acyl-homoserine lactone signal in the fire blight pathogen. J Bacteriol 187:3206–3213

Morales CQ, Posada J, Macneale E, Franklin D, Rivas I, Bravo M, Minsavage J, Stall RE, Whalen MC (2005) Functional analysis of the early chlorosis factor gene. Mol Plant Microbe Interact 18:477–486

Morozov SY, Solovyev AG (2003) Triple gene block: modular design of multifunctional machine for plant virus movement. J Gen Virol 84:1351–1366

Morton A, Davies DL, Blomquist CL, Barbara DJ (2003) Characteization of homologues of the apple proliferation immunodominant membrane protein gene from three related phytoplasmas. Mol Plant Pathol 4:109–114

Moy P, Qutob D, Chapman BP, Atkinson I, Gijzen M (2004) Patterns of gene expression upon infections of soybean plants by *Phytophthora sojae*. Mol Plant Microbe Interact 17: 1051–1062

Murata Y, Tamura N, Nakaho K, Mukaihara T (2006) Mutations in the *lrpE* gene of *Ralstonia solanacearum* affects Hrp pili production and virulence. Mol Plant Microbe Interact 19: 884–895

Mysore KS, Nam J, Gelvin SB (2000) An *Arabidopsis* histone H2A mutant is deficient in *Agrobacterium* T-DNA integration. Proc Natl Acad Sci USA 97:948–953

Nakamura M, Suprapta DN, Iwai H, Arai K (2001) Comparison of endopolygalacturonase activities of citrus and noncitrus races of *Geotrichum candidum* and cloning and expression of corresponding genes. Mol Plant Pathol 2:265–274

Narayanasamy P (2002) Microbial plant pathogens and crop disease management. Science Publishers, Enfield, CT, USA

Narayanasamy P (2005) Immunology in plant health and its impact on food safety. The Haworth Press, New York, USA

Nasser W, Faelen M, Hugouvieux-Cotte-Pattat N, Reverchon S (2001) Role of the nucleoid associated protein H-NS in the synthesis of virulence factors in phytopathogenic bacterium *Erwinia chrysanthemi*. Mol Plant Microbe Interact 14:10–20

Nasser W, Reverchon S, Vedel R, Boccara M (2005) PecS and PecT coregulate the synthesis of HrpN and pectate lyases, two virulence determinants in *E. chrysanthemi* 3937. Mol Plant Microbe Interact 18:1205–1214

Navarre DA, Wolpert JJ (1999) Victorin induction of an apoptotic/senescence-like response in oats. Plant Cell 11:237–249

Navarro L, Zipfel C, Rowland O, Keller I, Robatzek S, Boller T, Jones JD (2004) The transcriptional innate immune response to flg22: interplay and overlap with Avr gene-depedent defense responses and bacterial pathogenesis. Plant Physiol 135:1113–1128

Nelson EB (1991) Exudate molecules initiating fungal responses to seeds and roots. In: Keister DL, Cregan PB (eds) The rhizosphere and plant growth, Kluwer Academic Publishers, Dordrecht, The Nertherlands, pp 197–209

Nelson EB (2004) Microbial dynamics and interactions in the spermosphere. Annu Rev Phytopathol 42:271–309

Nelson RS, Citovsky V (2005) Plant viruses: invaders of cells and pirates of cellular pathways. Plant Physiol 138:1809–1814

Newman MA, Conrads-Strauch J, Scofield G, Daniels MJ, Dow JM (1994) Defense-related gene induction in *Brassica campestris* in response to defined mutants of *Xanthomonas campestris* with altered pathogenicity. Mol Plant Microbe Interact 7:553–562

Ngoko Z, Marasas WFO, Rheeder JP, Shephard GS, Wingfield MJ, Cardwell KF (2001) Fungal infection and mycotoxin contamination of maize in humid forest and western highlands of Cameroon. *Phytoparasitica* 29:352–360

Nicaise V, German-Retana S, Sanjuán R, Dubrana MP, Mazier M, Maisonneuve B, Candresse T, Caranta C, LeGall O (2003) The eukaryotic translation initiation factor 4E controls lettuce susceptibility to potyvirus *Lettuce mosaic virus*. Plant Physiol 132:1272–1282

Nicholson TP, Rudd BAM, Dawson M, Lazarus CM, Simpson TJ, Cox RJ (2001) Design and utility of oligonucleotide gene probes for fungal polyketide synthases. Chemical Biol 8:157–178

Nishiguchi M, Kikuchi S, Kiho Y, Ohno T, Meshi T, Okada Y (1985) Molecular basis of plant viral virulence: the complete nucleotide sequence of an attenuated strain of *Tobacco mosaic virus*. Nucleic Acids Res 13:5585–5590

Nishimura S, Kohmoto K, Kuwata M, Watanabe Y (1984) Studies on host-specific AF toxins produced by *Alternaria alternata* strawberry pathotype causing alternaria black spot of strawberry 1. Production of host-specific toxins and their biological activity. Ann Phytopathol Soc Jpn 50:600–609

Nishiuchi T, Masuda D, Nakashita H, Ichimura K, Shinozaki K, Yoshida S, Kimura M, Yamaguchi I, Yamaguchi K (2006) *Fusarium* phytotoxin trichothecenes have an elicitor-like activity in *Arabidopsis thaliana*, but the activity differed significantly among their molecular species. Mol Plant-Microbe Interact 19:512–520

Nissinen RM, Ytterberg AJ, Bogadanove AJ, van Wijk KJ, Beer SV (2007) Analysis of the secretomes of *Erwinia amylovora* and selected *hrp* mutants reveal novel type III secreted proteins and an effect of HrpJ on extracellular harpin levels. Mol Plant Pathol 8:55–67

Noël L, Thieme F, Nennstiel D, Bonas U (2001) cDNA-AFLP analysis unravels a genome-wide *hrpG*-regulon in the plant pathogen *Xanthomonas campestris* pv.*vesicatoria*. Mol Microbiol 41:1271–1281

Noël L, Thieme F, Nennstiel D, Bonas U (2002) Two novel type III-secreted proteins of *Xanthomonas campestris* pv. *vesicatoria* are encoded within the *hrp* pathogenicity island. J Bacteriol 184:1340–1348

Noël L, Thieme F, Gäbler J, Bültner D, Bonas U (2003) XopC and XopJ, two novel type III effector proteins from *Xanthomonas campestris* pv. *vesicatoria*. J Bacteirol 185:7092–7102

Nomura K, DebRoy S, Lee YH, Pumplin N, Jones J, He SY (2006) A bacterial virulence protein suppresses host innate immunity to cause plant disease. Science 313:220–223

Nomura K, Melotto M, He SY (2005) Suppression of host defense compatible plant–*Pseudomonas syringae* interactions. Curr Opin Plant Biol 8:361–368

Normura K, Nasser W, Kawagishi H, Tsuyumu S (1998) The *pir* gene of *Erwinia chrysanthemi* EC16 regulates hyperinduction of pectate lyase virulence genes in response to plant signals. Proc Natl Acad Sci USA 95:14034–14039

Noueiry AO, Lucas WJ, Gilbertson RL (1994) Two proteins of a plant DNA virus coordinate nuclear and plasmodesmatal transport. Cell 76:925–932

Noueiry AO, Lucas WJ, Gilbertson RL, Nomura K, Nasser W, Kawagishi H, Tsuyumu S (1999) The *pir* gene of *Erwinia chrysanthemi* EC16 regulates hyperinduction of pectate lyase virulence genes in response to plant signals. Proc Natl Acad Sci USA 95:14034–14039

Nurnberger T, Brunner F (2002) Innate immunity in plants and animals: emerging parallels between the recognition of general elicitors and pathogen-associated molecular patterns. Curr Opin Plant Biol 5:318–324

Nurnberger T, Brunner F, Kemmerling B, Piater L (2004) Innate immunity in plants and animals: striking similarities and obvious differences. Immunol Rev 198:249–266

Occhialini A, Cunnac S, Reymond N, Genin S, Boucher C (2005) Genome-wide analysis of gene expression in *Ralstonia solanacearum* reveals that the *hrpB* gene acts as a regulatory switch controlling multiple virulence pathways. Mol Plant Microbe Interact 18:938–949

Ochiai H, Inoue Y, Takeya M, Sasaki A, Kaku H (2005) Genome sequence of *Xanthomonas oryzae* pv.*oryzae* suggests contribution of large number of effector genes and insertion sequences to its race diversity. J Agri Res Quart 39:275–287

O'Donnell PJ, Jones JB, Antonine FR, Ciardi J, Klee H (2001) Ethylene-dependent salicylic acid regulates an expanded cell death response to a plant pathogen. Plant J 25:315–323

O'Donnell PJ, Schmelz E, Block A, Miersch O, Wasternack C, Jones JB, Klee HJ (2003) Multiple hormones act sequentially to mediate a susceptible tomato pathogen defense response. Plant Physiol 133:1181–1189

Oh CS, Kim JF, Beer SV (2005) The Hrp pathogenicity island of *Erwinia amylovora* and identification of three novel genes required for systemic infection. Mol Plant Pathol 6: 125–138

Oide S, Moeder W, Krasnoff S, Gibson D, Haas H, Yoshioka K, Turgeon BG (2006) *NPS6*, encoding a nonribosomal peptide synthetase involved in siderophore-mediated iron metabolism is a conserved virulence determinant of plant pathogenic Ascomycetes. Plant Cell 18: 2836–2853

Oka K, Akasmatsu H, Kodama M, Nakajima H, Kawada T, Otani H (2005) Host-specific AB-toxin production by germinating spores of *Alternaria brassiciola* is induced by a host-derived oligosaccharide. Physiol Mol Plant Pathol 66:12–19

Oka K, Okuba A, Kodama M, Otani H (2006) Detoxification of α-tomatine by tomato pathogens *Alternaria alternata* tomato pathotype and *Corynespora cassiicola* and its role in infection. J Gen Plant Pathol 72:152–158

Oku H, Shiraishi T, Ouchi S, Ishimura M, Matsueda R (1980) A new determinant of pathogenicity in plant disease. Naturwiss 67:310

Okumura K, Hyodo H, Kato M, Ikoma Y, Yano M (1999) Ethylene biosynthesis in sweet potato root tissue infected by black rot fungus (*Ceratocystis fimbriata*). Postharvest Biol Technol 17:117–125

Ormö M, Cubitt AB, Kallio K, Gross LA, Tsien RY, Remington SJ (1996) Crystal structure of the *Aequorea victoria* green fluorescent protein. Science 273:1392–1395

Otani H, Kohnobe A, Kodama M, Kohmoto K (1998) Involvement of host factors in the production of a protein host-specific toxin by *Alternaria brassicola*. In: Kohmoto K, Yodev OC (eds) Molecular genetics of host-specific toxin in plant disease, Kluwer Academic Publishers, Dordrecht, pp 63–69

Owens RA, Blackburn M, Ding B (2001) Possible involvement of the phloem lectin in long-distance viroid movement. Mol Plant Microbe Interact 14:905–909

Paape M, Solovyev AG, Erokhina TN, Minina EA, Schepetilnikov MV, Lesemann DE, Schiemann J, Morozov SY, Kellmann JW (2006) AT-4/1 an interactor of the *Tomato spotted wilt virus* movement protein belongs to a new family of plant proteins capable of direct intra and inter-cellular trafficking. Mol Plant Microbe Interact 19:874–883

Palani PV, Kasiviswanathan V, Chen JCF, Chen W, Hsu YH, Lin NS (2006) The arginine-rich motif of *Bamboo mosaic virus* satellite RNA-encoded P20 mediates self interaction, intercellular targeting and cell-to-cell movement. Mol Plant Microbe Interact 19:758–767

Pan SQ, Jin S, Boulton MI, Hawes M, Gordon MP, Nester EW (1995) An *Agrobacterium* virulence factor encoded by a Ti-plasmid gene or a chromosomal gene is required for T-DNA transfer into plants. Mol Microbiol 17:259–269

Panstruga R, Schulze-Lefert P (2002) Live and let live insights into powdery mildew disease and resistance. Mol Plant Pathol 3:495–502

Park G, Xue C, Zhao X, Kim Y, Orbach M, Xu JR (2006) Multiple upstream signals coverage on the adaptor protein Mst50 in *Magnaporthe grisea.* Plant cell 18:2822–2835

Park G, Xue C, Zheng L, Lam S, Xu JR (2002) MST 12 regulates infectious growth but not appressorium formation in rice blast fungus *Magnaporthe grisea.* Mol Plant Microbe Interact 15:183–192

Pascale M, Visconti A, Avantaggiato G, Prończuk M, Chelkowski J (1999) Mycotoxin contamination of maize hybrids after infection with *Fusarium proliferation.* J Sci Food Agric 79: 2094–2098

Pascholati SF, Deising H, Leite B, Anderson D, Nicholson RL (1993) Cutinase and nonspecific esterase activities in the conidial mucilage of *Colletotrichum graminicola.* Physiol Mol Plant Pathol 42:37–51

Pautot V, Holzer FM, Chaufaux J, Walling LL (2001) The induction of tomato leucine aminopeptidase genes (LapA) after *Pseudomonas syringae* pv. *tomato* infection is primarily a wound response triggered by coronatine. Mol Plant Microbe Interact 14:214–224

Paynter CD, Salisbury VC, Arnold DL, Jackson RW (2006) The use of bioluminescence for monitoring in planta growth dynamics of a *Pesudomonas syringae* plant pathogen. Eur J Plant Pathol 115:363–366

Pemberton CL, Whitehead NA, Sebaihia M, Bell KS, Hyman LJ, Harris SJ, Matlin AJ, Robson ND, Birch PRJ, Carr JP, Toth IK, Salmond GPC (2005) Novel quorum-sensing-controlled genes in *Erwinia carotovora* subsp *carotovora*: identification of a fungal elicitor homologue in a soft-rotting bacterium. Mol Plant Microbe Interact 18:343–353

Peng Q, Yang S, Charkowski AO, Yap MN, Steeber DA, Keen NT, Yang CH (2006) Population behavior analysis of *dspE* and *pelD* regulation in *Erwinia chrysanthemi* 3937. Mol Plant Microbe Interact 19:451–457

Pantaleo V, Grieco F, Castellano MA, Martelli GP (1999) Synthesis of infectious transcripts of *Olive latent virus 1*: genes required for RNA replication and virus movement. Arch Virol 144:1071–1079

Peng YH, Kadoury D, Gal-On A, Huet H, Wang Y, Raccah B (1998) Mutations in the HC-Pro gene of zucchini yellow mosaic potyvirus: effects on aphid transmission and binding to purified virions. J Gen Virol 79:897–904

Perbal MC, Thomas CL, Maule AJ (1993) *Cauliflower mosaic virus gene-1* product (P1) forms tubular structures which extend from the surface of infected protoplasts. Virology 195: 281–285

Perfect SE, Green JR, O'Connell RJ (2001) Surface characteristics of necrotrophic secondary hyphae produced by the bean anthracnose fungus *Colletotrichum lindemuthianum.* Eur J Plant Pathol 107:813–819

Perfect SE, Pixton KL, O'Connell RJ, Green JR (2000) The distribution and expression of a biotrophy-related gene CIH1 within the genes *Colletotrichum.* Mol Plant Pathol 1:213–221

Perino C, Gaudriceult S, Vian B, Barny MA (1999) Visualizaiton of harpin secretion in planta during infection of apple seedlings by *Erwinia amylovora.* Cell Microbiol 1:131–141

Petnicki-Ocwieja T, Schneider DJ, Tam VC, Chancey ST, Shan L, Jamir Y, Schechter LM, Janes MD, Buell CR, Tang S, Collmer A, Alfano JR (2002) Genomewide identification of proteins secreted by the Hrp type III protein secretion system of *Pseudomonas syringae* pv. *tomato* DC 3000. Proc Natl Acad Sci USA 99:7652–7657

Petnicki-Ocwieja T, van Dijk K, Alfano JR (2005) The *hrpK* operon of *Pseudomonas syringae* pv. *tomato* DC 3000 encodes two proteins secreted by the type III (Hrp) protein secretion system: HopB1 and HrpK, putative type III translocator. J Bacteriol 187:649–663

Pierson LS, and Pierson EA (1996) Phenazine antibiotic production in *Pseudomonas aureofaciens:* role in rhizosphere ecology and pathogen suppression. FEMS Microbiol Lett 136:101–108

Pietro AD, Roncero MI (1998) Cloning, expression and role in pathogenicity of *pg1* encoding the major extracellular endopolygalacturonase of the vascular wilt pathogen *Fusarium oxysporum*. Mol Plant Microbe Interact 11:91–98

Pitkin JW, Nikolskaya A, Ahn JH, Walton JD (2000) Reduced virulence caused by meiotic instability of the TOX2 chromosome of the maize pathogen *Cochliobolus carbonum*. Mol Plant Microbe Interact 13:80–87

Plisson C, Drucker M, Blanc S, German-Retana S, Le Gall O, Thomas D, Bron P (2003) Structural characterization of HC-Pro, a plant virus multifunctional protein. J Biol Chem 278: 23753–23761

Poinssot B, Vandelle E, Bentéjac M, Adrian M, Levis C, Brygoo Y, Garin J, Sicilia F, Coutos-Thévenot P, Pugin A (2003) The endopolygalacturonase 1 from *Botrytis cinerea* activates grapevine defense reactions unrelated to its enzymatic activity. Mol Plant Microbe Interact 16:553–564

Poplawsky AR, Chun W (1998) *Xanthomonas campestris* pv. *campestris* requires a functional *pigB* for epiphytic survival and host infection. Mol Plant Microbe Interact 11:466–475

Poplawsky AR, Chun W, Slater H, Daniels MJ, Dow JM (1998) Synthesis of polysaccharide extracellular enzymes and xanthomonad in *Xanthomonas campestris*: evidence for the involvement of two intercellular regulatory signals. Mol Plant Microbe Interact 11:68–70

Prats E, Gay AP, Mur LAJ, Thomas BJ, Carver TLW (2006) Stomatal lock-open, a consequence of epidermal cell death, follows transient suppression of stomatal opening in barley attacked by *Blumeria graminis*. J Exp Bot 57:2211–2226

Preiter K, Brooks DM, Penaloza-Vazquez A, Sreedharan A, Bender CL, Kunkel BN (2005) Novel virulence gene of *Pseudomonas syringae* pv. *tomato* strain DC 3000. J Bacteriol 187: 7805–7814

Preston JF, Rice JD, Ingram LO, Keen NT (1992) Differential polymerization mechanisms of pectate lyases secreted by *Erwinia chrysanthemi* EC 16. J Bacteriol 174:2039–2042

Proctor RH, Desjardins AE, Plattner RD, Hohn TM (1999) A polyketide synthase gene required for biosynthesis of fumonisin mycotoxins in *Gibberella fujikuroi* mating population A. Fungal Genet Biol 27:100–112

Proctor RH, Plattner RD, Brown DW, Seo JR, Lee YW (2004) Discontinuous distribution of fumonisin biosynthetic genes in *Gibberella fujikuroi* species complex. Mycol Res 108:815–822

Prusky D, Drori N, Kramer-Haimovich H, Rollins J (2003) External pH and nitrogen source affect secretion of pectate lyase by *C. gloeosporioides*. Phytopathology 93:S105(Abst.)

Pryce-Jones E, Carver T, Gurr SJ (1999) The roles of cellulose enzymes and mechanical force in host penetration by *Erysiphe graminis* f.sp. *hordei*. Physiol Mol Plant Pathol 55:175–182

Pugsley AP (1993) The complete general protein secretory pathway in Gam-negative bacteria. Microbiol Rev 57:50–108

Quayyum HA, Gijzen M, Traquair JA (2003) Purification of a necrosis-inducing host-specific protein from spore germination fluid of *Alternaria panax*. Phytopathology 93:323–328

Quidde T, Osbourn AE, Tudzynski P (1998) Detoxification of α-tomatine by *Botrytis cinerea*. Physiol Mol Plant Pathol 52:151–165

Quinones BC, Pujol CJ, Lindow SE (2004) Regulation of AHL production and its contribution to epiphytic fitness in *Pseudomonas syringae*. Mol Plant Microbe Interact 17:521–531

Rajeswari R, Gopaljee Jha, Sonti RV (2005) Role of in planta-expressed xylanase of *Xanthomonas oryzae* pv. *oryzae* in promoting virulence on rice. Mol Plant Microbe Interact 18:830–837

Quiñones B, Dulla G, Lindow SE (2005) Quorum sensing regulates exopolysaccharides production, motility and virulence in *Pseudomonas syringae*. Mol Plant Microbe Interact 18: 682–693

Rao ALN, Francki RIB (1982) Distribution of determinants for symptom production and host range on the three RNA components of *Cucumber mosaic virus*. J Gen Virol 61:197–205

Ramonell K, Berrocal-Lobo M, Koh S, Wan J, Edwards H, Stacey G, Somerville S (2005) Loss of function mutation in chitin responsive genes show increased susceptibility to the powdery mildew pathogen *Erysiphe cichoracearum*. Plant Physiol 138:1027–1036

Randall TA, Dwyer RA, Huitema E, Beyer K, Cvitanich C, Kelkar H, Ah Fong AMV, Gates K, Roberts S, Yatzkan E, Gaffney T, Law M, Testa A, Torto-Alalibo T, Zhang M, Zheng L, Mveller E, Windass J, Binder A, Birch PRJ, Gisi U, Govers F, Goco NA, Mauch F, van West P, Waugh ME, Yu J, Boller T, Kamoun S, Lam ST, Judelson HS (2005) Large-scale gene discovery in the oomycete *Phytophthora infestans* reveals likely components of phytopathogenicity shared with true fungi. Mol Plant Microbe Interact 18:229–243

Rantakari A, Virtaharju O, Vähämiko S, Taira S, Palva ET, Saarilahti HT, Romantschuk M (2001) Type III secretion contributes to the pathogenesis of the soft-rot pathogen *Erwinia carotovora*: partial characterization of the *hrp* gene cluster. Mol Plant Microbe Interact 14:962–968

Ray SK, Rajeswari R, Sharma Y, Sonti RV (2002) A high-molecular weight outer membrane protein of *Xanthomonas oryzae* pv. *oryzae* exhibits similarity to nonfibrial adhesions of animal pathogenic bacteria and is required for optimum virulence. Mol Microbiol 46:637–647

Redondo E, Krause-Sakate R, Yang SJ, Lot H, Le Gall O, Candresse T (2001) *Lettuce mosaic virus* (LMV) pathogenicity determinants in susceptible and tolerant lettuce varieties map to different region to the viral genome. Mol Plant Microbe Interact 14:804–810

Reeves PJ, Whitecombe D, Wharam S, Gibson M, Allison G (1993) Molecular cloning and characterization of 13 *out* genes from *Erwinia carotovora* subsp. *carotovora*: genes encoding members of a general secretion pathway (GSP) widespread in Gram-negative bacteria. Mol Microbiol 38:443–456

Reichel C, Beachy RN (1998) *Tobacco mosaic virus* infection induces severe morphological changes in the endoplasmic reticulum. Proc Natl Acad Sci USA 95:11169–11174

Reiss E, Bryngelsson T (1996) Pathogenesis-related proteins in barley leaves, induced by infection with *Drechslera teres* (Sacc) Shoem-by treatment with other biotic agents. Physiol Mol Plant Pathol 49:331–341

Requena A, Simón-Buela L, Salcedo G, Garcia-Arenal F (2006) Potential involvement of a cucumber homolog of phloem protein 1 in the long-distance movement of *Cucumber mosaic virus* particles. Mol Plant Microbe Interact 19:734–746

Restrepo S, Myers KL, del Pozo O, Martin GB, Hart AL, Buell CR, Fry WE, Smart CD (2005) Gene profiling of a compatible interaction between *Phytophthora infestans* and *Solanum tuberosum* suggests a role for carbonic anhydrase. Mol Plant Microbe Interact 18: 913–922

Reuhs BL, Kim JS, Matthysee AG (1997) Attachment of *Agrobacterium tumefaciens* to carrot cells and *Arabidopsis* wound sites is correlated with the presence of cell-associated acidic polysaccharide. J Bacteriol 179:5372–5379

Revers F, LeGall O, Candresse T, Maule AJ (1999) New advances in understanding the molecular biology of plant potyvirus interactions. Mol Plant Microbe Interact 12:367–376

Reyna NS, Yang Y (2006) Molecular analysis of the rice MAP kinase gene family in relation to *Magnaporthe grisea* infection. Mol Plant Microbe Interact 19:530–546

Rha E, Park HJ, Kim MO, Chung YL, Lee CW, Kim JW (2001) Expression of exopolygalacturonases in *Botrytis cinerea*. FEMS Microbiol Lett 201:101–109

Richael C, Gilchrist D (1999) The hypersensitive response a case of hold or fold? Physiol Mol Plant Pathol 55:5–12

Ricker KE, Bostock RM (1992) Evidence for release of the elicitor arachidonic acid and its metabolites from sporangia of *Phytophhtora infestans* during infection of potato. Physiol Mol Plant Pathol 41:61–72

Robert Baudouy J, Nasser W, Condemine G, Reverchon S, Shevchik VE, Hugouvieux-Cotte Pattat N (2000) Pectic enzymes of *Erwinia chrysanthemi* regulation and role in pathogenesis. In: Stacey G, Keen NT (eds) Plant–microbe interactions, vol.5. American Phytopathological Society, St. Paul, MN, USA, pp 221–268

Roberts AG, Oparka KJ (2003) Plasmodesmata and the control of symplastic transport. Plant Cell Environ 26:103–124

Roden JA, Belt B, Ross JB, Tachibana T, Vargas J, Mugett MB (2004b) A genetic screen to isolate type III effectors translocated into pepper cells during *Xanthomonas* infection. Proc Natl Acad Sci USA 101:16624–16629

Roden J, Eardley L, Hoston A, Cao Y, Mudgett MB (2004) Characterization of *Xanthomonas* AvrXv4 effector, a SUMO protease translocated into plant cells. Mol Plant Microbe Interact 17:633–643

Rogers LM, Kim YK, Guo WJ, González-Candelas L, Li DX, Kolattukudy PE (2000) Requirement of either a host- or pectin-induced pectate lyase for infection of *Pisum sativum* by *Nectria haematococca*. Proc Natl Acad Sci USA 97:9813–9818

Rojas CM, Ham JH, Schechter LM, Kim JF, Beer SV, Collmer A (2004) The *Erwinia chrysanthemi* EC16 *hrp/hrc* gene cluster encodes an active Hrp type III secretion system that is flanked by virulence genes functionally unrelated to the Hrp system. Mol Plant Microbe Interact 17: 644–653

Rojas MR, Zerbini FM, Allison RF, Gilbertson RL, Lucas WJ (1997) Capsid protein and helper component proteinase function as potyvirus cell-to-cell movement proteins. Virology 237: 283–295

Ronald PC, Staskawicz BJ (1988) The avirulence gene *avrBs1* from *Xanthomonas campestris* pv. *vesicatoria* encodes a 50-kD protein. Mol Plant Microbe Interact 1:191–198

Rossier O, Van den Ackerveken G, Bonas U (2000) HrpB2 and HrpF from *Xanthomonas* are type III-secreted proteins and essential for pathogenicity and recognition by the host plant. Mol Microbiol 38:828–838

Roper MC, Greve LC, Warren JG, Labavitch JM, Kirkpatrick BC (2007) *Xylella fastidiosa* requires polygalacturonase for colonization and pathogenicity in *Vitis vinifera* grapevines. Mol Plant Microbe Interact 20:411–419

Rose JK, Ham KS, Darvill AG, Albersheim P (2002) Molecular cloning and characterization of glucanase inhibitor proteins: coevolution of a counter defense mechanism by plant pathogens. Plant Cell 14:1329–1345

Rottem S (2002) Invasion of mycoplasmas into and fusion with host cells. In: Razin S, Hermann R (eds) Molecular biology and pathogenicity of mycoplasmas, Kluwer Academic Plenum Publishers, New York, pp 391–401

Roudet-Tavert G, German-Retana S, Delaunay T, Delécolle-B, Candresse T, Le Gall O (2002) Interaction between potyvirus helper component-proteinase and capsid protein in infected plants. J Gen Virol 83:1765–1770

Ruan Y, Kotraiah V, Straney DC (1995) Flavonoids stimulate spore germination in *Fusarium solani* pathogenic on legumes in a manner sensitive to inhibitors of cAMP dependent proein kinase. Mol Plant Microbe Interact 8:929–938

Rui O, Hahn M (2007) The Slt2-type MAP kinase Bmp3 of *Botrytis cinerea* is required for normal saprotrophic growth, conidiation, plant surface sensing and host tissue colonization. Mol Plant Pathol 8:173–184

Ruiz-Ferrer V, Boskovic J, Alfonso C, Rivas G, Llorca O, López-Abella D, López-Moya JJ (2005) Structural analysis of *Tobacco etch potyvirus* HC-Pro oligomers involved in aphid transmission. J Virol 79:3758–3765

Ryan RP, Fouhy Y, Lucey JF, Jiang B, He YQ, Feng JX, Tang JL, Dow JM (2007) Cyclic di-GMP signaling in the virulence and environmental adaptation of *Xanthomonas campestris*. Mol Microbiol: doi.10.1111/J.1365-2958. 2006.05531.x

Ryang BS, Kobori T, Matsumoto T, Kosaka Y, Ohki ST (2004) *Cucumber mosaic virus* 2b protein compensates for restricted systemic spread of *Potato virus Y* in doubly infected tobacco. J Gen Virol 85:3405–3414

Sadras VO, Quiroz F, Echarte L, Escande A, Pereyra VR (2000) Effect of *Verticillium dahliae* on photosynthesis, leaf expansion and senescence of field-grown sunflower. Ann Bot 86: 1007–1015

Sáenz P, Salvador B, Simón-Mateo C, Kasschau KD, Carrington JC, García JA (2002) Host-specific involvement of the HC protein in the long-distance movement of potyviruses. J Virol 76:1922–1931

Saito T, Meshi T, Takamatsu N, Okada Y (1987) Coat protein gene sequences of *Tobacco mosaic virus* encodes a host response determinant. Proc Natl Acad Sci USA 84:6074–6077

Salánki K, Carrére M, Jacquemond M, Balázs, Tepfer M (1997) Biological properties of pseudo-recombinant and recombinant strains created with *Cucumber mosaic virus* and *Tomato aspermy virus*. J Virol 71:3597–3602

Salisbury V, Pfoestl A, Wiesinger-Mayr H, Lewis R, Bowker KE, MacGowan AP (1999) Use of a clinical *Escherichia coli* isolate expressing *lux* genes to study the antimicrobial pharmacodynamics of moxifloxacin. J Antimicro Chemo 43:829–832

Sánchez-Torres P, González-Candeles L (2003) Isolation and characterization of genes differentially expressed during interaction between apple fruit and *Penicillium expansum*. Mol Plant Pathol 4:447–457

Sandrock RW, Van Etten HD (1998) Fungal sensitivity to and enzymatic degradation of the phytoanticipin α-tomatine. Phytopathology 88:137–143

Santen K, Marttilla S, Liljeroth E, Bryngelsson T (2005) Immunocytochemical localization of the pathogenesis-related PR-1 protein in barley leaves after infection by *Bipolaris sorokiniana*. Physiol Mol Plant Pathol 66:45–54

Sauvage C, Expert D (1994) Differential regulation by iron of *Erwinia chrysanthemi* pectate lyases: pathogenicity of iron transport regulatory (*cbr*) mutants. Mol Plant Microbe Interact 7:71–77

Schäfer W (1994) Molecular mechanisms of fungal pathogenicity to plants. Annu Rev Phytopathol 32:461–477

Schäfer P, Huckelhoven R, Kogel KH (2004) The white barley mutant Albostrians shows a susceptible but symptomless interaction phenotype with the hemibiotrophic fungus *Bipolaris sorokiniana*. Mol Plant Microbe Interact 17:366–373

Schechter LM, Roberts KA, Jamir Y, Alfano JR,Collmer A (2004) *Pseudomonas syringae* type III secretion system targeting signals and novel effectors studied with a Cya translocation reporter. J Bacteriol 186:543–555

Scheets K (1998) *Maize chlorotic mottle machlomovirus* and *Wheat streak mosaic rymovirus* concentrations increase in the synergistic disease corn lethal necrosis. Virology 242:28–38

Scheffer RP (1976) Host-specific toxins in relation to pathogenesis and disease resistance In: Heitefuss PH, Williams R (eds) Encyclopedia of plant physiology, Springer-Verlag, New York, USA, pp 247–269

Scheffer RP, Yoder OC (1972) Host-specific toxins and selective toxity. In: Wood RKS, Ballio A, Graniti A (eds) Phytotoxins in plant diseases, Academic Press, New York, pp 251–272

Schepetilnikov MV, Manske U, Solovyev AG, Zamyatnin Jr AA, Schiemann J, Morozov SY (2005) The hydrophobic segment of *Potato Virus X* TGNBp3 is a major determinant of the protein intracellular trafficiking. J Gen Virol 86:2379–2391

Scholthof KBG (1999) A synergism induced by satellite *Panicum mosaic virus*. Mol Plant Microbe Interact 12:163–166

Schornack S, Meyer A, Römer P, Jordan T, Lahaye T (2006) Gene-for-gene mediated reconigition of nuclear-targeted AvrBs3-like bacterial effector proteins. J Plant Physiol 163: 256–272

Schornack S, Peter K, Bonas U, Lahaye T (2005) Expression levels of avrBs3-like genes affect recognition specificity in tomato Bs4-but not in pepper Bs3-mediated perception. Mol Plant Microbe Interact 18:1215–1225

Schrammeijer B, Risseauw E, Pansegru W, Regensburg-Tuink TJG, Crosby WL, Hooykaas PJJ (2001) Interaction of the virulence protein VirF of *Agrobacterium tumefaciens* with plant homologs of the yeast Skp1 protein. Curr Biol 11:258–262

Schulze Gronover C, Kasulke D, Tudzynski P, Tudzynski B (2001) The role of G protein alpha subunits in the infection process of the gray mold fungus *Botrytis cinerea.* Mol Plant Microbe Interact 14:1293–1302

Schulze-Gronover C, Schorn C, Tudzynski B (2004) Identification of *Botrytis cinerea* up-regulated during infection and controlled by the G (alpha) subunit BCG1 using suppression subtractive hybridization (SSH). Mol Plant Microbe Interact 17:537–546

Schwach F, Vaistij FE, Jones L, Baulcombe DC (2005) An RNA-dependent RNA polymerase prevents meristem invasion by *Potato virus X* and is required for the activity but not the production of a systemic silencing signal. Plant Physiol 138:1842–1852

Scofield SR, Tobias CM, Rathjen JP, Chang JH, Lavelle DT, Michelmore RW, Staskawicz BJ (1996) Molecular basis of gene-for-gene specificity in bacterial speck disease of tomato. Science 274:2063–2065

Scholthof HB, Scholthof KBG, Kikkert M, Jackson AO (1995) *Tomato bushy stunt virus* spread is regulated by two nested genes that function in cell-to-cell movement and host-dependent systemic invasion. Virology 213:425–438

Scott-Craig JS, Panaccione DG, Cervone F, Walton JD (1990) Endopolygalacturonase is not required for pathogenicity for *Cochliobolus carbonum* on maize. Plant Cell 2:1191–1200

Sela-Buurlage MB, Epstein L, Rodriguez RJ (1991) Adhesion of ungerminated *Colletotrichum musae* conidia. Physiol Mol Plant Pathol 39:345–352

Selth LA, Dogra SC, Rasheed MS, Healy H, Randles JW, Rezaian M (2005) A NAC domain protein interacts with *Tomato leafcurl virus* replication accessory protein and enhances viral replication. Plant Cell 17:311–325

Seong K, Hou Z, Tracy M, Kistler HC, Xu JR (2005) Random insertional mutagenesis identifies genes associated with virulence in the wheat scab fungus *Fusarium graminearum*. Phytopathology 95:744–750

Shan L, Oh HS, Chen J, Guo M, Zhou J, Alfano JR, Collmer A, Jia X, Tang X (2004) The HopPtoF locus of *Pseudomonas syringae* pv. *tomato* DC3000 encodes a type III chaperone and a cognate effector. Mol Plant Microbe Interact 17:447–455

Shan L, He P, Zhou JM, Tang X (2000) A cluster of mutations disrupt the avirulence but not the virulence function of AvrPto. Mol Plant Microbe Interact 13:592–598

Shanmugam V, Viswanathan R, Raguchander T, Balasubramanian P, Samiyappan R (2002) Immunology of pathogen virulence and phytotoxin production in relation to disease severity: a case study in sheath blight of rice. Folia Microbiol 47:551–558

Shao F, Merritt PM, Bao Z, Innes RW, Dixon JEA (2002) *Yersinia* effector and a *Pseudomonas* avirulence protein define a family of cysteine proteases functioning in bacterial pathogenesis. Cell 109:575–588

Shapiro JA (1998) Thinking about bacterial populations as multicellular organisms. Annu Rev Microbiol 52:81–104

Shaw CH, Ashby AM, Brown A, Royal C, Loake GJ, Shaw C (1988) *virA* and *virG* are Ti-plasmid function required for chemotaxis of *Agrobacterium tumefaciens* toward acetosyringone. Mol Microbiol 2:413–417

Sheng J, Citovsky V (1996) *Agrobacterium*-plant cell DNA transport have virulence proteins with travel. Plant Cell 8:1699–1710

Shimizu R, Taguchi F, Marutani M, Mukhaihara T, Inagaki Y, Toyada K, Shiraishi T, Ichinose Y (2003) The ΔfliD mutant of *Pseudomonas syringae* pv. *tabaci*, which secretes flagellin monomers induces a strong hypersensitive reaction (HR) in nonhost tomato cells. Mol Genet Genom 269:21–30

Shintaku MH, Carter SA, Bao YM, Nelson RS (1996) Mapping nucleotides in the 126-kDa protein gene that control the differential symptoms induced by two strains of *Tobacco mosaic virus*. Virology 221:218–225

Shintaku MH, Palukaitis P (1990) Genetic mapping of *Cucumber mosaic virus* In: Pirone TP, Shaw JG (eds) Viral genes and plant pathogenesis, Springer-Verlag, New York, pp 156–164

Shintaku MH, Zhang L, Palukaitis P (1992) A single amino acid substitution in the coat protein of *Cucumber mosaic virus* induces chlorosis in tobacco. Plant Cell 4:751–757

Shiraishi T, Oku H, Yamashita M, Ouchi S (1978) Elicitor and suppressor of pisatin induction in spore germination fluid of pea pathogen, *Mycosphaerella pinodes*. Ann Phytopathol Soc Jpn 44:659–665

Shiraishi T, Saitoh K, Kim H, Kato T, Tahara M, Oku H, Yamada T, Ichinose T (1992) Two suppressors supprescins A and B, secreted by a pea pathogen, *Mycosphaerella pinodes*. Plant Cell Physiol 33:663–667

Shiraishi T, Yamada T, Ichinose Y, Kiba A, Toyoda K (1997) The role of suppressors in determining host-parasite specificities in plant cells. Internatl Rev Cytol 172:55–93

Siciliano F, Torres P, Sendín L, Bermejo C, Filippone P, Vellice G, Ramallo J (2006) Analysis of the molecular basis of *Xanthomonas axonopodis* pv. *citri* pathogenesis in *Citrus limon*. Electronic Biotechnol D01:10.2225/Vol 19-issue3-fulltext-20

Siewers V, Viaud M, Jimenez-Teja D, Collado IG Schulze-Gronover Ch, Pradier JM, Tudzynski B, Tudzynski P (2005) Functional analysis of the cytochrome p450 monooxygenase gene *cbot 1* of *Botrytis cinerea* indicates that botrydial is a strain-specific virulence factor. Mol Plant Microbe Interact 18:602–612

Siluè D, Tharreau D, Talbot NJ, Clergeot PH, Notteghem JC, Lebrum MH (1998) Identification and characterization of *apf1* in nonpathogenic mutant of the rice blast fungus *Magaporthe grisea* which is unable to differentiate appressoria. Physiol Mol Plant Pathol 53: 239–251

Singh VB, Kumar A, Kirubakaram SI, Ayyadurai N, Samish Kumar R, Sakthivel N (2006) Comparison of exopolysaccharides produced by *Xanthomonas oryzae* pv. *oryzae* strains BX01 and BX08 that varying degrees of virulence in rice (*Oryza sativa* L.). J Phytopathol 154: 410–413

Sjulin TM, Beer SV (1978) Mechanism of wilt induction by amylovorin in *Cotoneaster* shoots and its relation to wilting of shoots infected by *Erwinia amylolovora*. Phytopathology 68:89–94

Skadsen RW, Hohn TM (2004) Use of *Fusarium graminearum* transformed with *gfp* to follow infection patterns in barley and *Arabidopsis*. Physiol Mol Plant Pathol 64:45–53

Skamnioti P, Henderson C, Zhang Z, Robinson Z, Gurr SJ (2007) A novel role for catalase B in the maintenance of fungal cell-wall integrity during host invasion in the rice blast fungus *Magnaporthe grisea*. Mol Plant Microbe Interact 20:568–580

Slater H, Alvarez-Morales A, Barber CE, Daniels MJ, Dow JM (2000) A two-component system involving an HD-GYP domain protein links cell–cell signaling to pathogenicity gene expression in *Xanthomonas campestris*. Mol microbiol 38:986–1003

Smadja B, Latour X, Faure D, Chevalier S, Dessaux Y, Orange N (2004) Involvement of N-acylhomoserine lactones throughout plant infection by *Erwinia carotovora* subsp. *atroseptica* (*Pectobacterium atrosepticum*). Mol Plant Microbe Interact 17:1269–1278

Soellick TR, Uhrig JF, Butcher GL, Kellman JW, Schreier PH (2000) The movement protein NS_M of *Tomato spotted wilt tospovirus* (TSWV): RNA binding, interaction with the TSWV N protein and identification of interacting plant proteins. Proc Natl Acad Sci USA 97:2373–2378

Solovyev AG, Savenkov EJ, Agranovsky AA, Morozov SY (1996) Comparison of the genomic cis elements and coding regions in RNA components of the hordeiviruses *Barley stripe mosaic virus, Lychnis ringspot virus* and *Poa semilatent virus*. Virology 219:9–18

Sperandio V, Torres AG, Jarvis B, Nataro JP, Kaper JB (2003) Bacteria–host communication: the language of hormones. Proc Natl Acad Sci USA 100:8951–8956

Sreedharan A, Penaloza-Vazquez A, Kunkel BN, Bender CL (2006) CorR regulates multiple components of virulence in *Pseudomonas syringae* pv. *tomato* DC 3000. Mol Plant Microbe Interact 19:768–779

Stachel SE, Nester EW, Zambryski PC (1986) A plant cell factor induces *Agrobacterium tumefaciens vir* gene expression. Proc Natl Acad Sci USA 83:379–383

Stachel SE, Zambryski PC (1986) *virA* and *virG* control the plant induced activation of the T-DNA transfer process of *Agrobacterium tumefaciens*. Cell 46:325–333

Stahl DJ, Schäfer W (1992) Cutinase is not required for fungal pathogenicity on pea. Plant Cell 4:621–629

Stanley J, Townsend R, Curson SJ (1985) Pseudorecombinants between cloned DNAs of two isolates of *Cassava latent virus*. J Gen Virol 66:1055–1061

Stenger DC,Young BA, French R (2006) Random mutagenesis of *Wheat streak mosaic virus* HC. Pro: non-infectious interfering mutations in a gene dispensable for systemic infection of plants. J Gen Virol 87:2741–2747

Storms MMH (1998) The role of NS_M during *Tomato spotted wilt virus* infection. Doctoral Thesis Unviersity of Wageningen, Netherlands

Storms MMH, Kormelink R, Peters D, van Lent JWM, Goldbach RW (1995) The nosntructural NS_M protein of *Tomato spotted wilt virus* induces tubular structures in plant and animal cells. Virology 214:485–493

Stuurman N, Pacios Bras C, Schlaman HR, Wijfjes AH, Bloemberg G, Spaink HP (2000) Use of green fluorescent protein color variant expressed on stable broad-host range vectors to visualize rhizobia interacting with plants. Mol Plant Microbe Interact 13:1163–1169

Sugio A, Yang B, White FF (2005) Characterization of the *hrpF* pathogenicity peninsula of *Xanthomonas oryzae* pv. *oryzae.* Mol Plant Microbe Interact 18:546–554

Suksomtip M, Liu P, AndersonT, Tungpradabkul S, Wood DW, Nester EW (2005) Citrate synthase mutants of *Agrobacterium* are attenuated in virulence and display reduced *vir* gene induction. J Bacteriol 187:4844–4852

Suo Y, Leung DWM (2002) Accumulation of extracellular pathogenesis-related proteins in rose leaves following inoculation of *in vivo* shoots with *Diplocarpon rosae*. Scientia Horti 93: 167–178

Suzuki S, Oshima K, Kakizawa S, Arashida R, Jung HY, Yamaji Y, Nishigawa H, Ugaki M, Namba S (2006) Interaction between the membrane protein of a pathogen and insect microfilament complex determines insect-vector specificity. Proc Natl Acad Sci USA 103:4252–4257

Suzuki F, Sawada H, Azegami K, Tsuchiya K (2004) Molecular characterization of the *tox* operon involved in toxoflavin biosynthesis of *Burkholderia glumae*. J Gen Plant Pathol 70:97–107

Suzuki F, Zhu Y, Sawada H, Matsuda I (1998) Identification of proteins involved in toxin production by *Pseudomonas glumae*. Ann Phytopathol Soc Jpn 64:75–79

Sweigard JA, Chumley FG, Valent B (1998) Disruption of *Magnaporthe grisea* cutinase gene. Mol Gen Genet 232:181–190

Swords KM, Dahlbeck D, Kearney B, Roy M, Staskawicz BJ (1996) Spontaneous and induced mutations in a single open reading frame alter both virulence and avirulence in *Xanthomonas campestris* pv. *vesicatoria* avrBs2. J Bacteriol 178:4661–4669

Syller J (2006) The roles and mechanisms of helper component proteins encoded by potyviruses and caulimoviruses. Physiol Mol Plant Pathol 67:119–130

Szurek B, Rossier O, Hause G, Bonas U (2002) Type III-dependent translocation of the *Xanthomonas* AvrBs3 protein into the plant cell. Mol Microbiol 46:13–23

Tada Y, Hata S, Nakayashiki H, Tosa Y, Mayama S (2000) Signal mediators for phytoalexin production in defense response of oats elicited by victorin as a specific elicitor. J Gen Plant Pathol 66:185–190

Takamiya KI, Tsuchiya T, Ohta H (2000) Degradation pathway(s) of chlorophyll: what has gene cloning revealed? Trends Plant Sci 5:426–431

Takano Y, Kikuchi T, Kubo Y, Hamer JE, Mise K, Furusawa I (2000) The *Colletotrichum lagenarium* MAP-kinase gene *CMK1* regulates diverse aspects of fungal pathogenesis. Mol Plant Microbe Interact 13:374–383

Takano Y, Komeda K, Kojima K, Okuno T (2001) Proper regulation of cyclic AMP-dependent protein kinase is required for growth, conidiation and appressorium function in anthracnose fungus, *Colletotrichum lagenarium.* Mol Plant Microbe Interact 14:1149–1157

Talbot NJ (1998) Molecular variability of fungal pathogens using the rice blast fungus as a case study. In: Bridge P, Couteadudier Y, Clarkson J (eds) Molecular variability of fungal pathogens, CAB International, Wallingford, Oxon, UK

Talbot NJ, Ebbole NJ, Hamer JE (1993) Identification and characterization of Mpg1, gene involved in pathogenicity form the rice blast fungus *Maganporthe grisea.* Plant Cell 5:1575–1590

Tan KC, Sanchez P, Cooper RM, Oliver RP (2004) The disruption of a G (alpha) subunit sheds new light on the pathogenicity of *Stagonospora nodorum* on wheat. Mol Plant Microbe Interact 17:456–466

Tanaka A, Shiotani H, Yamamoto M, Tsuge T (1999) Insertional mutagenesis and cloning of the genes required for biosynthesis of the host-specific AK-toxin in the Japanese pear pathotype of *Alternaria alternata*. Mol Plant Microbe Interact 12:691–702

Tanaka A, Tsuge T (2000) Structural and functional complexity of the genomic region controlling AK-toxin biosynthesis and pathogenicity in Japanese pear pathotype of *Alternaria alternata*. Mol Plant Microbe Interact 13:975–986

Tang X, Xie M, Kim YJ, Zhou J, Klessig DF, Martin GB (1999) Overexpression of Pto activates defense responses and confers broad resistance. Plant Cell 11:15–30

Tenberge KB (1999) Host wall alterations by *Claviceps purpurea* during infection of rye: molecular cytology of a host–pathogen interaction. Scanning 21:111–112

ten Have A, Mulder W, Visser J, van Kan JAL (1998) The endopolygalacturonase gene *Bcpg1* is required for full virulence of *Botrytis cinerea*. Mol Plant Microbe Interact 11:1009–1016

ten Have A, Oude Breuil W, Wubben JP, Visser J, van Kan JAL (2001) *Botrytis cinerea* endopolygalacturonase genes are differentially expressed in various plant tissues. Fungal Genet Biol 33:97–105

ten Have A, Tenberge KB, Benen JAE, Tudzynski P, Visser J, van Kan JAL (2002) The contribution of cell wall degrading enzymes to pathogenesis of fungal pathogens. In: Kempken (ed) The Mycota XI agricultural applications, Springer-Verlag, Heidelberg, Germany, pp 341–358

Thines W, Webber RWS, Talbot NJ (2000) MAP-kinase and protein kinase A-dependent mobilization of triacylglycerol and glycogen during appressorium turgor generation by *Magnaporthe grisea*. Plant Cell 12:1703–1718

Thirumala-Devi K, Huitema E, Cakir C, Kamoun S (2006) Synergistic interactions of the plant cell death pathways induced by *Phytophthroa infestans* Nep1-like protein PiNPP1.1 and INF1 elicitin. Mol Plant Microbe Interact 19:854–863

Thompson JR, Doun S, Perry KL (2006) Compensatory capsid protein mutations in *Cucumber mosaic virus* confer systemic infectivity in squash (*Cucurbita pepo*). J Virol 80:7740–7743

Thwaites R, Spanu PD, Panopoulos, NJ, Stevens C, Mansfield JW (2004) Transcriptional regulation of components of the type III secretion system and effectors in *Pseudomonas syringae* pv. *phaseolicola*. Mol Plant Microbe Interact 17:1250–1258

Tian M, Benedetti B, Kamoun S (2005) A second Kazal-like protease inhibitor from *Phytophthora infestans* inhibits and interacts with the apoplastic pathogenesis-related protease P69B of tomato. Plant Physiol 138:1785–1793

Tian M, Huitema E, Da Cunha L, Torto-Alalibo T, Kamoun S (2004) A Kazal-like extracellular serine protease inhibitor from *Phytophthora infestans* targets the tomato pathogenesis-related protease P69B. J Biol Chem 279:26370–26377

Tian M, Win J, Song J, van der Hoorn R, van der Knaap E, Kamoun S (2007) A *Phytophthora infestans* cystatin like protein targets a novel tomato papain-like apoplastic protease. Plant Physiol 143:364–377

Timmer LW, Brown GE (2000) Biology and control of anthracnose diseases of citrus. In: Prusky D, Freeman S, Dickman MB (eds) Colletotrichum: host specificity pathology and host–pathogen interaction. APS Press, St. Paul, MN, USA, pp 300–316

Tinland B, Schoumacher F, Gloeckler V, Bravo-Angel AM and Hohn B (1995) The *Agrobacterium tumefaciens* virulence D2 protein is responsible for precise integration of T-DNA into the plant genome. EMBO J 14:3585–3595

Torto T, Li S, Styer A, Huitema E, Testa A, Gout NAR, van West P, Kamoun S (2003) EST mining and functional expression assays identify extracellular protein from *Phytophthora*. Genome Res 13:1675–1685

Trans-Kersten J, Guan Y, Allen C (2001) *Ralstonia solanacearum* needs motility for invasive virulence on tomato. J Bacteriol 183:3597–3605

Trans-Kersten J, Guam Y, Allen C (1998) *Ralstonia solanacearum* pectin methyl esterase is required for growth on methylated pectin, but not for bacterial virulence. Appl Environ Microbiol 64:4918–4923

Truesdell GM, Yang ZH, Dickman MB (2000) A G α-subunit gene from the phytopathogenic fungus *Colletotrichum trifolii* is required for conidial germination. Physiol Mol Plant Pathol 56:131–140

Trutnyeva K, Bachmaier R, Waigmann E (2005) Mimicking carboxy terminal phosphorylation differentially effects subcellular distribution and cell-to-cell movement of *Tobacco mosaic virus* movement protein. Virology 332:563–577

Tsai CH, Dreher TN (1993) Increased viral yield and symptom severity result from a single amino acid substitution in the *Turnip yellow mosaic virus* movement protein. Mol Plant Microbe Interact 6:268–273

Tsiamis G et al (2000) Cultivar-specific avirulence and virulence functions assigned to avrPphF in *Pseudomonas syringae* pv. *phaseolicola*, the cause of bean halo-blight disease. EMBO J 19:3204–3214

Tsien RY (1998) The green fluorescent protein. Annu Rev Biochem 67:509–544

Tsuda S, Kubota K, Kanda A, Ohki T, Meshi T (2007) Pathogenicity of *Pepper mild mottle virus* is controlled by the RNA silencing suppression activity of its replication protein but not the viral accumulation. Phytopathology 97:412–420

Tsuge S, Ochiai H, Inoue Y, Oku T, Tsuno K, Kaku H, Kubo Y (2004) Involvement of phosphoglucose isomerase in pathogenicity of *Xanthomonas oryzae* pv. *oryzae*. Phytopathology 94:478–483

Tsuji G, Fujii S, Tsuge S, Shiraishi T, Kubo Y (2003) The *Colletotrichum lagenarium* Ste12-like gene *CST1* is essential for appressorium penetration. Mol Plant Microbe Interact 16:315–325

Tudzynski P, Sharon A (2003) Fungal pathogenicity genes. In: Arora DK, Khachatourians GG (eds) Applied mycology and biotechnology, Volume 3-Fungal Genetics. Elsevier Science BV, Amsterdam, pp 187–212

Tuori RP, Wolpert TJ, Ciufetti LM (2000) Heterologous expression of functional Ptv ToxA. Mol Plant Microbe Interact 13:456–464

Tyler BM, Tripathy S, Zhang X, Dehal et al (2006) *Phytophthora* genome sequences uncover evolutionary origins and mechanisms of pathogenesis. Science 313:1261–1266

Ueki S, Citovsky V (2002) The systemic movement of a tobamovirus is inhibited by a cadmium-ion-induced glycine-rich protein. Nature Cell Biol 4:478–485

Uematsu T, Yoshimura D, Nishiyama K, Ibaraki T, Fujii H (1976) Pathogenic bacterium causing seedling rot of rice. Ann Phytopathol Soc Jpn 42:464–471

Uppalapati SR, Ishiga Y, Wangdi T, Kunkel BN, Anand A, Mysore KS, Bender CL (2007) The phytotoxin coronatine contributes to pathogen fitness and is required for suppression of salicylic acid accumulation in tomato inoculated with *Pseudomonas syringae* pv. *tomato* DC 3000. Mol Plant Microbe Interact 20:955–965

Uppalapati SR, Toyoda K, Yasuhiro I, Ichinose Y, Shiraishi T (2004) Differential regulation of MBP kinase by a glycoprotein elicitor and a polypeptide suppressor from *Mycosphaerella pinodes* in pea. Physiol Mol Plant Pathol 64:17–25

Uttaro AD, Cangelosi GA, Geremia PA, Nester EW, Ugalde RA (1990) Characterization of avirulent *exoC* mutants of *Agrobacterium tumefaciens*. J Bacteriol 172:1640–1646

Valette-Collet O, Cimerman A, Reignault P, Levis C, Boccara M (2003) Disruption of *Botrytis cinerea* pectin methylesterase gene *bcpme1* reduces virulence on several host plants. Mol Plant Microbe Interact 16:360–367

Vance CB (1999) Synergism: plant viruses. In: Granoff A, Webster RG (eds) Encyclopedia of virology, 2nd edn, Academic Press, San Diego, CA, pp 1694–1699

Vance VB, Berger PH, Carrington JC, Hunt AG, Shi XM (1995) 5′ proximal potyviral sequences mediate *Potato virus X*/ potyviral synergistic disease in transgenic tobacco. Virology 206: 583–590

Van Dijk K, Tam VC, Records AR, Petnicki-Ocwieja T, Alfano JR (2002) The ShcA protein is a molecular chaperone that assists in the secretion of the HopPsyA effector from the type III (Hrp) protein secretion system of *Pseudomonas syringae*. Mol Microbiol 44:1469–1481

Van Gijsegem F, Genin S, Boucher C (1995a) *hrp* and *avr* genes-key determinants controlling the interactions between plants and Gram-negative phytopathogenic bacteria. In: Singh US,

Singh RP (eds) Pathogens and host specificity in plant diseases: histochemical, biochemical genetic and molecular bases, vol I, Pergamon Press, UK, pp 273–292

Van Gijsegem F, Gough C, Zischek C, Niqueux E, Arlat M, Genin S, Berberis P, German S, Castello P, Boucher C (1995b) The *hrp* gene locus of *Pseudomonas solanacearum* which controls the production of a type III secretion system, encodes eight proteins related to components of the bacterial flagellar biogenesis complex. Mol Microbiol 15:1095–1114

Van Kan JAL, Van't Klooster JW, Wagemakers AM, Dees DCT, van der Vlugt- Bergmans CJB (1997) Cutinase A of *Botrytis cinerea* is expressed, but not essential during penetration of gerbera and tomato. Mol Plant Microbe Interact 10:30–38

Van Loon LC (1999) Occurrence and properties of plant pathogenesis-related proteins. In: Swapan KD, Muthukrishnan S (eds) Plant pathogenesis-related proteins in plants. CRC Press, Boca Raton, FL, USA, pp 21–47

Van Loon LC, van Strien EA (1999) The families of pathogenesis-related proteins, their activities and comparative analysis of PR-1 type proteins. Physiol Mol Plant Pathol 55:85–97

Vaux DL, Korsmeyer SJ (1999) Cell death in development. Cell 96:245–254

Veneault-Fourrey C, Laugé R, Langin T (2005) Nonpathogenic strains of *Colletotrichum lindemuthianum* trigger progressive bean defense responses during appressorium-mediated penetration. Appl Environ Microbiol 71:4761–4770

Venkatesh B. Babujee L, Liu H, Hedley P, Fujikawa T, Birch P, Toth, I, Tsuyumu S (2006) The *Erwinia chrysanthemi* 3937 PhoQ sensor kinase regulates several virulence determinants. J Bacteriol 188:3088–3098

Veronese P, Narasimhan ML, Stevenson RA, Zhu JK, Weller SC, Subbarao KU, Bressan RA (2003) Identification of a locus controlling *Verticillium* disease symptom response in *Arabidopsis thaliana*. Plant J 35:574–587

Viaud MC, Balhadere PV, Talbot NJ (2002) A *Magnaporthe grisea* cyclophlin acts as a virulence determinant during plant infection. Plant Cell 14:917–930

Viaud MC, Brunet-Simon A, Brygoo Y, Pradier JM, Levis C (2003) Cyclophilin A and calcineurin functions investigated by gene inactivation, cyclosporin A inhibition and cDNA arrays approaches in the phytopathogenic fungus *Botrytis cincerea*. Mol Microbiol 50:1451–1465

Viaud MC, Legeai F, Pradier JM, Brygo Y, Bitton F, Weissenbach J, Brunet-Simon A, Duclert A, Fillinger S, Fortini D, Gioti A, Giraud C, Halary S, Lebrum I, Le Pêcheur P, Samson D, Levis C (2005) Expressed sequence tages from phytopathogenic fungus *Botrytis cinerea*. Eur Plant Pathol 111:139–146

Voigt CA, Schäfer W, Salmon S (2005) A secreted lipase of *Fusarium graminearum* is a virulence factor required for infection of cereals. Plant J 42:364–375

Voinnet O (2003) RNA silencing bridging the gaps in wheat extracts. Trends Plant Sci 8:307–309

Voinnet O, Lederer C, Baulcombe DC (2000) A viral movement protein prevents spread of the gene silencing signal in *Nicotiana benthamiana*. Cell 103:157–167

Voinnet O, Pinto YM, and Baulcombe DC (1999) Suppression of gene silencing: a general strategy used by diverse DNA and RNA viruses of plants. Proc Natl Acad Sci USA 96:14147–14152

Vojnov AA, Slater H, Daniels MJ, Dow JM (2001) Expression of the *gum* operon directing xanthan biosynthesis in *Xanthomonas campestris* and its regulation in planta. Mol Plant Microbe Interact 14:768–774

Von Arnim F, Frischmuth T, Stanley J (1993) Detection and possible functions of *African cassava mosaic virus* DNA B gene products. Virology 192:264–272

Von Bargen S, Salchert K, Paape M, Piechulla B, Kellmann JW (2001) Interactions between the *Tomato spotted wilt virus* movement protein and plant proteins showing homologies to myosin, kinesin and DnaJ-like chaperones. Plant Physiol Biochem 39:1083–1093

Wang MB, Bian X-Y, Wu, L-X, Smith NA, Isenegger D, Wu RM, Masuta C, Vance CB, Watson JM, Rezaian A, Dennis ES, Waterhouse PM (2004c) On the role of RNA silencing in the pathogenicity and evolution of viroids and viral satellites. Proc Natl Acad Sci USA 101: 3275–3280

Wang K, Hadrami AE, Adam L, Daayf F (2004a) US-1 and US-8 genotypes of *Phytophthora infestans* differentially affect local, proximal and distal gene expression of phenylalanine ammonia lyase and 3-hydroxy, 3-methylglutaryl CoA reductase in potato leaves. Physiol Mol Plant Pathol 65:157–167

Wang LH, He YW, Gao YF, Wu JE, Dong YH, He CZ (2004b) A bacterial cell–cell communication signal with cross kingdom structural analogues. Mol Microbiol 51:903–912

Wang N, Lu SE, Wang J, Chen ZF, Gross DC (2006a) The expression of genes encoding lipidepsipeptide phytotoxins by *Pseudomonas syringae* pv. *syringae* is coordinated in response to plant signal molecules. Mol Plant Interact 19:257–269

Wang N, Lu SE, Yang Q, Sze SH, Gross DC (2006b) Identification of the *syr-syp* box in the promoter regions of genes dedicated to syringomycin and syringopeptin production by *Pseudomonas syringae* pv. *syringae* E301D. J Bacteriol 188:160–168

Wang P, Mathews DE, Van Etten HD (1992) Purification and characterization of cyanide hydratase from the phytopathogenic fungus *Gloeocercospora sorghi*. Arch Biochem Biophy 298: 569–575

Wang P, Van Etten HD (1992) Cloning and properties of a cyanide hydratase gene from the pathogenic fungus *Gloeocercospora sorghi*. Biochem Biophys Commun 187:1048–1054

Ward TJ, Bielawski JP, Kistler HC, Sullivan E, O'Donnell K (2002) Ancestral polymorphism and adaptative evolution in the trichothecene mycotoxin gene cluster of phytopathogenic *Fusarium*. Proc Natl Acad Sci USA 99:9278–9283

Weber E, Koebnik R (2005) Domain structure of HrpE, the Hrp pilus subunit of *Xanthomonas campestris* pv. *vesicatoria*. J Bacteriol 187:6175–6186

Weber E, Ojanen-Reuhs T, Huguet E, Hause G, Romantschuk M, Korhonen TK, Bonas U, Koebnik R (2005) The type III dependent Hrp pilus is required for productive interaction of *Xanthomonas campestris* pv. *vesicatoria* with pepper host plants. J Bacteriol 187:2458–2468

Weeds PL, Beever RE, Sharrock KR, Long PF (1999) A major gene controlling pathogenicity in *Botryostinia fuckeliana* (*Botrytis cinerea*). Physiol Mol Plant Pathol 54:13–35

Wehling MD, Guo M, Fu ZQ, Alfano JR (2004) The *Pseudomonas syringae* HopPtoV protein is secreted in culture and translocated into plant cells via type III protein secretion system in a manner dependent on the ShcV type III chaperone. J Bacteriol 186:3621–3630

Wei W, Opgenorth DC, Davis RE, Chang CJ, Summers CG, Zho Y (2006) Characterization of a novel adhesin-like gene and design of real-time PCR for rapid sensitive and specific detection of *Spiroplasma kunkelii*. Plant Dis 90:1233–1238

Weller DM, Saettler AW (1980) Colonization and distribution of *Xanthomonas phaseoli* and *Xanthomonas phaseoli* var *fuscans* in field-grown navy beans. Phytopathology 70:500–506

Wengelnik K, Bonas U (1996) HrpX, an AraC-type regulator, activates expression of five of the six loci in the *hrp* cluster of *Xanthomonas campestris* pv. *vesicatoria*. J Bacteriol 178:3462–3469

Wertz IE, Hanley MR (1996) Diverse molecular provocation of programmed cell death. Trends Biochem Sci 21:359–364

Weyman PD, Pan Z, Feng Q, Gilchrist DG, Bostock RM (2006) A circadian rhythm-regulated tomato gene is induced by arachidonic acid and *Phytophthora infestans* infection. Plant Physiol 140:235–248

Whalen MC, Wang JF, Carland FM, Heskell ME, Dahlbeck D, Minsavage GV (1993) Avirulence gene *avrRxv* from *Xanthomonas campestris* pv. *vesicatoria* specifies resistance on tomato line Hawaii 7998. Mol Plant Microbe Interact 6:616–627

White FF, Yang B, Johnson LB (2000) Prospects for understanding avirulence gene function. Curr Opin Plant Biol 3:291–298

Whitehead NA, Barnard AML, Slater H, Simpson NJL, Salmond GPC (2001) Quorum-sensing in gram-negative bacteria. FEMS Microbiol Rev 25:365–404

Whitehead NA, Byers JT, Commander P, Corbett MJ, Coulthurst SJ, Everson L, Harris AK, Pemberton CL, Simpson NJ, Slater H, Smith DS, Welch M, Williamson N, Salmond GP (2002) The regulation of virulence in phytopathogenic *Erwinia* species: quorum sensing, antibiotics and ecological considerations. Antonie Leeuwenhoek 81:223–231

Willis DK, Batra TM, Kinscherf TG (1991) Genetics of toxin production and resistance in phytopathogenic bacteria. Experimentia 47:765–771

Willis DK, Hrabak EM, Rich JJ, Barta TM, Lindow SE Panopoulos NJ (1990) Isolation and characterization of a *Pseudomonas syringae* pv. *syringae* mutant deficient in lesion formation on bean. Mol Plant Microbe Interact 3:149–156

Winans SC, Allenza P, Stachel SE, McBride KE, Nester EW (1987) Characterization of the *virE* operon of the *Agrobacterium* Ti plasmid pTiA6. Nucleic Acids Res 15:825–837

Wirawan IGP, Kang HW, Kojima M (1993) Isolation and characterization of a new chromosomal virulence gene of *Agrobacterium tumefaciens*. J Bacteriol 175:3208–3212

Wobbe KK, Akgoz M, Dempsey D' MA, Klessig DF (1998) A single amino acid change in *Turnip crinkle virus* movement protein P8 affects RNA binding and virulence on *Arabidopsis thaliana*. J Virol 72:6247–6250

Wolf S, Deom CM, Beachy RN, Lucas WJ (1989) Movement protein of *Tobacco mosaic virus* modifies plasmodesmatal size exclusion limit. Science 246:377–379

Wolpert TJ, Dunkle LD, Ciuffetti LM (2002) Host-selective toxins and avirulence determinants: what's in a name? Annu Rev Phytopathol 40:251–285

Wong SM, Thio SSC, Schintaku MH, Palukaitis P (1999) The rate of cell-to-cell movement in squash of *Cucumber mosaic virus* is affected by sequences of the capsid protein. Mol Plant Microbe Interact 12:628–632

Woods K, Hamilton CJ, Field RA (2004) Enzymatic liberation of lycotetraose from the *Solanum* glycoalkaloid α-tomatine. Carbohydrates Res 339:2325–2328

Woo HH, Hirsch AM, Hawes MC (2004) Altered susceptibility to infection by *Sinorhizobium meliloti* and *Nectria haematococca* in alfalfa roots with altered cell cycle. Plant Cell Reprod 22:967–973

Wubben JP, Mulder W, ten Have A, van Kan AL, Visser J (1999) Cloning and partial characterization of endopolygalacturonase genes from *Botrytis cinerea*. Appl Environ Microbiol 65: 1596–1602

Wubben JP, ten Have A, van Kan Jan AL, Visser J (2000) Regulation of endopolygalacturonase gene expression in *Botrytis cinerea* by galacturonic acid, ambient pH and carbon catabolite repression. Curr Genet 37:152–157

Wung CH, Hsu YH, Liou DY, Hung WC, Lin NS, Chang BY (1999) Identification of the RNA-binding sites of the triple gene block protein of *Bamboo mosaic potexvirus*. J Gen Virol 80:1119–1126

Xiao Y, Lu Y, Heu S, Hutcheson SW (1992) Organization and environmental regulation of the *Pseudomonas syringae* pv. *syringae* 61 *hrp* cluster. J Bacteriol 174:1734–1741

Xu GW, Gross DC (1988) Physical and functional analysis of the *syrA* and *syrB* genes involved in syringomycin production by *Pseudomonas syringae* pv. *syringae*. J Bacteriol 170:5680–5688

Xu HX, Mendgen K (1997) Targeted cell wall degradation at the penetration site of cowpea rust basidioporelings. Mol Plant Microbe Interact 10:87–94

Xu JR, Hamer JE (1996) MAP kinase and cAMP signaling regulate infection structure formation and pathogenic growth in rice blast fungus *Magnaporthe grisea*. Gene Develop 10:2696–2706

Xu JR, Urban M, Sweigard JA, Hamer JE (1997) The *CPKA* gene of *Magnaporthe grisea* is essential for appressorial penetration. Mol Plant Microbe Interact 10:187–194

Yabe K, Nakamura M, Hamasaki T (1999) Enzymatic formation of G-group aflatoxins and biosynthetic relationship between G-and G-group aflatoxins. Appl Environ Microbiol 65:3867–3872

Yakoby N, Beno-Moualen D, Keen NT, Dinoor A, Pines O, Prusky D (2001) *Colletotrichum gloeosporioides pelB* is an important virulence factor in avocado fruit–fungus interaction. Mol Plant Microbe Interact 14:988–995

Yakoby N, Freeman S, Dinoor A, Keen NT, Prusky D (2000) Expression of pectate lyase from *Colletotrichum gloeosporioides* in *C. magna* promotes pathogenicity. Mol Plant Microbe Interact 13:887–891

Yamada T, Hashimoto H, Shiraishi T, Oku H (1989) Suppression of pisatin phenylalanine ammonia-lyase mRNA, and chalcone synthase mRNA by a putative pathogenicity factor from the fungus *Mycosphaerella pinodes*. Mol Plant Microbe Interact 2:256–261

Yamauchi J, Takayanagi N, Komeda K, Takano Y, Okuno T (2004) cAMP-PKA signaling regulates multiple steps of fungal infection cooperatively with Cmk1 MAP kinase in *Colletotrichum lagenarium*. Mol Plant Microbe Interact 17:1355–1365

Yang Y, Ding B, Baulcombe DC, Verchot J (2000b) Cell-to-cell movement of the 25 K protein *Potato virus X* is regulated by three other viral proteins. Mol Plant Microbe Interact 13:599–605

Yang CH, Gavilanes-Ruiz M, Okinaka Y, Vedel R, Berthuy I, Boccara M, Chen JWT, Perna NT, Keen NT (2002) *hrp* genes of *Erwinia chrysanthemi* 3937 are important virulence factors. Mol Plant Microbe Interact 15:472–480

Yang B, Sugio A, White FF (2006) *Os8N3* is a host disease-susceptibility gene for bacterial blight of rice. Proc Natl Acad Sci USA 103:10503–10508

Yang B, White FF (2004) Diverse members of the AvrBs3/PthA family of type III effectors are major virulence determinants in bacterial blight disease of rice. Mol Plant Microbe Interact 17:1192–1200

Yang B, Zhu W, Johnson LB, White FF (2000a) The virulence factor AvrXa7 of *Xanthomonas oryzae* pv. *oryzae* is a type III secretion pathway-dependent nuclear-localized double-stranded DNA-binding protein. Proc Natl Acad Sci USA 97:9807–9812

Yanofsky MF, Porter SG, Young C, Albright LM, Gordon MP, Nester EW (1986) The virD operon from *Agrobacterium tumefaciens* encodes site-specific endonuclease. Cell 7:471–477

Yao J, Allen C (2006) Chemotaxis is required for virulence and competitive fitness of the bacterial wilt pathogen *Ralstonia solanacearum*. J Bacteriol 188:3697–3708

Yao CL, Conway WS, Ren RH, Smith D, Ross GS, Sams CE (1999) Gene encoding polygalacturonase inhibitor in apple fruit is developmentally regulated and activated by wounding and fungal infection. Plant Mol Biol 39:1231–1241

Yao N, Tada Y, Park P, Nakayshiki H, Tosa Y, Mayama S (2001) Novel evidence for apoptotic cell response and differential signals in chromatin condensation and DNA cleavage in victorin-treated oats. Plant J 28:13–26

Yip CK, Finlay BB, Strynadka NC (2005) Structural characterization of a type III secretion system filament protein in complex with its chaperone. Nat Struct Mol Biol 12:75–81

Yoneyama K, Kono Y, Yamaguchi I, Horikoshi M, Hirooka T (1998) Toxoflavin is an essential factor for virulence of *Burkholderia glumae* causing rice seedling rot disease. Ann Phytopathol Soc Jpn 64:91–96

Yoshikawa N, Saitou Y, Kitajima A, Chida T, Sasaki N, Isogai M (2006) Interference of long-distance movement of *Grapevine berry inner necrosis virus* in transgenic plants expressing a defective movement protein of *Apple chlorotic leaf spot virus*. Phytoapthology 96: 378–385

Yoshioka S, Okumura K, Hyodo H (2001) Ehtylene biosynthesis in sweet potato root tissue infected by *Ceratocystis fimbriata*. Acta Horti 553:139–141

Young GM, Schmiel DH, Miller VL (1999) A new pathway for the secretion virulence factors by bacteria: The flagellar export apparatus functions as a protein-secretion system. Proc Natl Acad Sci USA 96:6456–6461

Yu J, Chang PK, Cary JW, Bhatnagar D, Cleveland TE (1997) *avnA*, a gene encoding a cytochorme P-450 monooxygenase, is involved in the conversion of averntin to averufin in aflatoxin biosynthesis in *Aspergillus parasiticus*. Appl Environ Microbiol 63:1349–1356

Yu J, Mohamed SM, Bhatnagar D, Cleveland TE (2003a) Substrate-induced lipase gene expression and aflatoxin production in *Aspergillus parasiticus* and *Aspergillus flavus*. J Appl Microbiol 95:1334–1342

Yu W, Murfett J, Schoelz JE (2003b) Differential induction of symptoms in *Arabidopsis* by P6 of *Cauliflower mosaic virus*. Mol Plant Microbe Interact 16:35–42

Yu J, Wayadande AC, Fletcher J (2000) *Spiroplasma citri* surface protein P89 implicated in adhesion to cells of the vector *Circulifer tenallus*. Phytopathology 90:716–722

Yun MH, Torres PS, Oirdi ME, Rigano LA, Gonzalez-Lamothe R, Marano MR, Castagnaro AP, Dankert MA, Bouarab K, Vojnov (2006) Xanthan induces plant susceptibility by suppressing callose deposition. Plant Physiol 141:178–187

Zahiri AR, Babu MR, Saville BJ (2005) Differential gene expression during teliospore germination in *Ustilago maydis*. Mol Genet Genom 273:394–403

Zamyatnin AA Jr, Solovyev AG, Savenkov EI, Germundsson A, Sandgreen M, Valkonen JPT, Morozov SY (2004) Transient coexpression of individual genes encoded by the triple gene block of *Potato mop-top virus* reveals requirements for TGBp1 trafficking. Mol Plant Microbe Interact 17:921–930

Zelinger E, Hawles CR, Gurr SJ, Dewey FM (2004) An immunocytochemical and ultra structural study of the extracellular matrix produced by germinating spores of *Stagonospora nodorum* on natural and artificial surfaces. Physiol Mol Plant Pathol 65:125–135

Zeyen RJ, Bushnell WR, Carver TLW, Robbins MP, Clark TA, Boyles DA, Vance CP (1995) Inhibiting phenylalanine-ammonia lyase and cinnamyl alcohol dehydrogenase suppress Mla 1 (HR) but not Mlo5 (non-HR) barley powdery mildew resistances. Physiol Mol Plant Pathol 47:119–140

Zhang F, Simon AE (2003) Enhanced viral pathogenesis associated with a virulent mutant virus or a virulent satellite RNA correlates with reduced virion accumulation and abundance of free coat protein. Virology 312:8–13

Zhang L, Xu J, Birch RG (1999) Engineered detoxification confers resistance against a pathogenic bacterium. Nature Biotechnol 17:1021–1024

Zhao Y, Blumer SE, Sundin GW (2005) Identification of *Erwinia amylovora* genes induced during infection of immature pear tissue. J Bacteriol 187:8088–8103

Zhao Y, He S-Y, Sundin GW (2006) The *Erwinia amylovora avrRpt2(EA)* gene contributes to virulence on pear and AvrRpt 2 (EA) is recognized by *Arabidopsis* RPS2 when expressed in *Pseudomonas syringae*. Mol Plant Microbe Interact 19:644–654

Zhao Y, Owens R, Hammond R (2001a) Use of a *Potato virus X* vector in a whole plant assay to demonstrate nuclear targeting of *Potato spindle tuber viroid*. J Gen Virol 82:1491–1497

Zhao Z, Sagulenko E, Ding Z, Christie PJ (2001b) Activities of virE1 and VirE1 secretion chaperone in export of the multifunctional VirE2 effector via an *Agrobacterium* type IV secretion pathway. J Bacteriol 183:3855–3865

Zhen XH, Li YZ (2004) Ultrastructural changes and location of β-1,3-glucanase in resistant and susceptible cotton callus cells in response to treatment with toxin of *Verticillium dahliae* and salicylic acid. J Plant Physiol 161:1367–1377

Zheng L, Campbell M, Murphy L, Lam S, Xu JR (2000) The *BMP1* gene is essential for pathogenicity in the gray mold fungus *Botrytis cinerea*. Mol Plant Microbe Interact 13:724–732

Zhou T, Fan ZF, Li HF, Wong SM (2006) *Hibiscus chlorotic ringspot virus* p27 and its isoforms affect symptom expression and potentiate virus movement in kenaf (*Hibiscus cannabinus* L.). Mol Plant Microbe Interact 19:948–957

Zhu Y, Green L, Woo YM, Owens RA, Ding B (2001) Cellular basis of *Potato spindle tuber viroid* systemic movement. Virology 279:69–77

Zhu W, MaGbanua MM, White FF (2000) Identification of two novel *hrp*-associated genes in the *hrp* gene cluster of *Xanthomonas oryzae* pv. *oryzae*. J Bacteriol 182:1844–1853

Zhu Y, Qi Y, Xun Y, Owens R, Ding B (2002) Movement of *Potato spindle tuber viroid* reveals regulatory points of phloem-mediated RNA traffic. Plant Physiol 130:138–146

Ziemienowicz A, Tinland B, Bryant J, Gloeckler V, Hohn B (2000) Plant enzymes but not *Agrobacterium* VirD2 mediate T-DNA ligation. Mol Cell Biol 20:6317–6322

Zipfel C, Felix G (2005) Plants and animals: a different taste for microbes? Curr Opin Plant Biol 8:353–360

Zipfel C, Kunze G, Chinchilla D, Caniard A, Jones JDG, Boller T, Felix G (2006) Perception of the bacterial PAMP EF-Tu by the receptor EFR restricts *Agrobacterium* – mediated transformation. Cell 125:749–760

Zipfel C, Robtatzek S, Navarro L, Oakeley EJ, Jones JD, Felix G, Boller T (2004) Bacterial disease resistance in *Arabidopsis* through flagellin perception. Nature 428:764–767

Zolobowska L, van Gijsegem F (2006) Induction of lateral root structure formation on petunia roots: a novel effect of GMI 1000 *Ralstonia solanacearum* infection impaired in Hrp mutants. Mol Plant Microbe Interact 19:597–606

Zorreguieta A, Germia RA, Cangelosi GA, Nester EW, Ugalde RA (1988) Identification of the product of *Agrobacterium tumefaciens* virulence gene. Mol Plant Microbe Interact 1:121–127

Zou L, Wang X, Xiang Y, Zhang B, Li YR, Xiao Y, Wang J, Walnslay AR, Chen G (2006) Elucidation of the *hrp* clusters of *Xanthomonas oryzae* pv. *oryzicola* that control the hypersensitive response in nonhost tobacco and pathogenicity in susceptible host rice. Appl Environ Microbiol 72:6212–6224

Zupan J, Muth TR, Draper O, Zambryski P (2000) The transfer of DNA from *Agrobacterium tumefaciens* into plants: a feast of fundamental insights. Plant J 23:11–28

Zuppini A, Navazio L, Sella L, Castiglioni C, Favaron F, Mariani P (2005) An endopolygalacturonase from *Sclerotinia sclerotiourm* induces calcium-mediated signaling and programmed cell death in soybean cells. Mol Plant Microbe Interact 18:849–855

Chapter 3
Molecular Ecology and Epidemiology

Abstract During the past two decades, significant advancements have been made in the field of molecular ecology and epidemiology, resulting in accumulation of large amount of genomic data. Application of molecular techniques has been demonstrated to be essential for rapid, reliable and reproducible results for detection, identification and differentiation of microbial plant pathogens. The specificity of reactions of molecular techniques has been shown to be a clear advantage over conventional methods. Establishment of the identity of the microbial pathogen is the basic requirement of epidemiological investigations. Equally important is the possibility of quantification of pathogen(s) present in plants, air, soil and water by molecular techniques adding credibility to the conclusions arrived at. The effects of host plant, pathogen and environment on the incidence and spread of the diseases have been assessed by using molecular methods. The interactions between the pathogens, particularly viruses and their vectors has been studied in detail in some pathosystems. Molecular methods have opened up the possibility of having an insight into the disease development from different sources of infection which may or may not exhibit symptoms of infection.

Molecular ecology including epidemiology, over the past decades, has emerged as a new and rapidly expanding branch of biological sciences. Interdisciplinary approach is needed basically for several scientific disciplines including biology, ecology, epidemiology, molecular evolution and genetics. Molecular techniques are employed to study the complex patterns and processes of biological diversity at different levels of organization from genes to organisms, populations, species and ecosystems, but all experiments are oriented towards the genome of the organisms. A wide range of organisms including microbial pathogens has been studied by applying molecular ecological principles. It has been demonstrated that it has the potential to provide answer to specific key questions left unresolved in earlier ecological studies, in addition to revealing a wide array of additional features of the pathogen ecology. Molecular epidemiology aims to throw light on those processes that may result in a successful state of disease in a host plant, primarily to enable identification of a weak link in the biotic and molecular interactions or processes that may be capitalized upon to achieve disease control.

P. Narayanasamy, *Molecular Biology in Plant Pathogenesis and Disease Management: Disease Development, Vol. 2*,

Agricultural crops are affected by different kinds of microbial pathogens, each one requiring specific set of conditions for population build up to reach levels required for infection and spread. The interactions among plant host, pathogen and environmental conditions will significantly influence the disease intensity and extent of disease spread, resulting in varying magnitude of qualitative and quantitative losses. Among the plant factors, the levels of resistance/susceptibility of the crop variety or plant species may be most important. The resistance of the cultivar may be genetically manipulated by transfer of suitable gene(s) from wild relatives or other plant species through breeding methods. Genetic modification through biotechnological procedures is considered as an effective alternative. However, the usefulness and feasibility of this approach appears to be limited. The pathogen factors include the virulence (aggressiveness), variability in pathogenic potential within a morphologic species and host range. The environmental factors that may decisively influence are rainfall, temperature, agricultural practices etc. The temporal and spatial interactions of these factors may favor or reduce the incidence and spread of diseases. The viruses and phytoplasmas depend entirely on vectors for their spread from infected plants to healthy plants, whereas some of the fungal and bacterial pathogens are also spread by vectors. In such cases, the vector population and their efficiency may have a significant effect on the spread of these diseases. In general, traditional methods involving isolation in culture of the microbial pathogens and the assessment of their populations have been applied. The molecular techniques that may provide results based on experiments for detection and quantification of microbial plant pathogens more rapidly and accurately are discussed in this chapter.

When a pathogen species or its variants appear in a location, the identity of the pathogen has to be established as early as possible. It is also important to quantify the pathogen population at various stages of disease development. Further, the effects of various ecological factors on the pathogen population build up have to be assessed. The nucleic acid-based and immunological techniques have been demonstrated to be suitable for specific, reliable and rapid identification of microbial pathogens. However, their use was limited, because of the cost of the techniques and non-availability of adequate technical manpower. Following concerted efforts of researchers, many techniques have been simplified for routine tests and for large scale application in the field conditions. Furthermore, huge amount of genomic data are available for the development of ecological genomics that can merge with epidemiology. The information about the population genetics of pathogens has remarkably benefited from epidemiological investigations. It is possible to track disease outbreaks by studying simultaneously how pathogen frequencies change within and among populations and how pathogen populations grow and spread (Zwankhuizen et al. 1998). The genetic sequences and gene expression studies have yielded information that can help predictions about epidemic features and outcomes and understanding host resistance and pathogen evolution. In addition, modeling plant disease epidemics and pathogen evolution requires thorough understanding of both plant and pathogen characteristics that influence the dynamics between a pathogen and its host. The simultaneous use of molecular, cellular, organismal, population and ecological approaches can be expected to provide the information on various aspects of host–pathogen interaction. Population changes in pathogens, their reproduction and dispersal depend on this interaction which has a bearing on the dynamics of resistance evolution (Antonovics 2003).

A basic need in epidemiology is the availability of improved diagnostic systems and information of pathogen genomics will significantly provide a range of tools for choosing the suitable ones. Various physical, immunological and nucleic acid-based techniques are presented in Volume 1, Chapter 2 to enable the researchers to apply the effective test(s) to study the characteristics of the target pathogen(s). The possibility of determining precisely and rapidly, the abundance of pathogens in a range of environmental settings is the critical requirement for constructing models of the risk of invasion by the pathogen(s) in question. Such investigations require the techniques to detect and identify the desired pathogens precisely. Further, the detection of the pathogens in asymptomatic plants, seeds and asexually propagative materials such as tubers, corms and setts assume great importance in eliminating primary sources of infections that may be otherwise, unwittingly introduced into fields. The molecular diagnostic assays also help in evaluation of host or pathogen when infection is still at very low levels, opening the way for developing effective disease management systems that can eradicate the pathogen(s) before they become well-established in a field or region. With refinement for successful diagnosis of gene expression, it may be possible to assess features such as quorum sensing status (von Bodman et al. 2003).

The development of systems to effectively detect a wider range of targets in a single assay is considered as a significant advancement in disease diagnosis. Conventional PCR assays have been effectively multiplexed to detect up to seven targets in viruses or viroids (Bertolini et al. 2001; Ito et al. 2002; Ragozzino et al. 2004). In order to overcome the limitation of real-time PCR for increasing the number of targets that can be detected, the PCR array was developed. This method is essentially a technique for separating, running and resolving simplex PCR, by using small channels to transport reagents from the sample-loading ports to each of the 48 reaction wells in which lyophilized primers and probes are resuspended. It is possible to generate real-time PCR data for up to 48 different assays (potentially 48 target pathogens) from eight individual samples in a single run (Gallagher et al. 2005). The Thru-Hole® technology (BioTrove, Woburn, MA, USA) permits a total of 3072 separate assays to be performed simultaneously on a single OpenArray® plate (Mumford et al. 2006). These techniques will be helpful to determine the qualitative and quantitative variations of pathogen population over a period of time and geographical locations.

The DNA microarray is a technique that allows the resolution of specific hybridization events between the nucleic acid in a sample and known nucleic acid probes bound in a solid phase. It is analogous essentially to a dot-blot in reverse where the probe rather than the sample is bound to the solid phase. The DNA capture probes (or spots) for each of the genes/pathogens to be detected are immobilized onto a solid support in a spatially separated and individually recognizable manner. Nucleic acid from the sample is extracted and labeled and their labeled nucleic acid (target) is then hybridized to the array. Then the array is scanned to identify the hybridization and to resolve the presence of the gene/pathogen by the pre-defined position of the DNA capture probe on array. Potato viruses (Boonham et al. 2003) and cucurbit-infecting tobamoviruses (Lee et al. 2003) were detected either individually or in mixed infections at sensitivity levels similar to that of TAS-ELISA assessments. The microarrays could detect a range of virus isolates and strains and discriminated at the species level by using synthetic oligonucleotide-based arrays. The serotypes and subgroups of *Cucumber mosaic virus* (CMV) that differed by only 8% in an amplified PCR product, larger than 700 bp were discriminated by employing oligonucleotide-based array (Deyong et al. 2005).

Microarray techniques have been employed for the discrimination of closely-related *Fusarium* spp. causing Fusarium head blight (FHB) disease of wheat and barley. These pathogens produce mycotoxins that can enter the food chain inducing mycotoxicoses in human beings and animal. The estimation of populations of mycotoxin-producing species of *Fusarium* is essential for mycotoxin management (Nicolaisen et al. 2005). Based on amplification of the 16S region, microarray methods have been developed for the detection and identification of bacterial pathogens also (Fessehaie et al. 2003). The use of microarrays may pave the way for diagnostic protocols for detection and discrimination of large number of microorganisms for studying the community characteristics of systems such as disease-suppressive soils, the phyllosphere and endophytic communities of disease-inducing agents. Detection of pathogenic or beneficial microbes (biocontrol agents) either in soil or irrigation water may prove to be one of the important application of array technique relevant to epidemiology.

3.1 Viral Pathogens

3.1.1 Molecular Biology of Virus Infection

Molecular ecology, during the past ten years or more has been shown to be useful for elucidating a wide variety of factors involved in the emergence of plant viruses especially occurring in the tropics, that have been the principal cause(s) of serious crop losses. Some crop virus diseases that have increased in incidence, geographical distribution or host range or changed pathogenesis or newly evolved have been investigated. Molecular analysis of viral isolates can provide information that facilitates an understanding of virus epidemiology.

A viral pathogen to be successful, at first, should be able to multiply, a disease process that requires "replication-susceptible" host plant(s) followed by a host permissive to invasion of cells adjacent to the primary infection sites and ultimately to systemic spread. Whitefly-transmitted geminiviruses have become serious pathogens of several agronomic and horticultural crops in tropical and subtropical regions. Host-specific factors are required to initiate replication and to redirect host cells to synthesize begomoviral DNA, resulting in a direct influence of these factors on viral host range. The host range may in part be controlled by the ability of the viral Rep gene(s) to influence the expression of the required number of host genes. *Bemisia tabaci* is the only known vector of begomoviruses worldwide, suggesting a highly conserved mechanism for the whitefly-mediated transmission of begomoviruses (Bird and Maramorosch 1978; Rosell et al. 1999). Two *Cassava mosaic geminivirus* (CMG) species, *African cassava mosaic virus* (ACMV) and *East African cassava mosaic virus* (EACMV) were recognized based on serology and genome sequencing (Hong et al. 1993; Swanson and Harrison 1994). Later, the appearance of a cassava geminivirus outbreak revealing severe symptoms of infection, indicated that this CMG was due to an interspecies recombination between ACMV and EACMV (Deng et al. 1997; Zhou et al. 1997). The recombinant virus contained an EACMV-like genome, but the capsid protein sequence had a typical ACMV fragment of 459 nucleotides (nts). This recombinant isolate was designated EACMV-UG (Fauquet and Stanley 2003; Fauquet et al. 2005). Sequence comparisons of many other geminiviruses have indicated that recombination appears to be a common phenomenon playing a very important role in their evolution and emergence.

Mixed ACMV and EACMV-UG infections were frequently noticed leading to the CMD pandemic. ACMV and EACMV-like viruses interact synergistically. Accumulation of the viral DNA-A and DNA-B for each virus dramatically increased reaching higher levels (20–50 folds) than the levels attained in single infections. This synergism was considered as a key factor in the spread of the CMD pandemic in East and Central Africa (Harrison et al. 1997; Legg 1999). The molecular basis of the biological mechanism of synergism between these CMGs was studied. A differential and combined action of two geminiviral suppressors of gene silencing is likely to facilitate the rapid accumulation of viral DNAs in plants infected by both viruses. ACMV and EACMV-UG possess two posttranscriptional gene silencing (PTGS) suppressors AC4 and AC2 respectively. ACMV employs AC4 protein (for enhancing pathogenicity) as an efficient PTGS suppressor. On the other hand, the AC2 protein from EACMV-UG may function as transcriptional activator protein. Thus in cassava plants infected by both CMGs, the AC4 provides an immediate PTGS suppression. It can benefit both viruses to enhance their replication to reach high concentrations. This situation may aid in production of more AC2 that can suppress PTGS more effectively for the remaining period of plant's life. These findings indicate the importance of the control of gene silencing of the host plant for geminiviruses and in turn, the importance of the molecular studies in understanding the ecological phenomena (Fargette et al. 2006).

Genetic diversity of plant viruses has a significant influence on the development of virus epidemics. A survey of the genetic diversity of CMV-associated begomoviruses across the major cassava growing areas in Kenya was undertaken. Representative members of the population were characterized by sequence analysis. The full-length sequences of 109 components (68 DNA-A and 41 DNA-B) were determined, representing isolates of EACMV and *East African cassava mosaic Zanzibar virus*, in addition to a novel begomovirus species *East African cassava mosaic Kenya virus*. The DNA-B components were much less diverse than their corresponding DNA-A components. Distinct geographical distributions were associated with all virus species, emphasizing the need to prevent both movement of viruses between these regions and the importation of the disease from adjacent countries and islands in the Indian Ocean that would undoubtedly encourage further diversification (Bull et al. 2006).

Following the development of appropriate molecular diagnostic techniques, the importance and diversity of *Banana streak virus* (BSV) belonging to the genus *Badnavirus*, family *Caulimoviridae* was recognized only recently and especially in tropical crops. Due to the tedious process required for obtaining viral DNA from banana plants, the absence of an indicator host and the failure of mechanical transmission hampered molecular studies on BSV. In addition, badnaviruses are highly variable at both genomic and serological levels affecting significant progress. However, many BSV sequences of the polymerase coding region of ORF3 were determined. Experiments combining different molecular tools such as PCR, molecular hybridization and in situ hybridization, revealed that the *Musa* genome contained integrated BSV sequences that can cause episomal infection under certain conditions (Harper et al. 1999; Ndowora et al. 1999). BSV integrated sequences may have probably arisen from "illegitimate" recombination between host and viral DNA. The integrated sequence of *Banana streak OL virus* (BSOLV) found in *Musa* sp. cv. Obina l'Ewai contained a rearranged but complete virus genome (Ndowora et al. 1999). These potentially activable sequences have been observed in only some B genome-containing *Musa*, although there are other badnavirus sequences found in both *Musa* A and B genomes that have not been associated with disease (Geering et al. 2005a, b). The range of BSV-like sequence variability estimated with a fragment of the RT gene was found to be very high compared to the variability of other badnaviruses. Further, no obvious correlation could be seen between a particular BSV clade or individual sequence to specific *Musa* genotypes, cultivars or disease severity and symptoms. The difficulty in distinguishing between episomal sequence derived from a classical viral evolutionary process and those derived from a long history of coevolution inside the banana genome, poses obstacles to have a clear understanding of the molecular epidemiology of BSV.

Epidemiological experiments may commence with models within individuals, predicting infection levels based on the expression of particular gene(s) and then expand on these to predict infection rates in plant populations based on gene expression rates in individuals. These genetic structure of populations of *Cucumber mosaic virus* (CMV) satellite RNA (sat-RNA) and its evolution were analyzed during the course of a CMV epidemic in tomatoes in Eastern Spain. RNase protection assay (RPA) was employed to characterize 62 variants of CMV-sat-RNA from three epidemic episodes in 1987–1991. Sixty haplotypes were differentiated by RPA patterns. High genetic variability of CMV-sat-RNA resulted in very heterogeneous populations. The high diversity of the population is maintained through time by the continuous generation of variants by mutation, counterbalanced by negative selection. Because of this, replacement of certain haplotypes might occur from year to year. The sequential accumulation of mutations in CMV-sat- RNA may result in fast genetic divergence reaching possibly the upper permitted threshold level (Aranda et al. 1993). In a further study, molecular epidemiology of CMV and its sat-RNA in locations where CMV was endemic in vegetable crops was taken up. The genetic structure of CMV population (about 300 isolates) representing 17 outbreaks (subpopulations) in different crops, regions and years was compared. The RPA for genetic analyses of CMV isolates was performed using cRNA probes representing RNA1, RNA2 and two ORFs in RNA3. The results indicated that heterologous genetic combinations were not favored. About 30% of CMV isolates were supporting sat-RNA. Molecular analyses of CMV-sat-RNA isolates revealed high genetic diversity, due both to the accumulation of point mutations and to recombination. Further, CMV-sat-RNA might have spread epidemically on the extant virus population from an original reservoir in Eastern Spain (Garcia-Arenal et al. 2000).

By applying single-strand conformation polymorphism (SSCP) and nucleotide sequence analyses of genomic regions of 2b, CP, MP and 3′ nontranslated region of RNA3, the structure and genetic diversity of a California CMV population was determined. The California CMV population showed low genetic diversity and composed of one to three predominant haplotypes for specific genomic regions. Phylogenetic analysis suggested that genetic exchange by reassortment contributed to the evolution of the California CMV population. Based on the analysis of various population genetic parameters and distribution of synonymous and nonsynonymous mutations, it is considered that different coding regions and even different parts of coding regions were under different evolutionary constraints, including a short region of the 2b gene (Lin et al. 2004).

Papaya ringspot virus (PRSV) belonging to *Potyvirus* genus is an important viral pathogen of papaya and cucurbits. Two biotypes with similar biological properties, but differing in their ability to infect papaya have been recognized. The biotype PRSV-P can infect papaya, whereas PRSV-W has natural host range within *Cucurbitaceae* and it is unable to infect papaya (Purcifull 1984; Tomlinson 1987). The epidemiological investigations showed that the pattern of disease incidence and spread of PRSV-P was similar to that of other aphid-transmitted nonpersistent viruses and this biotype was not found in intercropped cucurbits. There is evidence to indicate the evolution of PRSV-P from PRSV-W, presumably by mutation (Bateson et al. 1994). Very close sequence similarity of CP-coding regions of P and W isolates within Australia has been observed. The CP sequences of several isolates of PRSV-P from different countries and some isolates of PRSV-W were compared. Nucleotide and amino acid divergence of up to 14% and 10% respectively was noted between these biotypes. In order to assess the relative impact of mutation on the structure of PRSV populations, the CP sequences of isolates of both biotypes of PRSV from Vietnam, Thailand, India and the Philippines were determined. They were analyzed along with 28 PRSV sequences already published. In Thailand, variation and diversity in the isolates were similar. Phylogenetic analyses of PRSV showed that this virus might have originated in Asia, particularly in the Indian subcontinent, since PRSV population in this location showed greater diversity and might have been probably existed for the longest period of time. The results indicated that mutation together with local and long-distance movement contributed to population variation, confirming that populations of the PRSV-P have evolved on several occasions from PRSV-W populations (Bateson et al. 2002).

3.1.2 Molecular Determinants of Virus Transmission

Transmission of plant viruses from infected plant species to the healthy plants of the same or different species is a vital factor affecting the development of epidemics. Plant-feeding organisms (vectors) insects, mites, nematodes or fungi function as a means of transportation of viruses from plant to plant. Vector transmission consists of three steps: acquisition, retention and inoculation of the virus by the vector. The mechanisms of the virus-vector interaction vary widely depending on the nature of the virus. In the case of non-circulative (mechanical or nonpersistent) viruses, two distinct molecular strategies regulate the virus-vector interaction: (i) the capsid strategy in which the virus coat protein (CP) interacts directly with the vector and (ii) the helper strategy in which at least one nonstructural viral gene product, the helper component (HC) is needed for effective transmission of the virus (Foissart et al. 2002).

The transmissibility of viruses belonging to the genus *Cucumovirus* entirely depends on the characteristics of CP. On the other hand, viruses belonging to the genera *Potyvirus* and *Caulimovirus*, the characteristics of both CP and the virus-encoded nonstructural accessory factors such as HC are important for aphid transmission. The CP of *Cucumber mosaic virus* has multiple functions such as encapsidation of viral genome, involvement in systemic movement and host range and transmission determinant. Two domains in the CP amino acids are involved in aphid transmission (Palukaitis et al. 1992). The vector specificity and variations in the transmission rates by different aphid species may be due to changes in the amino acids of CP. The amino acid changes that significantly alter the transmission rates with a particular aphid species may not have any effect on transmission of CMV by another aphid species (Perry et al. 1998). The ease of binding and/ or release of virus particles in the mouth parts or foregut of the vector or instability of virus particles during the transmission of noncirculative viruses was assessed by Lopez-Moya (2002).

The helper strategy adopted by the genera *Potyvirus* and *Caulimovirus* can be explained by the bridge hypothesis. The nonstructural HC protein molecules may be capable of engaging two types of interactions viz., one with the virus CP and the other with a putative receptor in the vector's mouthparts, creating a reversible "molecular bridge" allowing virus retention in the vector. This hypothesis implies that loss of transmissibility may be due to either from mutations in the HC rendering it nonfunctional or from defects in the CP abolishing the HC–CP interaction. The phenomenon termed as HC-transcomplementation involving the sequential acquitistion of HC and virus particles was suggested to be in operation. The HC acquired first by the vector assists the transmission of virus particles present in the same cell, in other cells or even other host plants probed by the vector (Foissart et al. 2002). Drucker et al. (2002) provided evidence that the transmissible complex of *Cauliflower mosaic virus* (CaMV) was not formed in the infected plant cells. But it was formed in the insect vector, after sequential acquisition of HC and virus particles. The results indicated that this sequential acquisition may possibly result in HC-transcomplementation being prominent upon natural CaMV vector transmission.

The potyvirus HC is a virus-encoded protein around 50-kDa. The HC is specific for viruses in the genus *Potyvirus* and some important residues for transmission have been identified in the HC (Atreya et al. 1992; Atreya and Pirone 1993). It was hypothesized that the HC adsorbs virus particle to unidentified structures in the stylets of vectors and act as a bridge. Retention sites for virus particles are located in the cuticular linings of aphid mouthparts (Ammar et al. 1994). By employing radiolabeled *Tobacco etch virus* (TEV) particles, the presence of retention sites for combination of active HC and transmissible virus particles could be detected (Wang et al. 1996, 1998). The virus-encoded helper factor was shown to be the 18-kDa product of gene II (P18) in the case of CaMV (Armour et al. 1983) Transmission-active CaMV P18 factor was associated with viroplasms present in host cells. Further, P18 formed stable complexes with microtubules (Blanc et al. 1996). In addition to P18, a second helper protein P15 encoded by gene III, was also required for vector transmission of CaMV (Leh et al. 1999).

The virus family *Luteoviridae* includes the genera *Polerovirus* [*Potato leafroll virus* (PLRV) and *Cereal yellow dwarf virus* (CYDV)], *Luteovirus* [*Barley yellow dwarf virus* (BYDV), *Soybean dwarf virus* (SbDV)] and *Enomovirus* [*Pea enation mosaic virus* (PEMV)] which are transmitted by different aphid species in a nonpropagative-circulative manner. The uptake and transport of the virus through the gut is accomplished by an active receptor-mediated endocystosis process. The putative receptors mediating this process are yet to be identified (Li et al. 2001; Gray and Gildow 2003). After passing through the gut, virus particles have to move from the hemolymph through the accessory salivary gland to be transmitted to healthy plants during feeding. In order to study the genetic regulation of *Polervirus* and *Luteovirus* transmission in the aphid *Schizaphis graminum*, crosses between two genotypes of *S. graminum* with varying ability to transmit the RPV strain of CYDV and SGV strain of BYDV were made. The hybrid genotypes were used to determine the genetic control of transmission of the viruses. The amount of RPV ingested by an aphid and/or moving across the hindgut into the hemocoel was quantified using real-time PCR. The virus was detected in the hemolymph of all aphid species and genotypes though the concentration of the virus much less compared to other body structures. Virus contents of hemolymph were similar in most aphid species and genotypes irrespective of their efficiency of transmission (Fig. 3.1). Only one genotype (F2-C2) consistently had a significantly lower concentration of virus in the hemolymph. A major gene or linked genes may regulate the virus transmission in *S. graminum*. However, individual hybrid genotypes differed significantly in their ability to transmit each virus, indicating that in addition to a major gene, minor genes may also influence the transmission of each virus independently. Gut and salivary gland-associated barriers of virus transmission were identified. But hemolymph factors do not seem to be important in determining the transmission phenotype. The results indicated that aphid transmission of luteoviruses was genetically regulated in the insect, whereas the tissue-specific barriers to virus transmission were not genetically linked. This study appears to be the first report indicating the existence of multiple genetically-controlled barriers to circulative transmission by aphids and that these different barriers could be isolated using a genetic strategy (Burrows et al. 2006).

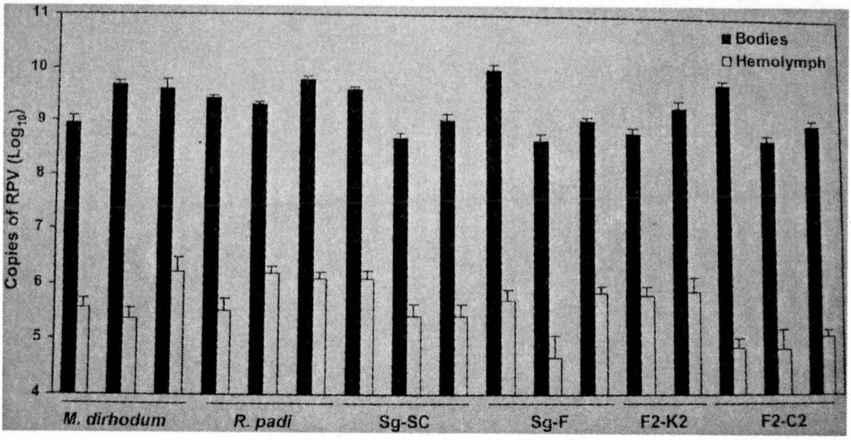

Fig. 3.1 Detection of *Cereal yellow dwarf virus*-RPV in the extracts of bodies and hemolymph from six aphid genotypes by real-time RT-PCR assay; Filled columns: Extracts of bodies of the different genotypes of aphids; Open columns: Hemolymphs from the different genotypes of aphids; Error bars represent the SE of the mean. (Courtesy of Burrows et al. 2006; The American Phytopathological Society, St. Paul, MN, USA)

Aphis glycines was able to transmit potyviruses *Bean yellow mosaic virus* (BYMV) and *Soybean mosaic virus* (SbMV) from soybean to soybean more efficiently than *Myzus persicae*. The luteovirus *Soybean dwarf virus* (SbDV) was widely distributed in red and white clover, although it was not observed in soybean in Kentucky. The transmission experiments showed that *A. glycines* could not transmit SbDV. An RT-PCR assay developed in this investigation, detected SbDV in single individuals of *A. glycines*, using a pair of primers designed to amplify a 372 bp PCR fragment in the coding region of SbDV coat protein. Although *A. glycines* was not a vector of SbDV, the virus could be detected in 100% of the aphids tested by RT-PCR, after an acquisition period of 24–48 h and also even after additional inoculation feeding on soybean plants for 24–48 h. It is, therefore, cautioned that monitoring SbDV incidence is essential, because of the potential generation of *A. glycines*-transmissible SbDV variants, as indicated by the RT-PCR assay (Wang et al. 2006).

The transmission of begomoviruses by *Bemisia tabaci* exclusively suggested the operation of a highly conserved mechanism of these viruses for the *B. tabaci*-mediated transmission of these viruses (Rosell et al. 1999). The evidences supporting this suggestion are (i) the extreme conservation in the amino acid sequence of the coat protein (CP) shared among all begomoviruses as compared to wide divergence of CP sequences of leafhopper-transmitted geminiviruses by different vector species; (ii) replacement of a begomovirus CP with that of a leafhopper-transmitted virus rendering the begomovirus to become leafhopper-transmissible (Briddon et al. 1990) and (iii) CP deletion mutants with the replacement of CP sequences of nonwhitefly-transmissible virus resulting in failure of transmission by whitefly (Noris et al. 1998). As *B. tabaci* is reported to be associated with more than 1000 crops and weed species, genetic markers have been used to separate populations of the vector (Rosell et al. 1997). Cassava plants severely infected by CMD appear to interact synergistically with *B. tabaci* populations (Omongo 2003) leading to enhanced colonization of CMD-infected plants by whiteflies. Higher concentration of whitefly eggs on symptom-free leaves of diseased plants and enhancement of rates of fecundity on CMD-infected plants may favor greater spread of the viruses associated with the pandemic situations (Colvin et al. 2004).

Tomato chlorosis virus (ToCV) belonging to genus *Crinivirus*, family *Closteroviridae* is unique among whitefly-transmitted viruses, because of its ability to be transmitted by four whitefly vectors viz., *B. tabaci* biotype A (BTA), biotype B (BTB), *Trialeurodes abutilonea* and *T. vaporariorum*. *T. abutilonea* and BTB were highly efficient transmitters and *T. vaporariorum* and BTA were less efficient vectors. ToCV persisted for up to 5 days in *T. abutilonea*, 2 days in BTB and only one day in BTA and *T. vaorariorum*. A portion of the CP coding region of five geographically diverse ToCV isolates was compared and found to be highly conserved. Further, the sequences of the HSP7h of the isolates occurring in various geographical locations showed high similarity (97%–100%). This suggests that relatively little diversity exists among ToCV isolates indicating that they all might have originated from a single source. Possibly all isolates with known sequence information may represent a single introduction into tomato followed by inadvertent dispersal of ToCV throughout the world. As there are different vector species, once introduced, ToCV becomes established in crops, weeds and ornamental hosts (24 plant species) found in different countries and potentially can move to new areas through vectors. A prudent approach is to index vegetable and ornamental crops for the presence of ToCV and other criniviruses to develop an effective management system (Wintermantel and Wisler 2006).

The nematodes belonging to the families Longidoridae and Trichodoridae are ectoparasites feeding on the plant roots and they are involved in the transmission of plant viruses. The transmission of viruses by nematodes has been shown to be semipersistent and nonpropagative in nature. The virus particles are retained by binding to specific sites on the surface of the nematode oesophagus. The specificity of transmission of a virus by a nematode species is considered to be due to the ability or inability of the virus to be retained within the mouthparts of the nematode. The nematode-transmitted viruses belonging to the genera *Nepovirus* and *Tobravirus* have two genomic RNA1 and RNA2. The smaller RNA2 encodes the viral CP that determines the transmission properties of both nepoviruses and tobraviruses (Harrison and Murrant 1977; Ploeg et al. 1993). The RNA2 encoded polyprotein of *Grapevine fan leaf virus* (GFLV) is cleaved into three proteins 2A (28 kDa), 2B (38 kDa) and 2C (56 kDa). The 2C protein forms the viral CP (Gaire et al. 1999). Although the GFLV and *Arabis mosaic virus* (ArMV) are serologically related, they are transmitted by two different nematode species, *Xiphinema index* and *X. diversicaudatum*. The CP is the sole determinant of transmission specificity (Belin et al. 1999, 2001). The RNA2 of tobraviruses particularly *Tobacco rattle virus* (TRV) is extremely diverse encoding only CP in some isolates and encoding additional proteins in other isolates. The transmission genes are located in RNA2

(Vassilakos et al. 2001). The RNA2 of all tobraviruses contains CP gene and two other genes *2b* and *2c*. The CP is not the sole determinant of transmission of tobraviruses as in the case of nepoviruses (MacFarlane et al. 1996; MacFarlane 2003).

3.2 Fungal Pathogens

Models of disease foci and the patterns in which these foci expand in time and space have been developed in epidemiological studies. In agricultural systems, the spatial pattern of host genotypic resistance manipulated through use of intercrops or mixed genotypes (multilines) within a crop species has been successfully adopted as disease management strategy (Garrett and Mundt 1999; Zhu et al. 2000; Mundt 2002). The genomics approach in epidemiology will be useful to explore multiple spatial and temporal scales as well as levels of details in genomic status. It is important to determine the type of scale of variation required to include in prediction of epidemics for a given pathosystem. Microarray analyses can provide information in optimal scales of variation in expression data that may be included in predictive models. However, at present the cost of such analyses may be a limiting factor for large scale application. Within a host plant, the local phenotype might be at the scale of a leaf or cell. Local gene expression may be at the point of infection as exemplified by the presence of green island of host tissue formed around the pustule (sorus) formed by a rust pathogen. Regional gene expression within an individual plant might be the expression in tissues adjacent to infection. Further, competition between two pathogens may influence differentially depending on the time of infection and type of plant tissue infected (Al-Naimi et al. 2005).

The detection of mRNA by RT-PCR assay provides a better assessment of cell viability than the DNA-based methods which do not differentiate dead and living cells. An RT-PCR procedure was developed to determine viable *Mycosphaerella graminicola* populations in wheat presymptomatically and monitor disease development during the epidemic period. By using a specific primer set E1/STSP2R, only single fragment was amplified from total RNA of *M. graminicola* and infected wheat leaves, but not from those of healthy leaves. In addition, no amplification was detected with total RNA of five other common fungal pathogens infecting wheat, indicating the specificity of the assay. This diagnostic test was applied successfully to monitor the disease development in inoculated and naturally infected plants, using the same primer set (E1/STS2R) developed from a sequence of β-tubulin genes. The specificity and sensitivity (5 ng of total RNA) of this test was high enough to detect *M. graminicola* presymptomatically in inoculated plants as well as in the field samples at 4 dai, when no visible symptoms appeared. After the production of symptoms, the increase in band intensity of the amplified RT-PCR products in the gel was in general agreement with the results of visual disease assessment (Guo et al. 2005).

Expression foci in which host individuals share altered gene expression patterns, may form around inoculation sources in host populations. These expression foci may have properties related to those of disease foci. In response to exposure to pathogens and other microorganisms, a wide range of gene expression may be evident. The variation of host response may vary from increased resistance [systemic acquired resistance (SAR)/induced systemic resistance (ISR)]. The infection efficiency of *Puccinia striiformis* decreased by 44%, if plants were earlier exposed to an "inducer race" of the pathogen (Calonec et al. 1996). Waves of different expression patterns surrounding initial sources of infection may be observed with increasing distances from a primary inoculum source. Furthermore, the past presence of pathogens through on-going expression in host population can be studied by using genomic tool. Such an estimate of "footprints" of the pathogen may provide support to studies on long-term epidemics and changes in host resistance over a period of time (Blain et al. 2002).

It has been difficult to have an understanding of factors governing long-term disease dynamics due to nonavailability of long time series data, since survey are generally short-term responses to problems of immediate concern. Detection of ancient DNA in historical samples has provided critical information on infectious diseases of humans and animals (Zink et al. 2002) and plants (May and Ristaino 2004). The long-term dynamics of *Phaeosphaeria nodorum* and *Mycosphaerella graminicola* causing Septoria blotch diseases of wheat populations were investigated by using PCR assays and an archive of samples from the long-term Broadbalk winter wheat experiment run continuously at Rothamsted since autumn 1843. Quantitative real-time PCR assays were applied to quantify the fungal pathogens and wheat DNAs present in a set of samples covering a 160-year period of wheat production. The relative abundance of DNA of these two fungal pathogens in the sample, reflected the relative importance of the two wheat diseases they caused as determined in disease surveys undertaken in UK. Curiously, there was a very strong correlation between the changes in the atmospheric pollution, as measured by SO_2 emissions and the changes in the ratio of the pathogens over the 160-year period. A positive correlation between abundance of *P. nodorum* DNA and SO_2 emissions and a negative correlation between abundance of *M. graminicola* DNA and SO_2 emission were observed. There appears to be links between pathogen infection and SO_2 in the atmosphere (Bearchell et al. 2005).

Four different diseases of citrus are due to *Alternaria* spp. *A. alternata* causes Alternaria brown spot of tangerines, Alternaria leaf spot of rough lemon and Alternaria black rot of many citrus fruits. The causal agents can be differentiated using pathogenicity tests, toxin assays or genetic markers. Mancha foliar disease of Mexican lime is due to *A. limicola*. Studies of the population genetics, host specificity and ecology of *A. alternata* from different ecological niches on citrus have revealed host-specific forms of the pathogen which induce disease on different citrus species, existence of three phylogenetic lineages of the fungus which cause brown spot worldwide and closely related nonpathogenic isolates which colonize healthy citrus tissues. The brown spot and leaf spot pathogens produce different host-specific toxins (HSTs) (Kohmoto et al. 1979, 1993) and they are referred as tangerine and rough lemon pathotypes (Nishimura and Kohmoto 1983). Populations of the brown spot pathogen sampled from Minnesota and Orlando tangelo in Florida were screened with RAPD markers and the pathogen was moderately differentiated genetically in both hosts (Peever et al. 2000). Both RAPD markers and sequence data from an endoPG showed that the isolates from Florida and Colombia were distinct from isolates sampled in other countries.The endoPG data grouped the isolates into three phylogenetic lineages that correlated to the country of origin. The Florida strains were significantly more virulent than others. This enhanced virulence of Florida strains may be due to the presence of more copies of genes controlling toxin biosynthesis (ACTT) resulting in production of higher concentrations of ACT-toxin by them (Akimitsu et al. 2003).

White pine blister rust is caused by *Cronartium ribicola*, while *C. quercuum* f.sp. *fusiforme* is the incitant of fusiform rust disease. The DNA markers were employed to generate multilocus haplotypes (MLHs) to improve the understanding of pine rust genetics and to apply that knowledge to investigate the epidemiology of these diseases. Well differentiated populations were identified in *C. ribicola* in eastern and western North America. The North American population structure was consistent with historical records showing that multiple introductions of several million seedlings from nurseries in Germany, France and Holland were responsible for the establishment of the pathogen in the eastern North America. In contrast, the incidence of the disease in western North America is attributed to a single introduction from a French nursery. *C. quercuum* f.sp. *fusiforme* contained four metapopulations that were genetically differentiated. Correlation between microsatellite markers and either a longitudinal or latitudinal gradient was observed. There seemed to be a high level of genetic variability within and among these metapopulations which is consistent with some local adaptation and limited gene flow. A main difference between these pathosystems is that *C. quercuum* f.sp. *fusiforme* is native, whereas *C. ribicola* is an exotic pathogen. Possibly *C. quercuum* f.sp. *fusiforme* has co-evolved with its pine hosts and adapted locally to host genotypes and environments. This pathogen has a very large genetic diversity in contrast to the narrow genetic basis of *C. ribicola* as revealed by RFLP, RAPD and PCR-SSCP assays (Hamelin et al. 2005).

Several types of multilocus DNA fingerprinting methods such as RFLPs and RAPDs have been employed to assess genetic diversity of rust pathogens infecting wheat crops. Selectively amplified microsatellites (SAMs) have revealed significantly higher levels of genetic diversity between isolates of *Puccinia graminis* f.sp. *tritici* and *P. graminis* f.sp. *avenae.* Microsatellites or multiple sequence repeats (SSRs) consisting of tandem repeats of short (1–6 bp) have been examined for their potential to assess the genetic diversity of microbial plant pathogens. The SSRs have been found to be useful molecular markers, since they are codominant, highly abundant, distributed randomly throughout most eukaryotic genomes and tend to be highly polymorphic due to variation in the number of repeat units (Weber and May 1989; Enkerli et al. 2001). The PCR assays have been developed to amplify SSR loci, using a single primer specific to DNA flanking only one side of the microsatellite and a primer that anchors to the repeat motif. These markers designated sequence-tagged microsatellites (STMs) are generally developed using SAM analysis and STM profiling (STMP) (Hayden and Sharp 2001a, b). The STMP technique was employed to develop 110 single-locus microsatellites of *P. graminis* f.sp. *tritici.*

Low microsatellite polymorphisms were exhibited among 10 pathogenically diverse *P. graminis* f.sp. *tritici* isolates from various regions in Australia over a period of 70 years. Two polymorphic loci were detected, each revealing two alleles. Limited cross species amplification was observed for other wheat rust pathogens *P. triticina* (leaf rust) and *P. striiformis* f.sp. *tritici* (stripe). On the other hand, six polymorphic microsatellite loci with a total of 15 alleles were detected in a historical group of 16 isolates of *P. graminis* f.sp. *avenae* collected in a survey during 1999. The results revealed the dynamic population structure of *P. graminis* f.sp. *avenae* in Australia that may have a role in the epidemics of oat stem rust disease (Keiper et al. 2006).

The ascospores and/or conidia of *Monilinia fructicola* can infect blossoms and young fruits, causing blossom blight and fruit brown rot diseases of stone fruits. The importance of spore inoculum in initiating the disease may vary depending on crop growth stages. Accurate assessment of inoculum potential in orchards and spore density in the air to estimate dynamics of inoculum may be useful for prediction and risk assessment of disease development. This information is important for determining the critical time periods for efficient and timely fungicide sprays. A real-time PCR procedure was developed to quantify the dynamics of spore density of *M. fructicola* from spore trap samples collected from stone fruit orchard in California. A linear relationship between the number of spores of *M. fructicola* counted using the compound microscope and the corresponding number of spores determined with the real-time PCR assay, the r^2 value being 0.60 at $P = 0.002$. The spore density in five of six orchards/year combination ranged from 0.0 to 0.05 spores per ml, while the sixth orchard showed much higher value and wider range of fruit and blossom infections, compared with other orchards. This real-time PCR protocol could detect as few as 20 spores of *M. fructicola* per spore trap tape segment/slide from the field and detect DNA concentration corresponding to only two spores per sample from pure culture of *M. fructicola*. This procedure has the potential for quantitative determination of spore inoculum potential and for different epidemiological investigations (Ma et al. 2003; Luo et al. 2007).

The mode of dispersal of fungal pathogens, a major epidemiological factor, may be through air, rain water, soil and in some cases through insects. The oomycetes producing sporangia during asexual reproduction require free water or high relative humidity during dispersal of zoospores. Splash droplets generated by rain drop impaction on plants have been found to be the dispersal agent of *Phytophthora infestans* causing late blight disease of tomato and potato. The traditional method involving collection of droplets carrying spores on glass is tedious and time-consuming, in addition to the requirement of adequate skill for specific spore recognition. In order to overcome these drawbacks, a new approach for analysis of splash-dispersed *P. infestans*, an engineered fluorescent strain transformed with the gene encoding the green fluorescent protein (GFP) was developed. The use of fluorescent sporangia facilitated the detection and counting of exclusively viable sporangia which are of epidemiological importance by image processing. This study reveals a new application for the GFP marker in plant disease epidemiology in laboratory conditions. However, the principal limitation of this approach is the requirement of a transformed strain of the pathogen to be investigated (Saint-Jean et al. 2005).

3.3 Bacterial Pathogens

The DNA sequencing provides more complete genetic information that can be used to address longstanding questions of bacterial and other microbial population structure, host specificity and phylogenetics. The identity of the pathogen up to species level may be efficiently established by determining the sequence of specific gene(s). The number and types of resistance genes in natural populations of plant hosts may be determined by identifying sequences associated with resistance (*R*) genes. Likewise, estimates of the number and types of effector proteins in a single bacterial pathogen such as the number of gene products transferred into plant cells by the type III secretion system (TTSS) of *Pseudomonas syringae* pv. *tomato* DC3000 can be obtained (Collmer et al. 2002; Alfano and Collmer 2004; Chang et al. 2005). The effector proteins may function as virulence proteins and also as targets of plant disease resistance proteins. Evaluation of genomes of xylem-limited, fastidious bacterial pathogen *Xylella fastdiosa* (*Xf*) was done to determine the missing sequences by comparison to other less fastidious bacterial species (Van Sluys et al. 2002). Genetic and pathogenic variation of *Xf* is an important factor affecting disease epidemiology, since the almond leaf scorch disease (ALSD) induced by this bacterial pathogen has reemerged threatening almond production in the San Joaquin valley of California. The strains of *Xf* are differentiated by pure culture-dependent DNA polymorphism-based methods such as RAPD and RFLP techniques. A conserved genomic region may be useful for species recognition, while a single nucleotide polymorphism (SNP) within the conserved region may be used to differentiate genotypes below the species level. Based on SNPs in the 16S rRNA gene (16S rDNA) of *Xf* strains (G- and A-types) were found to co-exist simultaneously in the same infected almond orchard. The concept of mixed genotype infection could affect the epidemiological investigations based on the assumption that one genotype causes ALSD in one location and consequently the disease management strategy (Chen et al. 2005).

Erwinia amylovora (*Ea*) causes the fire blight disease in pome fruits and other rosaceous plants in many countries around the world. By applying pulse field gel electrophoresis (PFGE) technique, strains of *Ea* present in Europe were classified into six major pattern types, Pt1–Pt6 (Zhang et al. 1997, 1998). Further detailed characterization is necessary for epidemiological studies. The *Ea* strains carry a low copy number of plasmid of 29-kb (pEa29) and different strains exhibited variations in fragment size of an approximately one-kb *Pst* I fragment. Three types were recognized based on the variations in fragment size among strains present in Central Europe (Lecomte et al. 1997). The fragment size variation was due to a short sequence DNA repeat (SSR) of eight nucleotides that are present 3–15 times in the one-kb fragment. The SSR units in 104 Austrian isolates of *Ea* were determined by DNA sequencing and this formed the basis for the classification of different *Ea* strains. The stability of SSR numbers was assessed to determine the suitability of this marker for strain differentiation. The number of SSR unit was stable under laboratory and certain stress conditions. Hence, this SSR marker was used for differentiation of strains from SSR-3 to SSR-15. A large number of *Ea* strains from

Austria were screened for their SSR numbers for epidemiological identification purposes. The strains with very high or low SSR numbers could be traced under field conditions. The occurrence of strains with low and very high SSR numbers was recorded in an orchard during the first year of fire blight incidence. The orchard was probably infected from the West (Switzerland, strains with 5 and 6 SSRs) and from the North (Austria, strains with SSRs of 12, 13 and 14). The SSR data showed that there was no continuous propagation, but several "imports" of infected plant materials probably accelerated the pathogen spread to the eastern provinces of Austria (Ruppitsch et al. 2004).

Burkholderia (Pseudomonas) cepacia was reported to cause onion slipper skin disease by Walter Burkholder of Cornell University in 1949. But this phytopathogen has emerged as a human pathogen in the past two decades, causing numerous outbreaks, especially among cystic fibrosis (CF) patients. This bacterium has a particular predilection for lung patients with CF and has emerged as an important opportunistic human pathogen in hospitalized and immunocompromised patients. The isolate from infected onion caused the disease in onion on inoculation. On the other hand, the environmental isolates tested or strains from CF lung did not induce the disease in onion. *B. cepacia* has the ability to propagate as an environmental microbe and as an opportunistic pathogen. This may be due to the presence of an unusually large (more than four times that of *Mycoplasma influenzae* and two times that of *Escherichia coli*) complex and variable genome. Numerous insertion sequences are present in the genome which is divided into 1–4 circular replicons. Such genome structure may account for the nutritional versatility and adaptability of *B. cepacia*. This pathogen is inherently resistant to several antibiotics, can metabolize diverse substrates and is found in the soil and moist environments. *B. cepacia* has been shown to be antagonistic to many plant pathogens and it can degrade toxic compounds such as pesticides and herbicides. Its application as a biocontrol agent has attracted intense interest from agroindustries. Abundant caution has to be exercised, while promoting *B. cepacia* as a biocontrol agent because of the possible danger involved (Holmes et al. 1998).

Bacterial blight of onion (BBO) caused by *Xanthomonas axonopodis* pv. *allii (Xaa)* infects also garlic, leek and Welsh onion. The association of seed-borne inoculum with development of BBO epidemic was considered as a possibility (Roumagnac et al. 2004). By using molecular techniques, the seed contamination with *Xaa* in experimental plots was assessed and the relationship between disease status of leaves and inflorescences and seed contamination was explored. This study indicated that seed-borne inoculum associated with naturally contaminated seeds was likely to be the primary source of outbreaks in a tropical environment. The onion plants in experimental plots established in an area where no other commercial crops were present, were inoculated with a rare bacterial haplotype that has a unique AFLP fingerprint and has never been recorded at the experimental site or in its vicinity, so that occurrence of exogenous inoculum in the experimental plots could be quantified by AFLP analysis. Based on AFLP analysis, most strains of *Xaa* isolated from leaf or floral stem lesions had the fingerprints identical to that of the inoculated LMG 16528 strain. But a single haplotype different from

Table 3.1 Detection of exogenous inoculum of *Xanthomonas axonopodis* pv. *allii (Xaa)* using AFLP technique

Experimental plot	Total no. of plants infected	No. of isolates of *Xaa* recovered	Percentage of plants infected with inoculated strain	Percentage of plants with alternative haplotype	Disease foci with exogenoous inoculum (No.)
I	371	74	48	26	5
II	281	46	41	5	2
III	3,137	125	118	7	3
IV	4,293	133	121	12	2

Source: Humeau et al. (2006).

the inoculated strain was also detected during the 2-year investigation. This new CFBP 6369-like haplotype was detected in two to five foci per plot, making a total of 12 foci (Table 3.1). Seed contamination with *Xaa* was detected by nested PCR assay and also by using semiselective NCTM 1 medium which showed similar levels of seed contamination. A significant positive correlation between the incidence of *Xaa* and seed contamination was seen. The systematic AFLP analysis showed the presence of a haplotype different from the inoculated one in all experimental plots. The occurrence of exogenous inoculum stressed the need to monitor the genotype of the pathogen populations for an accurate assessment of the biological significance of the putative inoculum sources and for a sustainable disease management system. This study seems to be the first of its kind demonstrating the usefulness of a genotyping method (AFLP and rep-PCR) for assessing the epidemiological significance of multiple inoculum sources on a pathosystem (Humeau et al. 2006).

The three races causing bacterial spot of tomatoes and peppers were reclassified as *Xanthomonas euvesicatoria* (71), *X. vesicatoria* (T2) and *X. perforans* (T3). Strains of *X. perforans* (T3) became predominant by progressive elimination of the other two species. T3 strains produced at least three bacteriocin-like compounds (BCN-A, BCN-B and BCN-C) toxic to T1 strains. These bacteriocin-like compounds provided the T3 strains with a competitive advantage over T1 strains. Further, certain combinations of bacteriocins were more effective in reducing the populations of T1 stains. The relative importance of the bacteriocin-like compounds was assessed by constructing different mutant forms of wild-type T3 strain. Mutants expressing BCN-B and either BCN-A or BCN-C reduced T1 populations less than mutants expressing only BCN-A or BCN-C. The triple knockout mutant, Mut-ABC also was more competitive than the T1 strain for survival in vivo (Hert et al. 2005).

The plant pathogens belonging to the class Mollicutes are cell wall-less bacteria existing exclusively in the phloem sieve cells. They depend on the phloem-feeding leafhoppers for spread from infected plants to healthy plants, while the vectors feed on the plant tissues. The symptom of corn stunt disease caused by *Spiroplasma kunkelii* is not likely to be recognized until close to flowering time. Early diagnosis and disease forecasting are essentially required to contain the disease spread. Several diagnostic methods have been employed earlier to detect the

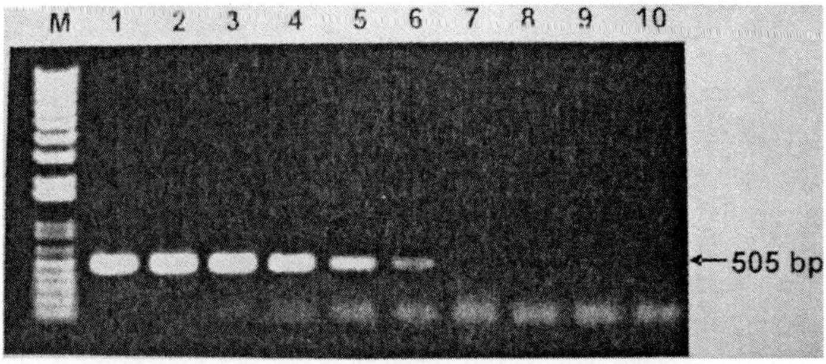

Fig. 3.2 Detection of *Spiroplasma kunkelii* using conventional PCR assay for amplification of targeted gene sequences of spiroplasma pathogen on 2.5% agarose gel electrophoresis. Note the generation of amplicons (505-bp) with the specific primer set CSSF1/CSSR1; Lane M: 1 kb- Plus DNA ladder (standard); Lanes 1–10: Different concentrations of DNA template from 50×10 to 50×10^{-9} ng. (Wei et al. 2006; The American Phytopathological Society, St. Paul, MN, USA)

presence of *S. kunkelii* in corn stunt-affected plants and leafhopper vectors carrying the pathogen. A real-time PCR format employing the primers designed based on the sequences of a unique gene encoding adhesin-like protein. The primer sets CSSF1/CSSR1 or sk104F/sk104R amplified the target DNA fragment in *S. kunkelii*. The limit of detection of the assay was as low as 5 ng DNA of the pathogen and the assay was more sensitive (100 folds) than the conventional PCR (Fig. 3.2). The usefulness of the real-time PCR developed in this study was demonstrated under field conditions also. Both the infected corn plants and the inoculative leafhoppers could be identified rapidly and reliably. This diagnostic procedure has the potential for use in epidemiological and breeding programs (Wei et al. 2006).

3.4 Genomics and Disease Resistance

Management of crop diseases through disease resistance is considered as the most important and feasible strategy. Cultivars with durable resistance are preferred to cultivars exhibiting other forms of resistance to diseases caused by microbial pathogens. An understanding of the evolution of host and pathogen genes affecting the interactions between host and pathogen is essential to locate the sources of disease resistance. Different *R* gene loci are apparently evolving in different ways. Some of them may be under strong diversifying selection, while others are not. It will be very useful for crop improvement, if durability of resistance can be predicted based on genomic analysis. The genomic approaches for identifying the genes coding for disease effector proteins of bacterial and fungal pathogens and the functional genomic approaches to obtain desired knockout mutants greatly advanced the knowledge of the functions of the genes involved in pathogenesis and disease resistance (**? ?**). The evolution of disease resistance in plants increasingly becomes clear following the isolation and sequential analysis of several resistance genes. Molecular analyses of resistance proteins and their corresponding avirulence proteins have revealed direct physical interaction in certain pathosystems (Jia et al. 2000; Deslandes et al. 2003). The genes responsible for quantitative traits may be identified by applying genomic tools, making it possible to determine whether resistance governed by quantitative traits is really more durable. Comparison of agricultural and unmanaged (natural) systems, the latter being under the selection pressure and former under selection pressure mediated by human decision-making will be important to understand the roles that plant pathogens might have played in structuring plant communities in various agroecosystems (Garrett et al. 2006).

References

Akimitsu K, Peever TL, Timmer LW (2003) Molecular, ecological and evolutionary approaches to understanding Alternaria diseases of citrus. Mol Plant Pathol 4:435–446

Alfano JR, Collmer A (2004) Type III secretion system effector proteins: double agents in bacterial disease and plant defense. Annu Rev Phytopathol 42:385–414

Al-Naimi FA, Garrett FA, Bockus WW (2005) Competition, facilitation and niche differentiation in two foliar pathogens. Oecologia 143:449–457

Ammar ED, Järlfors U, Pirone TP (1994) Association of potyvirus helper component protein with virions and the cuticle lining the maxillary food canal and foregut of an aphid vector. Phytopathology 84:1054–1059

Antonovics J (2003) Toward community genomics? Ecology 84:598–601

Aranda MA, Fraile A, Garcia-Arenal F (1993) Genetic variability and evolution of the satellite RNA of *Cucumber mosaic virus* during natural epidemics. J Virol 67:5896–5901

Armour SL, Mercher U, Pirone TP, Lyttle DG, Essember RC (1983) Helper component for aphid transmission encoded by region III of *Cauliflower mosaic virus* DNA. Virology 129:25–30

Atreya CD, Atreya PL, Thornbury DW, Pirone TP (1992) Site-directed mutations in the potyvirus HC-Pro gene affect helper component activity, virus accumulation and symptom expression in infected tobacco plants. Virology 191:106–111

Atreya CD, Pirone TP (1993) Mutational analysis of the helper component-proteinase gene of a potyvirus: effects of amino acid substitutions, deletions and gene replacement on virulence and aphid transmissibility. Proc Natl Acad Sci USA 90:11919–11923

Bateson MF, Henderson J, Chaleeprom W, Gibbs AJ, Dale JL (1994) *Papaya ringspot virus*: isolate variability and origin of PRSV type P in Australia. J Gen Virol 75:3547–3553

Bateson MF, Lines RE, Revill P, Challeprom W, Ha CC, Gibbs AJ Dale JL (2002) On the evolution and molecular epidemiology of the potyvirus *Papaya ringspot virus*. J Gen Virol 83:2575–2585

Bearchell SJ, Fraaije BA, Shaw MW, Fitt BDL (2005) Wheat archive links long-term fungal pathogen population dynamics to air pollution. Proc Natl Acad Sci USA 102:5438–5442

Belin C, Schmitt C, Demangeat G, Komar V, Pinck L, Fuchs M (2001) Involvement of RNA2 encoded proteins in the specific transmission of *Grapevine fan leaf virus* by its nematode vector *Xiphinema index*. Virology 291:161–171

Belin C, Schmitt C, Gaire F, Walter B, Demangeat G, Pinck L (1999) The nine C-terminal residues of the *Grapevine fan leaf virus* movement protein are critical for systemic virus spread. J Gen Virol 80:1347–1356

Bird J, Maramorosch K (1978) Viruses and virus diseases associated with whiteflies. Adv Virus Res 22:55–110

Bertolini E, Olmos A, Martinez MC, Gorris MT, Cambra M (2001) Single-step multiplex RT-PCR for simultaneous and colorimetric detection of six RNA viruses in olive trees. J Virol Methods 96:33–41

Blain JP, Coutos-Thevenot P, Marion D, Ponchet M (2002) From elicitins to lipid-transfer proteins: a new insight in cell signaling involved in plant defence mechanisms. Trends Plant Sci 7:293–296

Blanc S, Schmidt I, Vantard M, Scholthof HB, Kuhl G, Esperandieu P, Cerutti M, Louis C (1996) The aphid transmission factor of *Cauliflower mosaic virus* forms a stable complex with microtubules in both insect and plant cells. Proc Natl Acad Sci USA 93:15158–15163

Boonham N, Walsh K, Smith P, Madagan K, Graham I, Barker I (2003) Detection of potato viruses using microarray technology: towards a generic method for plant viral disease diagnosis. J Virol Methods 108:181–187

Briddon RW, Pinner MS, Stanley J, Markham PG (1990) *Geminivirus* coat protein gene replacement alters insect specificity. Virology 177:85–94

Bull SE, Briddon RW, Serubombwe WS, Ngugi K, Markham PG, Stanley J (2006) Genetic diversity and phylogeography of cassava mosaic viruses in Kenya. J Gen Virol 87: 3053–3065

Burrows ME, Caillaud MC, Smith DM, Benson EC, Gildow FE, Gray SM (2006) Genetic regulation of polerovirus and luteovirus transmission in the aphid *Schizaphis graminum*. Phytopathology 96:828–837

Calonec A, Goyeau H, de Vallavielle-Pope C (1996) Effects of induced resistance on infection efficiency and sporulation of *Puccinia striiformis* on seedlings in varietal mixtures and on field epidemics in pure stands. Eur J Plant Pathol 102:733–741

Chang JH, Urbach JM, Law TF, Arnold LW, Hu A Gombar S, Grant SR, Ausubel FM, Dangl JL (2005) A high-throughput, near saturating screen for type III effector genes from *Pseudomonas syringae*. Proc Natl Acad Sci USA 102:2549–2554

Chen J, Groves R, Civerlo EL, Viveros M, Freeman M, Zheng Y (2005) Two *Xylella fastidiosa* genotypes associated with almond leaf scorch disease on the same location in California. Phytopathology 95:708–714

Collmer A, Lindberg M, Petnicki-Ocwieja T, Schneider DJ, Alfano JR (2002) Genomic mining type III secretion system effectors in *Pseudomonas syringae* yields new picks for all TTSS prospectors. Trends Microbiol 10:462–469

Colvin J, Omongo C, Maruthi M, Otim-Nape G, Thresh J (2004) Dual begomovirus infections and high *Bemisia tabaci* populations: two factors driving the spread of a cassava mosaic disease. Plant Pathol 53:577–584

Deng D, Otim-Nape G, Sangare A, Ogwal S, Beachy R, Fauquet C (1997) Presence of a new virus closely associated with cassava mosaic outbreak in Uganda. Afr J Root Tuber Crops 2:23–28

Deslandes L, Olivier J, Peeters N, Feng DX, Khounlotham M, Boucher C, Somssich I, Genin S, Marco Y (2003) Physical interaction between RRS1-R, a protein conferring resistance to bacterial wilt and PopP2, a type III effector targeted to the plant nucleus. Proc Natl Acad Sci USA 100:8024–8029

Deyong Z, Willingmann P, Heinze C, Adam G, Pfunder M, Frey B, Frey JE (2005) Differentiation of *Cucumber mosaic virus* isolates by hybridization to oligonucleotides in microarray format. J Virol Methods 123:101–108

Drucker M, Froissart R, Hebrard E, Uzest M, Ravallec M, Esperandieu P, Mani JC, Pugniere M, Roquet F, Fereres A, Blanc S (2002) Intracellular distribution of viral gene products regulates a complex mechanism of *Cauliflower mosaic virus* acquisition by its aphid vector. Proc Natl Acad Sci USA 99:2422–2427

Enkerli J, Widmer F, Gessler C, Keller S (2001) Strain-specific microsatellite markers in the entomopathogenic fungus *Beauveria brongniartii*. Mycol Res 105:1079–1087

Fargette D, Konaté G, Fauquet C, Muller E, Petershmitt M, Thresh JM (2006) Molecular ecology and emergence of tropical viruses. Annu Rev Phytopathol 43:235–260

Fauquet C, Mayo M, Maniloff J, Desselberger U, Ball L (eds) (2005) Virus taxonomy. VIIIth report of the international committee on taxonomy of viruses. Elsevier/Academic, London

Fauquet C, Stanley J (2003) *Geminivirus* classification and nomenclature: progress and problems. Ann Appl Biol 142:165–189

Fessehaie A, De Boer SH, Levesque AC (2003) An oligonucleotide array for the identification and differentiation of bacteria pathogenic on potato. Phytopathology 93:262–269

Foissart R, Michalakis Y, Blanc S (2002) Helper component-transcomplementation in the vector transmission of plant viruses. Phytopathology 92:576–579

Gaire F, Schmitt C, Stissi-Garaud C, Pinck L, Ritzenthaler C (1999) Protein 2A of *Grapevine fan leaf nepovirus* is implicated in RNA2 replication and colocalises to the replication sites. Virology 264:25–36

Gallagher WM, Bergin OE, Rafferty M, Kelly ZD, Nolan IM, Fox EJ, Cullhane AC, Mc Ardle L, Fraga MF, Hughes L (2005) Multiple markers for melanoma progression regulated by DNA methylation: insights from transcriptomic studies. Carcinogenesis 26:1856–1867

Garcia-Arenal F, Escriu F, Aranda MA, Alonso-Prados JL, Malpica JM, Fraile A (2000) Molecular epidemiology of *Cucumber mosaic virus* and its satellite RNA. Virus Res 71:1–8

Garrett KA, Hulbert SH, Leach JE, Travers SE (2006) Ecological genomics and epidemiology. Eur J Plant Pathol 115:35–51

Garrett KA, Mundtt CC (1999) Epidemiology in mixed host populations. Phytopathology 89: 984–990

Geering A, Olszewski N, Harper G, Lockhart B, Thomas J (2005a) Banana contains a diverse array of endogenous badnaviruses. J Gen Virol 86:511–520

Geering A, Pooggin M, Olszewski N, Lockhart B, Thomas J (2005b) Characterization of *Banana streak Mysore virus* and evidence that its DNA is integrated in the B genome of cultivated *Musa*. Arch Virol 150:787–796

Gray SM, Gildow FE (2003) Luteovirus-aphid interactions. Annu Rev Phytopathol 41:539–566

Guo JR, Schnieder F, Beyer M, Verreet JA (2005) Rapid detection of *Mycosphaerella graminincola* in wheat using reverse transcription-PCR assay. J Phytotopathol 153:674–679

Hamelin RC, Allaire M, Bergeron MJ, Nocole MC, Lecours N (2005) Molecular epidemiology of white pine blister rust: recombination and spatial distribution. Phytopathology 95:793–799

Harper G, Osuji. J, Heslop-Harrison J, Hull R (1999) Integration of *Banana streak badnvirus* into the *Musa* genome: molecular and cytogenetic evidence. Virology 255:207–213

Harrison BD, Murant AF (1977) Nematode transmissibility of pseudo-recombinant isolates of *Tomato black ring virus*. Ann Appl Biol 86:209–212

Harrison B, Zhous X, Otim-Nape G, Liu Y, Robinson D (1997) Role of a novel type of double infection in the geminivirus-induced epidemic of severe cassava mosaic in Uganda. Ann Appl Biol 131:437–448

Hayden MJ, Sharp PJ (2001a) Sequence-tagged microsatellite profiling (STMP): a rapid technique for developing SSR markers. Nucleic Acids Res 29:e43

Hayden MJ, Sharp PJ (2001b) Targeted development of informative microsatellite (SSR) markers. Nucleic Acids Res 29:e44

Hert AP, Roberts PD, Momol MT, Minsavage GV, Tudor-Nelson SM, Jones JB (2005) Relative importance of bacteriocin-like genes in antagonism of *Xanthomonas perforans* tomato race 3 to *Xanthomonas euvesicatoria* tomato race 1 strains. Appl Environ Microbiol 71: 3581–3588

Holmes A, Govan J, Goldstein R (1998) Agricultural use of *Burkholderia (Pseudomonas) cepacia*: a threat to human health. Emerg Infect Dis 4:221–227

Hong Y, Robinson D, Harrison B (1993) Nucleotide sequence evidence for the occurrence of three distinct whitefly-transmitted geminiviruses in cassava. J Gen Virol 74:2437–2443

Humeau L, Roumagnac P, Picard Y, Robéne-Soustrade I., Chiroleu F, Gagnevin L, Pruvost O (2006) Quantitative and molecular epidemiology of bacterial blight of onion in seed production fields. Phytopathology 96:1345–1354

Ito T, Ieki H, Ozaki K (2002) Simultaneous detection of six citrus viroids and *Apple stem grooving virus* from citrus plants by multiplex reverse transcription-polymerase chain reaction. J Virol Methods 106:235–239

Jia Y, McAdams SA, Bryan GT, Hershey HP, Valent B (2000) Direct interaction of resistance and avirulence gene products confers rice blast resistance. Eur Mol Biol Org J 19:4004–4014

Keiper FJ, Haque MS, Hayden MJ, Park RF (2006) Genetic diversity in Australian populations of *Puccinia graminis* f.sp. *avenae*. Phytopathology 96:96–104

Kohmoto K, Itoh Y, Shimomura N, Kondoh Y, Otani H, Kodama M, Nishimura S, Nakatsuka S (1993) Isolation and biological activities of two host-specific toxins from the tangerine pathotype of *Alternaria alternata*. Phytopathology 83:495–502

Kohmoto K, Scheffer RP, Whiteside JO (1979) Host-selective toxins from *Alternaria citri*. Phytopathology 69:667–671

Leach JE, Vera Cruz CM, Bai J, Leung H (2001) Pathogen fitness penalty as a predictor of durability of disease resistance genes. Annu Rev Phytopathol 39:187–224

Lecomte P, Manceau C, Paulin JP, Keck M (1997) Identification of PCR analysis on plasmid pEA 29 of isolates of *Erwinia amylovora* responsible for an outbreak in Central Europe. Eur J Plant Pathol 103:91–98

Lee GP, Min BE. Kim CS, Choi SH, Harn CH, Kim SU, Ryu KH (2003) Plant virus cDNA chip hybridization for detection and differentiation of four cucurbit-infecting Tobamoviruses. J Virol Methods 110:19–24

Legg J (1999) Emergence, spread and strategies for controlling the pandemic of *Cassava mosaic virus* disease in East and Central Africa. Crop Prot 18:627–637

Leh V, Jacquot E, Geldreich A, Herman T, Leclerc D, Cerutti M, Yot P, Keller M, Blanc S (1999) Aphid transmission of *Cauliflower mosaic virus* requires the PIII protein. EMBO J 18: 7077–7085

Li C, Cox-Foster D, Gray SM, Gildow F (2001) Vector specificity of *Barley yellow dwarf virus* (BYDV) transmission: Identification of potential cellular receptors binding BYDV-MAV in the aphid *Sitobion avenae*. Virology 286:125–133

Lin HX, Rubio L, Smythe AB, Falk BW (2004) Molecular population genetics of *Cucumber mosaic virus* in California: evidence for founder effects and reassortment. J Virol 78:6666–6675

Lopez-Moya JJ (2002) Genes involved in insect-mediated transmission of plant viruses. In: Khan JA, Dijkstra J (eds) Plant viruses as molecular pathogens. CBS Publishers and Distributors, New Delhi, India, pp 31–60

Luo Y, Ma Z, Reyes H C, Morgan D, Michailides TJ (2007) Quantification of airborne spores of *Monilinia fructicola* in stone fruit orchards of California using real-time PCR. Eur J Plant Pathol 118:145–154

Ma Z, Luo Y, Michailides TJ (2003) Nested PCR assays for detection of *Monilinia fructicola* in stone fruit orchards and *Botryospheria dothidea* from pistachios in California. J Phytopathol 151:312–322

MacFarlane SA (2003) Molecular determinants of the transmission of plant viruses by nematodes. Mol Plant Pathol 4:211–215

MacFarlane SA, Wallis CV, Brown DJF (1996) Multiple genes involved in the nematode transmission of *Pea early browning virus*. Virology 219:417–422

May KJ, Ristaino JB (2004) Identity of the mtDNA haplotype(s) of *Phytophthora infestans* in historical specimens from the Irish Potato Famine. Mycol Res 108:471–479

Mumford R, Boonham N, Tomlinson J, Barker I (2006) Advances in molecular phytodiagnosis-new solutions for old problems. Eur J Plant Pathol 116:1–19

Mundt CC (2002) Use of multiline cultivars and cultivar mixtures for disease management. Annu Rev Phytopathol 40:381–410

Ndowora T, Dahal G, La Flaeur D, Harper G, Hull R (1999) Evidence that *Badnavirus* infection in *Musa* can originate from integrated pararetroviral sequences. Virology 255: 214–220

Nicolaisen M, Justesen AF, Thrane U, Skouboe P, Holmstrom K (2005) An oligonucleotide microarray for the identification and differentiation of trichothecene producing and non-producing *Fusarium* species occurring on cereal grain. J Microbiol Methods 62:57–69

Nishimura S, Kohmoto K (1983) Host-specific toxins and chemical structures from *Alternaria* species. Annu Rev Phytopathol 21:87–116

Noris E, Vaira AM, Caciagli P, Masenga V, Gronenborn B, Accotto GP (1998) Amino acids in the capsid protein of *Tomato yellow leaf curl virus* that are crucial for systemic infection, particle formation and insect transmission. J Virol 72:10050–10057

Omongo C (2003) Cassava whitefly, *Bemisia tabaci* behavior and ecology in relation to the spread of the cassava mosaic pandemic in Uganda. Doctoral Thesis, Univ Greenwich, Greenwich

Palukaitis P, Roosink M J, Dietzgen RG, Francki RIB (1992) *Cucumber mosaic virus*. Adv Virus Res 41:281–348

Peever TL, Olsen L, Ibáñez A, Timmer LW (2000) Genetic differentiation and host specificity among populations of *Alternaria* spp. causing brown spot of grapefruit and tangerine x grapefruit hybrids in Florida. Phytopathology 90:407–414

Perry KL, Zhang L, Palukaitis P (1998) Amino acid changes in the coat protein of *Cucumber mosaic virus* differentially affect transmission by the aphids *Myzus persicae* and *Aphis gossypii*. Virology 242:204–210

Ploeg AT, Ribinson DJ, Brown DJF (1993) RNA2 of *Tobacco rattle virus* encodes the determinants of transmissibility by trichodorid vector nematodes. J Gen Virol 74:1463–1466

Purcifull DE, Edwardson JR, Hiebert E, Gonsalves D (1984) *Papaya ringspot virus*. CMI/AAS description of plant viruses, No. 292, CAB International, Wallingford, UK

Ragozzino E, Faggioli F, Barba M (2004) Development of a one tube-one step RT-PCR protocol for the detection of seven viroids in four genera: *Apscaviroid, Hostuviroid, Pelamoviroid and Pospiviroid.* J Virol Methods 121:25–29

Rosell RC, Bedford ID, Frohlich DR, Gill RJ, Brown JK, Markham PG (1997) Analysis of morphological variation in distinct populations of *Bemisia tabaci* (Homoptera: Aleyrodidae). Ann Entomol Soc Am 90:575–589

Rosell, RC, Torres-Jerez I, Brown JK (1999) Tracing the geminivirus-whitefly transmission pathway by polymerase chain reaction in whitefly extracts, saliva and hemolymph and honey dew. Phytopathology 89:239–246

Roumagnac P, Pruvost O, Chiroleu C, Hughes G (2004) Spatial and temporal analyses of bacterial blight of onion caused by *Xanthomonas axonopodis* pv. *allii*. Phytopathology 94:138–146

Ruppitsch W, Stöger AR, Keck M (2004) Stability of short sequence repeats and their application for the characterization of *Erwinia amylovora* strains. FEMS Microbiol Lett 234:1–8

Saint-Jean S, Testa A, Kamoun S, Madden LV (2005) Use of a green fluorescent marker for studying splash dispersal of sporangia of *Phytophthora infestans*. Eur J Plant Pathol 112:391–394

Swanson M, Harrison B (1994) Properties, relationships and distribution of cassava mosaic geminiviruses. Trop Sci 34:15–25

Tomlinson JA (1987) Epidemiology and control of virus diseases of fruits and vegetables. Ann Appl Biol 110:661–681

Van Sluys MA, Monteiro-Vitrorello CB, Camargo LEA, Menck CFM, da Silva ACR., Ferro JA, Oliveira MC, Setubal JC, Kitajima JP, Simpson AJ (2002) Comparative genomic analysis of plant-associated bacteria. Annu Rev Phytopathol 40:169–189

Vassilakos N, Vellios EK, Brown EC, Brown DJF, MacFarlane SA (2001) Tobravirus 2b protein acts in trans to facilitate transmission by nematodes. Virology 279:478–487

von Bodman SB, Bauer WD, Coplin DL (2003) Quorum sensing in plant pathogenic bacteria. Annu Rev Phytopathol 41:455–482

Wang RY, Ammar ED, Thornbury DW, Lopez-Moya JJ, Pirone TP (1996) Loss of potyvirus transmissibility and helper component activity correlate with non-retention of virions in aphid stylets. J Gen Virol 77:861–867

Wang RY, Kritzman A, Hershman DE, Ghabrial SS (2006) *Aphis glycines* as a vector of persistently and nonpersistently transmitted viruses and potential risks for soybean and other crops. Plant Dis 90:920–926

Wang RY, Powell G, Hardie J Pirone TP (1998) Role of the helper component in vector-specific transmission of potyviruses. J Gen Virol 79:1519–1524

Weber JL, May PE (1989) Abundant class of human DNA polymorphisms which can be typed using the polymerase chain reaction. Amer J Hum Genet 44:388–396

Wei W, Opgenorth DC, Davis RE, Chang CJ, Summers CG, Zhao Y (2006) Characterization of a novel adhesin-like gene and design of a real-time PCR for rapid, sensitive and specific detection of *Spiroplasma kunkelii*. Plant Dis 90:1233–1238

Wintermantel WM, Wisler GC (2006) Vector specificity, host range and genetic diversity of *Tomato chlorosis virus*. Plant Dis 90:814–819

Zhang Y, Geider K (1997) Differentiation of *Erwinia amylovora* strains by pulsed field gel electrophoresis. Appl Environ Microbiol 63:4421–4426

Zhang Y, Merghi M, Bazzi C, Geider K (1998) Genomic analysis by pulsed field gel electrophoresis of *Erwinia amylovora* strains from the Mediterranean region including Italy. J Plant Pathol 80:225–232

Zhou X., Liu Y, Calvert L, Munoz C, Otim-Nape G, Robinson D, Harrison B (1997) Evidence that DNA-A of a geminivirus associated with severe cassava mosaic disease in Uganda has arisen by interspecific recombination. J Gen Virol 78:2101–2111

Zhu Y, Chen H, Fan J, Wang Y, Li Y, Chen J, Fan J, Yang S, Hu L, Leaung H, Mew CW, Teng PS, Wang Z, Mundt CC (2000) Genetic diversity and disease control. Nature 406:718–722

Zink AR, Reischl U, Wolf H, Nerlich AG (2002) Molecular analysis of ancient microbial infections. FEMS Microbiol Lett 213:141–147

Zwankhuizen MJ, Govers F, Zadoks JC (1998) Development of potato late blight epidemics: disease foci, disease gradients and infection sources. Phytopathology 88:754–763

Glossary

ABC transporters A large family of proteins embedded in plasma membrane characterized by a highly conserved ATP-binding domain; they constitute a class of membrane transporter proteins capable of transferring sugar molecules, inorganic ions, polypeptides, certain anticancer drugs and antibiotics

Acquired resistance Enhancement of resistance of plants following infection or treatment with biotic or abiotic resistance inducers

Actin Structural proteins are present abundantly in eukaryotic cells that can interact with many other proteins; these proteins polymerize to form cytoskeletal filaments

Activator Transcription factor that stimulates transcription

Active site The region of an enzyme that binds substrate molecules and catalyzes an enzymatic reaction

Active transport Export of an ion or small molecule across a membrane against its concentration gradient driven by the hydrolysis of ATP

Alarm signal In response to pathogenic infection, chemical(s) produced, may send signal to host cell proteins/genes leading to synthesis of compounds inhibitory to the invading pathogen

Algorithm A logical description of the manner in which a problem can be solved and that can be used as a specification for how to write a computer program; a computational procedure that employs a combination of simple operations to process, analyze and present pictorially data about sequences of DNA, RNA, proteins or other molecules

Alignment This is an arrangement of two or more molecular sequences one below the other in such a way that regions that are identical or similar are placed immediately below one another

Allele It is an alternate form of a gene occupying a given locus on the same chromosome which controls expression (of product) in different ways

Allozyme An allele of the gene may be involved in the production of an enzyme with properties slightly different from that of the enzyme produced by the gene concerned

Amino acid profile Quantitative delineation of how much of each amino acid is present in a particular protein

Aminoacyl-tRNA synthetase Enzyme that catalyzes the linkage of an amino acid to a specific tRNA to be used for protein synthesis

Amplicon A specific sequence of DNA that is produced from the template fragment by reactions such as polymerase chain reaction assay

Amplification Synthesis of additional copies of a DNA sequence found either as chromosomal or extrachromosomal DNA

Analog A compound or molecule that is a structural derivative of a 'parent' compound; this term also refers to a molecule that may be structurally similar, but not identical to another, and may exhibit many or some of the biological functions of the other

Ankyrin a protein capable of binding spectrin and links the actin cytoskeleton to the plasma membrane

Annealing The process by which the complementary base pairs in the strands of DNA combine

Annotated gene A gene is said to be annotated when it has been recognized from a large segment of genome sequence and if its cellular role is known to some extent

Antibody A protein (immunoglobulin) formed in an animal circulatory system in response to introduction of an antigen (usually a protein or nucleoprotein)

Anticodon The sequence of three bases on tRNA that combines with the complementary triplet codon on mRNA

Antigen A protein or live or inactivated microbe capable of inducing the production of a specific antibody in an animal system and capable of reacting with that antibody in a specific manner

Antisense technology A molecular technique that uses a nucleic acid sequence complementary to mRNA so that the two bind and the mRNA is effectively neutralized

Apoplast Space outside of the plasma membrane of cells constituted by cell walls and conducting cells of the xylem in which aqueous phase of intercellular solutes is present

Apoptosis It is a natural phenomenon encountered in animal cells for the elimination of cells involving a highly regulated, energy-dependent process by which a cell brings about its own death (programmed cell death in plants)

Attacins These proteins constitute a class of antimicrobial peptides produced in the insect hemolymph; synthesis of outer membrane proteins of Gram negative bacteria is inhibited by attacins

Avidin A protein found in egg white and bacteria like *Streptomyces* that binds with very high affinity to the vitamin biotin (vitamin H)

Avirulence Opposite of virulence indicating the inability of a pathogen to infect a particular species or cultivar of a plant, presumably because of the presence of avirulence gene(s) whose product(s) alerts the host resistance mechanism

Avirulence conferring enzyme (ACE) The avirulence gene of the fungal pathogen coding for the enzyme that is recognized by the resistant cultivar carrying corresponding resistance gene The rice resistance gene *Pi33* recognizes *Magnaporthe grisea* isolates carrying *ACE 1* gene

***Avr* gene** Pathogen gene that is responsible for the inability to infect a given plant species or cultivar

Avr protein The protein coded for by the *avr* gene, capable of functioning as an elicitor of defense responses in plants

Bacterial artificial chromosome (BAC) DNA molecules that are used as cloning vectors, derived from plasmids. This type of vector can be employed for cloning large inserts (of about 150,000 bp) of DNA in bacteria

Base pair (bp) Association of two complementary nucleotides in a DNA or RNA molecule stabilized by hydrogen bonding between their base components: adenine pairing with thymine or uracil and guanine pairing with cytosine

Biofilm Bacteria or fungi may be embedded in polysaccharide matrix that plays a role in attachment to host surface, colonization and invasion

Bioinformatics Computer based analysis of data on biological sequencing of the genome of an organism to predict gene function, protein and RNA structure, genome organization and molecular bases in relation to responses of plants to microbial pathogens

Biolistic This term has been coined from the words 'biologic' and 'ballistic'; refers to process involving the use of pellets coated with the desired genes that are fired from a gun into seeds or plant tissues in order to get plants expressing these transgenes

Bioluminescence Emission of light from a living organism

BLAST (**B**asic **L**ocal **A**lignment **S**earch **T**ool) This program is used widely for searching sequence databases for entries that are similar to a specified sequence in question

Bootstrapping A statistics protocol used to estimate the reliability of a result (such as construction of phylogenetic tree) that involves sampling data with replacement from the original set of data

Calmodulin A small cytosolic regulatory protein that binds four Ca^{2+} ions; the Ca^{2+}/calmodulin complex can interact with many proteins resulting in inhibition or activation of those proteins

Capsid Protein coat of viruses that encloses the viral genome

***Cauliflower mosaic virus* 35S promoter (CaMV 35S)** This promoter consisting of DNA sequences of the virus is employed very frequently in genetic engineering to control expression of an inserted gene (synthesis of desired protein in a plant)

cDNA A complementary DNA strand to an RNA strand and synthesized from RNA strand using reverse transcriptase enzyme. It is also known as copy DNA

cDNA clone A DNA molecule synthesized from a mRNA sequence via sequential use of reverse transcription and DNA polymerase

cDNA library A collection of cDNA clones which represent all the genetic information expressed by a given cell or by a given tissue type is known as a cDNA library

Cecropins Antimicrobial proteins isolated from insects capable of forming pores in and causing lysis of bacterial cell membrane (see humoral immunity)

Chaperon protein A protein molecule attached to and to facilitate the export of an effector protein secreted through a type III secretion system of bacteria. It may protect the effector protein from coming in contact with other proteins. The chaperones may also assist with correct folding as the protein molecule emerges from the cell's ribosome

Chemotaxis Capacity of a cell or organism to move toward or away from certain chemicals

Chromosome walking A technique employed for determining the location of and sequencing a given gene within the genome of the organism, by sequencing specific DNA fragments that overlap and span collectively

cis-acting protein This protein has the exceptional property of acting only on the molecule of DNA from which it was expressed

Cistron Viral sequences of nucleotides of DNA/RNA that code for particular protein and are considered to be analogous to genes of other organisms

Clade A clade consists of all the species descending from an internal node of genealogical tree

Clone Genetically identical individuals produced asexually from one individual

Cloning of DNA Introduction of a segment of DNA from one species into the DNA of another species leading to the formation of many copies of the hybrid DNA by replication

Codon A sequence of three nucleotide (triplet) bases on mRNA that interacts with the anticodon on tRNA and specifies the incorporation of a specific amino acid into a polypeptide during translation process

Codon usage The frequency of occurrence of each codon in a gene or genome, especially the relative frequencies with which synonymous codons are used

Colony hybridization A technique used in situ hybridization to identify bacterial colonies carrying inserted DNA that is homologous with some particular sequence (used as probe)

Complementary sequences Two nucleic acid sequences that may form an exactly matching double strand as a result of A-T and G-C pairing. Complementary sequences run in opposite directions

Constitutive genes These genes code for certain protein products that are required at all times for general cell maintenance and metabolism. These genes are expressed as a function of the interaction of RNA polymerase with the promoter without additional regulation. They are also known as household genes

Contigs Overlapping DNA segments that as a collection, form a longer and gapless segment of DNA

Cosmid It is a larger insert cloning vector, useful for analysis of highly complex genomes; forms an important component of many genome mapping projects; cosmid vectors are constructed by incorporating the *cos* sites from the phage λ

Cross-protection Enhancement of resistance in plants following infection by one virus or its strain(s) to reinfection by the same virus or its strains/related viruses

Cytoplasmic resistance Resistance attributed to the genetic material present in the cell cytoplasm

Cytoskeleton A three-dimensional network of fibrous elements comprising primarily of microtubules, actin microfilaments and intermediate filaments found in the cytoplasm of eukaryotic cells; the cytoskeleton offers structural support for the cell and permits directed movement of organelles, chromosomes and even the cell

Cytosol Unstructured aqueous phase of cytoplasm excluding organelles, membranes and insoluble cytoskeletal components

Dalton A mass unit used to indicate the size of a biomolecule; one dalton is equivalent to molecular mass of a hydrogen atom

Data mining A computerized program used to search for relationships between and overall patterns among the data available within a database

Defensins Defense-related, cysteine-rich, antimicrobial peptides present in the plasma membrane of host plant species; they constitute a group designated defensins capable of providing resistance against microbial plant pathogens

Denaturation The conformation of a protein or nucleic acid may be drastically altered because of disruption of various bonds due to heating or exposure to chemicals; this may result in loss of biological function(s)

Dendrogram A branching diagram that shows the relative sequence similarity between many different proteins or genes to indicate the phylogenetic relationships; typically horizontal lines indicate the degree of differences in sequences, while vertical lines are used for clarity to separate branches

DNA fingerprinting/ DNA profiling This technique involves the use of restriction enzymes and electrophoresis to determine the differences and similarities in the DNA of individual organisms

DNA library This represents a collection of cloned DNA molecules/fragments of entire genome or DNA copies of all mRNAs produced by a cell type (cDNA library) inserted into suitable cloning vector

DNA ligase The enzyme that seals breaks in DNA strand and also catalyzes the formation of the final phosphoester bond in DNA replication

DNA microarray Oligonucleotides or fragments of cDNAs are printed on a glass slide or membrane filter at a high density, permitting simultaneous analysis of thousands of genes by hybridization of the microarray with fluorescent probes

Domain A portion of polypeptide chain that folds into a compact globular units of the protein forming the basic unit of the tertiary structure and remains distinct even within the tertiary structure of the protein; a discrete part of a protein with its own distinct function

Downstream A relative direction of DNA sequence, as the DNA is usually written with 5′ end to the left; downstream would be to the right of a reference point – eg. the start codon is downstream of the protmoter

Ectopic A gene inserted in an unnatural location

Effectors These molecules can manipulate host cell structure and function, thereby facilitating infection (virulence factors/toxins) and /or triggering defense responses (avirulence factors/elicitors)

Effector proteins Bacterial virulence determinants injected into host cells via type III secretion system (TTSS) of the bacterial pathogen

Elongation factor A group of nonribosomal proteins required for continued translation of mRNA (protein synthesis) following initiation; guiding the elongation phase of transcription or translation, during protein synthesis

Endoplasmic reticulum (ER) An extensive network of membrane-enclosed tubules and sacs involved in protein sorting and processing as well as in lipid synthesis

Epitope A specific group of atoms on an antigen molecule that is recognized by the specific antibody produced against the antigen concerned; it is also called as antigenic determinant

Episome An independent genetic element (DNA) that is present inside the bacterial species in addition to the normal bacterial cell genome. The episome can replicate either as an autonomous unit or as one integrated into host genome

Exon A segment of an eukaryotic gene that contains a coding sequence and this part of gene sequence is transcribed into an mRNA and translated to give rise to a specific domain of the protein

Expressed sequence tags (ESTs) These are partial sequences of cDNA clones in an expressed cDNA library; useful for identifying all unique sequences (genes) to determine their functions; a profile of mRNAs allowing cloning of a large number of genes being expressed in a cell population

Expression vector A modified plasmid or virus that carries a gene or cDNA into suitable host cell in which it directs synthesis of the encoded protein; expression vector can be used to screen DNA libraries for a gene of interest or to produce large amounts of a protein from its cloned gene

Expressivity The intensity of realization of the effect of a gene in a phenotype; the degree to which a particular effect is expressed by individuals

Extracellular matrix Secreted proteins and polysaccharides that fill the spaces between cells and bind cells and tissues together

Flagellin These proteins associated with flagella of bacteria, are capable of functioning as a receptor system for general elicitors

Flanking sequence A segment of DNA molecule that either precedes or follows the region of interest on the molecule

Flux The flow of intermediates in metabolism; the rate at which substrates enter and exit a pathway

Footprinting This is a technique employed for identifying protein-binding regions of DNA or RNA based on the ability of a protein bound to a region of DNA or RNA to protect that region from digestion

Free radical A molecule that has one or more unpaired electrons

Functional genomics Functions and interactions of genes or groups of genes belonging to host plants, pathogens or both are genetically examined

Gene cloning Characterization of gene functions by isolating and multiplying individual gene or sequences by insertion into bacteria (*Escherichia coli*); the sequences are duplicated as the bacteria multiplies by fission

Gene expression Conversion of genetic information within a gene into an actual protein or cell process; an overall process applied to assess the information encoded in a gene that can be converted into an observable phenotype (production of a protein)

Gene expression profiling Identification of specific genes that are 'switched on' in a cell at particular point of time or process, enabling the precise definition of the phenotypic condition of that cell

Gene flow Movement of genes (under examination) through specific process, from one population to another population geographically separated apart

Gene knockouts Genetically modified individuals containing either a defective gene or lacking a gene

Gene silencing Interrupting or suppressing the activity of desired gene(s), resulting in the loss of coordination for production of specific proteins

Gene targeting Insertion of antisense DNA molecules into selected cells of the organism in order to block the activity of undesirable genes such as oncogenes

Genetic code The correspondence between nucleotide triplets and amino acids in proteins; the sequence of bases in mRNA specify the amino acid sequence of a polypeptide, read in triplets (codons), based on a set of rules

Genome sequencing Reading of all nucleotides present in the entire genomic DNA of the organism in an orderly way

Genomic library Contains a collection of recombinant DNA clones that collectively constitute the genome of the organism (see DNA library)

G protein A family of membrane–bound cell-signaling proteins regulated by guanine nucleotide binding

Green fluorescent protein (GFP) A protein from jellyfish possessing the property of fluorescing; useful as a marker in fluorescence microscopy

GUS gene Production of β-glucuronidase (GUS protein) in certain organisms such as *Escherichia coli* is controlled by this gene; it is commonly used as a marker gene for genetically engineered plants

Haplotype A collection of alleles in an individual that are located on one chromosome; alleles within a haplotype are, often, inherited as a single unit from one generation to another; it also refers to a group of genomic variations found repeatedly in many individuals within a population

Harpins Proteins coded by *hrp* genes present in the type III secretion system (TTSS) of phytobacterial pathogens; these proteins can induce resistance in susceptible plants

Heterotrimeric G protein A guanine nucleotide-binding consisting of three subunits

High-throughput methods Large number of genes or gene products can be studied using these partially automated protocols

Homologs The individuals considered to have sequences that are evolutionarily related by descent from a common ancestor; proteins or genes exhibiting similarity (homology)

Homology A sequence of amino acid in two or more proteins that are identical to each other; nucleic acid homology indicates the presence of complementary strands that can hybridize with each other

Horizontal gene transfer Incorporation of a 'foreign' gene acquired from an unrelated species into the genome of another organism

Host-specific toxins (HSTs) Some pathogens elaborate toxins that can induce all primary symptoms as the pathogen itself; production of HSTs is governed by specific genes which are expressed in susceptible plants

Hot spots Events such as mutations with unusual high frequency may occur in certain sites in genes, termed as hot spots

Humoral immune response In response to infection, rapid production and secretion of soluble blood serum components occurs in the animal body

Humoral immunity By injecting viable nonpathogenic or inactivated phytopathogenic bacteria into insects, formation of new proteins in the hemolymph of these insects is induced leading to the development of humoral immunity; these proteins have antibacterial properties; cecropins and attacins are formed in the hemolymph of *Hyalophora cecropia*

Hybridization probes DNA or RNA molecules that are complementary to a region in DNA; useful for detecting specific gene(s) in fingerprinting

Hybridoma A hybrid cell line produced by fusing a myeloma (capable of multiplying indefinitely) with a lymphocyte (capable of producing antibodies); the hybridoma provides continuous supply of specific immunoglobulins (antibodies)

Hypersensitive response (HR) A protective or defense response of plants to infection or inoculation with pathogen in which the initially infected cells and the adjacent ones express the phenomenon of programmed cell death (PCD) resulting in self destruction to cordon off the infected area and to restrict the spread of pathogen to other cells/tissues (see programmed cell death)

Idiotype The specific site of antibody molecule capable of combining with the specific site in the antigen (epitope) is said to have an idiotype (for that epitope), serving as an identifying characteristic of an epitope

Indel An insertion or deletion occurring in a protein or nucleic acid sequence; frequently it may be difficult to find out whether a change in a sequence is due to a deletion in that sequence or an insertion of a related sequence

Indexing Testing the plants or seeds or propagative plant materials for the presence of microbial pathogens by biological and/ molecular techniques

Induced systemic resistance (ISR) Colonization of roots of plants by plant growth promoting rhizobacteria (PGPRs) induces systemic resistance to pathogens infecting tissues/organs far away from the roots of plants; this type of resistance to disease is referred to as induced systemic resistance

Inducible promoter The promoter in which start or increase of promotion occurs usually in the presence of a pathogen or toxin or toxic metabolites of the pathogen leading to initiation of defense-related activity

In silico Experimental process performed on a computer and not by bench research

Intergenic sequence DNA sequence between two genes; sometimes called as 'junk DNA'

Intron The sequence for protein synthesis is absent in the intron; this noncoding sequence of the gene interrupting exons is cut out of the mRNA (splicing) prior to translation

Isoforms Different forms of the same protein whose amino acid sequences differ slightly and whose general activities are similar; isoforms may be encoded by different genes or by a single gene whose primary transcript undergoes alternative splicing

Isozymes Different forms of an enzyme involved in the same or similar reactions, but need different optima for their activity

Karyopherin A family of nuclear transport proteins that function as an importin, exportin or occasionally both; each karyopherin binds to a specific signal sequence in a protein (cargoprotein) to be imported or exported

Killer toxin (KT) A proteinaceous toxin isolated from microorganisms like yeast has antimicrobial properties

Knockout gene Selective inactivation of a desired gene by replacing it with a non-functional (disrupted) allele in an otherwise normal organism

Lectins Plant proteins capable of binding to specific carbohydrates; they can be used to detect the carbohydrates in situ

Leucine-rich repeats (LRR) Segments of amino acids containing multiple copies of leucine present repeatedly in a protein; these proteins are known as LRR proteins

Ligand A molecule other than an enzyme substrate that binds tightly and specifically to a macromolecule, usually a protein forming a macromolecule-ligand complex

Linkage On the same chromosome two different loci governing two different traits may be inherited together; closer the loci greater are the chances of linkage

Linker A short segment of ds-DNA that can be ligated onto a second fragment of DNA to facilitate the cloning of that fragment. Linkers contain a restriction site so that they can be ligated to produce the desired sticky ends for ligation

Locus The locatable position of a gene on a chromosome; all alleles of a particular gene occupy the same locus

Marker-assisted selection (MAS) A known resistance gene or chromosomal sequence closely linked to a gene that is used as a genetic marker to select the progenies of crosses or genotypes containing the selected marker by DNA testing; this procedure is very useful in hastening the development of cultivars with built-in resistance to crop diseases

Messenger RNA (mRNA) An RNA molecule serving as a template for protein synthesis

Metabolome It represents the entire metabolic content of a cell or organism; the complete set/complement of all metabolites and other molecules involved in or produced during a cell's metabolism

Microarray A glass slide or silicon chip onto which several DNA probes are deposited for simultaneous determination of gene expression levels of many genes in the same tissue sample

MicroRNA (miRNA) A naturally occurring short non-coding RNA that can act to regulate gene expression

MicRNAs (messenger-RNA-interfering complementary RNAs) These complementary RNA molecules bind to the RNA transcripts of specific genes resulting in blockage of their translation; they are also called antisense RNAs

Molecular cloning Insertion of a desired DNA fragment into a DNA molecule (vector) that can replicate independently in a host cell

Molecular markers DNA sequence (s) that is characteristic of the DNA segment/gene is used for comparing or detecting the similarities of related organisms or genotypes (see marker assisted selection)

Monocistronic Messenger RNAs encoding a single polypeptide chain

Monoclonal antibody An antibody secreted by a clonal line of B lymphocytes

Monophyletic A group of species on a phylogenetic tree sharing a common ancestor that is not shared by species outside the group; a clade is a monophyletic group

Motif A sequence of amino acids or nucleotides that perform a particular role and is often conserved in other species or molecules

Movement proteins Virus coded-specific proteins involved in the movement of plant viruses in the host plants

Mutagen A chemical capable of inducing a high frequency of mutations

Nonhost resistance Resistance exhibited by a plant species on which the pathogen in question is unable to establish infection; the plant species is termed as a nonhost and the interaction is called as incompatible

Ontogenic resistance The level of resistance of plants may vary with the growth stages of host plant species; plants are highly susceptible to viruses in the early growth stages (seedling), but they develop resistance progressively as the plants become older

Open reading frame (ORF) Region of a gene which contains a series of triplet coding for amino acids without any termination codon is known as ORF; the sequences contained in the ORF may be potentially translatable into a protein

Operon One or more genes may be present in this gene unit; they specify a polypeptide and an operator regulates the structural gene

Orthologs Sequences from different organisms (species) that are evolutionarily related by descent from a common ancestral sequence and later diverged from one another as a result of speciation

Pathogenicity factors The factors (genes) of pathogen origin, are essential for initiation of infection and further colonization of plants

Pathogenesis-related (PR) proteins PR proteins are coded for by the host plant, but induced only in pathological or related conditions. They are produced postinfectionally during plant-pathogen interactions; they have different roles in the development of resistance to the diseases caused by pathogens; these proteins are classified into 14 families; PR genes/PR proteins are widely used as marker genes/ proteins to study the defense mechanisms of plants

Peptide mapping Following partial hydrolysis of a protein, a characteristic two dimensional pattern may be recognized (on paper or agar gel) due to the separation of a mixture of peptides

Phytoalexins These are low-molecular antimicrobial compounds that accumulate at the site of infection by incompatible pathogens. Several biosynthetic enzymes are involved in the production of phytoalexins and hence highly coordinated signal transduction events are required

Phytoanticipins Toxic compounds naturally present in the plants that can affect the development of pathogens adversely

Plasmid Represents an independent, stable, self-replicating piece of extrachromosomal DNA in bacterial cells; it does form a component of normal cell genome, but does not get integrated into the host chromosome

Point mutation A mutation resulting in a change in only one nucleotide in a DNA molecule

Polycistronic Coding region representing more than one gene may be present in the mRNA; it codes for two or more polypeptide chains; RNA genomes of plant viruses (such as *Tobacco mosaic virus*) are polycistronic

Posttranscriptional modification A set of reactions that occur to change the structure of newly synthesized polypeptides

Primer A short nucleic acid sequence containing free 3′-hydroxyl group that forms base pair with a complementary template strand and functions as the starting point for addition of nucleotides to copy the template strand

Probe Defined RNA or DNA fragment radioactively or chemically labeled that is used to detect specific nucleic acid sequences by hybridization

Programmed cell death (PCD) Death of cells at the site of initiation of infection by the pathogen as the early response of the host; these cells have a predetermined function amounting to suicide to prevent further spread of the pathogen to other cells/tisssues (see hypersensitive response)

Promoter A region of DNA upstream of a gene that can act as a binding site for a transcription factor and ensure the transcription of the gene concerned

Proteasome A large protease complex that degrades proteins tagged by ubiquitin

Proteome The entire range of proteins expressed in a specified cell

Proteomics Comprehensive analysis of all cell proteins

Pseudogene It is the nonfunctional copy of the gene

Quorum sensing The ability of bacterial or fungal propagules to sense the concentration of certain signal molecules in their environment prior to activation of infection process

Reactive oxygen species (ROS) Intermediate and radical species generated from oxygen such as superoxide or hydrogen peroxide as resistance response in plants

Recognition factors Specific receptor molecules or structures on the host plant surface useful for recognizing the presence of a pathogen

Recombinant DNA A hybrid DNA formed by incorporation of DNA segment from one species into another species

Recombinant protein A polypeptide synthesized by transcription of the hybrid DNA and translation of the mRNA

Regulatory proteins These proteins can bind to DNA and influence the action of RNA polymerase thereby acquire the ability to control the rate of protein synthesis

Reporter gene It is a gene inserted into the DNA of a cell capable of reporting the occurrence of signal transduction or gene expression

Repression of gene function Inhibition of transcription or translation by binding of a product of a gene (repressor protein) to a specific site in the DNA or RNA molecule

Restriction endonucleases Hydrolytic enzymes capable of catalyzing the cleavage of phosphoester bonds at specific nucleotide sequences in DNA

Restriction site This is a specific nucleotide segment of (base pairs) in a DNA molecule that can be recognized and cleaved by the restriction endonuclease enzyme employed

Reverse genetics A method of analysis of gene function by introducing mutations into a cloned gene

Reverse transcription Synthesis of DNA from an RNA template using appropriate reverse transcriptase

RNA-induced silencing complex (RISC) Large multiprotein complex associated with a short ss-RNA that mediates degradation or translational repression of a complementary or near complementary mRNA

RNA interference (RNAi) Degradation of mRNAs by short complementary double-stranded RNA molecules

RNA processing Refers to various modifications that are made in RNAs within the nucleus

RNA splicing A process that results in removal of introns and joining exons in pre-mRNA

Sequencing The process used to determine the sequential arrangement of amino acids/nucleotides in protein /nucleic acid molecule

Serotypes Groups of an organism differentiated based on the variations in the serological reactions with different antibodies; monoclonal antibodies are frequently used for characterization of serotypes of plant pathogens

Signaling Communication established between and within cells of an organism

Signal molecules Host factors (molecules) that respond to the presence or initiation of infection by a pathogen and transmit the signal to and activate proteins or genes in the tissues away from the site of pathogen entry leading to restriction of pathogen spread/disease development

Signal transduction Reception, conversion and transmission of 'chemical message' by a cell

Single nucleotide polymorphism (SNP) Variation detected in individual nucleotides within a DNA molecule; SNPs usually occur in the same genomic location in different individuals

siRNAs Specific short sequences of dsRNA of less than 30-bp in length that can trigger degradation of mRNA containing the same sequence (present in siRNA) within the cell as part of process known as RNA interference (RNAi)

Site-directed mutagenesis A laboratory protocol employed to modify the amino acid sequence of a protein

Somaclonal variation Variability detected in different calli for various characteristics, including disease resistance; the calli exhibiting resistance to disease(s) may be regenerated into whole plants that are tested for the level of resistance to disease(s)

Sticky ends Exposed complementary single strands of DNA can bind (stick) to complementary single strands in another DNA molecule, producing a hybrid piece of DNA

Synteny Two genetic loci are presumed to be linked to the same chromosome, whether or not linkage has been demonstrated, as against the asyntenic loci that are linked to different chromosomes

Systemic acquired resistance (SAR) Resistance to diseases caused by microbial pathogens can be induced by biotic and abiotic inducers of resistance in treated plants; a set of genes referred to as 'SAR genes' is activated by the inducers resulting in the systemic resistance in various organs/parts of plants

Tat pathway The twin-arginine translocation (Tat) system is involved in physiological functions of bacterial pathogens; the Tat pathway operates in the inner membrane of Gram negative bacteria

TATA box A conserved sequence in the promoter of several eukaryotic protein-coding genes where the tranascription initiation complex assembles

Transactivating protein A protein that 'switches on' a cascade of genes /gene regulation

Transactivation Activation of transcription by binding a transcription factor to the DNA regulatory sequence in question

Transcripts During transcription of a gene, various segments of mRNA, known as transcripts, are formed

Transcription The genetic information contained in one strand of DNA is used as a template and transcribed through the enzyme-catalyzed process to specify and produce a complementary mRNA strand; the genetic information in the DNA is rewritten into mRNA

Transcription factor A protein that binds to a regulatory region of DNA, often upstream of the coding region and influences the rate of gene transcription

Transcriptome The entire set of all gene transcripts (mRNA segments derived from transcription process) in a specified organism; provides the knowledge of their roles in the growth, structure, health and disease of the organism concerned

Transduction Bacterial genes may be transferred from one bacterium to another by means of bacterial virus (bacteriophage)

Transformation This a process by which free DNA may be transferred directly into a competent recipient cell; plant cells may be transformed with genes from different kinds of organism to enhance the level of resistance to microbial pathogen(s)

Transgenic An organism carrying in its genome one or more DNA sequences (transgene) from another organism

Translation Synthesis of a polypeptide chain from an mRNA template

Transposon A DNA sequence (segment or molecule) capable of replicating and inserting one copy (of itself) at a new location in the genome

Trypsin inhibitor (TI) protein This protein is constitutively expressed in mature maize kernels of resistant maize genotype at higher levels compared to susceptible leaves

Tubulin A family of globular cytoskeletal proteins that polymerize to form microtubules

Two-hybrid system The yeast or bacterial system that is employed for detecting specific protein-protein interaction; the protein of interest is used as a 'bait' to 'fishout' proteins that may bind to it (referred to as 'prey')

Ubiquitin A small protein present in all eukaryotic cells that has an important role in 'tagging' other protein molecules; the tagged protein molecules are said to be 'ubiquitinated'

Upstream A relative direction of nucleic acids often used to describe the location of a protmoter relative to the start transcription site; the start codon is upstream of the stop codon

Vector An agent used to carry foreign DNA in recombinant DNA technology; natural carrier of microbial pathogens, commonly viruses from infected plant to healthy plant resulting in the spread of the pathogen/disease under natural conditions

Virulence factors Genes or factors essential for and contribute to the virulence of the pathogen; may not be required for the growth and reproduction of the pathogen

Yeast two-hybrid system An experimental technique of detecting the protein-protein interactions in yeast cells

Zinc finger domain A kind of DNA-binding domain consisting of loops containing cysteine and histidine residues that bind zinc ions

Index